Farming

compiled by

Sarah Johnson

Themes in Environmental History, 6

'Themes in Environmental History' is a series of readers for students and researchers. Each volume aims to cover a prominent subject in the discipline, combining theoretical chapters and case studies. All chapters have been previously published in the White Horse Press journals *Environment and History* and *Environmental Values*.

1. *Bioinvaders* (2010) ISBN 978-1-874267-55-3
2. *Landscapes* (2010) ISBN 978-1-874267-60-7
3. *Indigenous Knowledge* (2012) ISBN 978-1-874267-68-3
4. *Animals* (2014) ISBN 978-1-874267-80-5
5. *Trees* (2015) ISBN 978-1-874267-88-1
6. *Farming* (2016) ISBN 978-1-874267-89-8

British Library Cataloguing in Publication Data
A catalogue record for this book is available from the British Library

ISBN   978-1-874267-89-8 (PB)

Contents

Publisher's Introduction

At its simplest, farming is, as defined by Barbara Orland, 'the link between the landscape and nutrition'. It is the complex set of factors by which food production results from human 'knowing and struggling with the natural environment', such that, beyond foraging and hunting, human beings consciously manipulate the landscape and its organisms to feed themselves. In modern industrialised societies, via a long process of enclosure and mechanisation, the directness of the link is lost and between ecology and nutrition stands the apparatus of the market. The stories of farming, very broadly, are of evolution from subsistence to industrialisation, a process shaped by relations of property and labour and one which has often severely challenged ecological and cultural equilibrium. Ian Rotherham has recently outlined the land-use changes that have shaped the 'eco-cultural' landscape of Britain (and, indeed, those of many European countries), with often devastating consequences for natural and social orders, as farming has evolved from the communal, subsistence-based activity of predominantly rural populations to large-scale business in an increasingly urbanised society whose population, for the most part, lives remotely from the land.[1]

This volume considers several traditional farming systems in diverse parts of the world and their ecological and social effects, foregrounding the key characteristics of subsistence, rotation (of cropping or grazing areas) and mobility and low population density. It then addresses the colonial era – showing how external powers effected the transformation or displacement of indigenous agriculture by resource-hungry, market-driven farming businesses, echoing in the imperial arena the transformations that had already radically altered rural environments in more developed nations.[2] The third section complicates this narrative by exploring cases in which the emphasis is on changing environment and changing attitudes, rather than on a simple narrative about power – thus, we see the effects of cash-driven farming on virgin environments in Brazil and Madagascar; and, in Australia, the evolution of settler farming to nature tourism, changing attitudes to bush burning and attempts to align profitability with conservation techniques aimed at reversing or halting soil degradation. This section highlights the co-evolution of agriculture and human attitudes and values. The fourth section returns to the trajectory of industrialisation, investigating how the inputs into agricultural systems change with attempts to maximise and standardise outputs – with ever-greater dependency on chemicals (and associated costs in terms of energy and money). The fifth section contains two papers that chal-

---

1   Ian D. Rotherham. *Eco-history. An Introduction to Biodiversity and Conservation.* (Cambridge: The White Horse Press, 2014).
2   Of course, change was not necessarily always exogenous: forces like industrialisation or colonialism hastened and scaled-up the transformation of agriculture globally, but far-reaching changes have also begun at the local level (Orland) and, most often, have been the result of adaptation to local circumstances (Moon, Griggs).

lenge the ascendancy of industrialised monoculture, exploring 'counter-cultural' farming movements of the late twentieth century. This volume uses disparate (though often connected) cases that represent the most exciting scholarship on farming published in *Environment and History* and *Environmental Values*, not to delineate an overarching narrative but to highlight the adaptativeness of human agriculture, the sustained ingenuity required to wrest a living from the soil for a global population that has spiralled into – increasingly greedy – billions. The questions that the later papers throw up as to the sustainability of global food-ways from the mid-twentieth century onwards (for example the drenching of crops in petrochemicals, the selective breeding of ever-more-productive, ever-less-viable animal varieties, the increased feeding of nutritious crops to meat animals) represent a key problem the twenty-first century must face.

COMMUNALITY AND SUBSISTENCE: TRADITIONAL SYSTEMS

Dominguez et al. discuss the agdal systems of the Moroccan High Atlas, 'seasonal prohibitions' that limit access to resources to ensure their annual recovery from human exploitation. These systems, legislated by tribal assemblies, have the 'aim of ensuring the continuity of the local agro-pastoral system', which includes the production of livestock but also associated cultural values. The territory and the legislation that governs it is perceived as sacred, its governance overseen by saints, and pastoral activities are accompanied by religious festivals that bring together communities who, by virtue of their transhumant activities, spend long periods apart. In line with many traditional systems, the agdal is communal, separating ownership and usage rights, allows access to pasture by all and 'guarantees a certain degree of social justice among herders', as well as playing 'a role in identity building and the feeling of belonging to a certain group that confers rights and duties over those territories'. Furthermore, the rotational opening and closing of agdals creates an 'ecological mosaic', regarded as highly desirable in terms of biodiversity.

This traditional resource-management system ensures sustainability of subsistence, of ecology and of cultural values but is, of course, remote from the market economy. It may be termed an endogenous system and other articles in the volume show analogous systems working well – both for human populations and the environment – until powerful external factors intervene. Thus, Orland describes how the equilibrium in Swiss Alpine cattle farming between 'natural resources, economic interests and cultural values' and 'production methods' was compromised by the demands of national and international trade as early as the sixteenth century. In cultivation, as opposed to pastoralism, the story has been the same in diverse parts of the world: Oliveira and Winiwarter show that the indigenous slash-and-burn systems of Brazil, characterised by intercropping 'to reduce plant pathogens and weeds' and using fire to release soil nutrients were sustainable, indeed productive, and ecologically neutral under conditions of low population density and high mobility. However, when colonial powers arrived, the prior conditions of communality and subsistence were replaced by a market-directed model of monoculture in pursuit of cash returns. Caillavet outlines a similar change in the Ecuadorian Andes, where a 'raised field / flooded ditch complex' aided subsistence by ironing out hydraulic irregularities of drought and flood, raising temperatures and resisting erosion. However, this system 'did not correspond to [Spanish colonial] production priorities' and systematic drainage efforts by the colonisers resulted in its abandonment. Caillavet links these efforts to more than economics, however, suggesting that the colonisers were prejudiced against irrigated agriculture, associating it with disease and the despised culture of 'moriscos'; they were also discomfited by the blending of habitation and farming and the 'incongruous configuration of the landscape' resulting from the 'continuities and inter-relations between lakes, the lake rims with their reeds ('totora'), and the flooded fields whose ditches link up in turn with rivers and lakes', leaving 'no clear borderline between dry and wet, fishing and agriculture'. To the colonisers, this was tantamount to 'barbarism', regardless of whether it worked well as an agricultural system. Caillavet argues that this unease is why colonial documentation elides all details of indigenous agricultural systems, rendering them historically almost invisible.

Both Caillavet and Dominguez et al. suggest that preserving or redeploying traditional systems that guarantee subsistence, if not surplus, 'can be of great interest for a renewed approach to development' (Dominguez et al.), though Caillavet warns that 'retrieval of ancient techniques is not a viable option if they cannot be sustained under present-day economic conditions'. As Suh points out, the principles of permaculture owe much to traditional Eastern agricultural systems informed by Taoism, Buddhism and Confucianism, but practitioners struggle to scale-up permaculture farming to a form that is economically viable beyond mere subsistence without importing 'energy sources, resulting in the extensive use of machines that consume non-renewable fossil fuels'. However, Dominguez et al., Orland and Clement all address ways in which traditional systems can retain their integrity and be made to bear economic fruit in the modern age: Dominguez et al. propose that the agdal system could aid landscape conservation, in that the current eco-cultural landscape is a substantial asset in terms of tourism; Orland notes that farming in the Alps attracts large subsidies because it preserves flora, fauna and landscape, as well as supporting the culturally-valued stereotype of dairy products as 'a source of health, physical strength, and the spirit of freedom of the Swiss people' and the herder who produces them as an embodiment of national virtues; and Clement points out that the 'dehesas' created under an age-old traditional system of transhumant woodland pasturage survive in the modern age because they provide habitat for 'lucrative breeding of brave bulls for corridas and ... black pigs ... to produce a fine ham called Jamón de Pata Negra'.

Introduction

RESOURCE MINING IN DEVELOPING SYSTEMS

A number of authors here (Clement, Oliveira and Winiwarter, Orland) draw attention to the beneficial effects of traditional farming on landscape, creating biodiverse 'ecological mosaics'.[3] However, it is worth noting that farmers' primary aim was never to shape the landscape according to aesthetic or conservation interests; rather, the landscapes that resulted reflected the limits of possibility in meeting (even surpassing) the demands of subsistence. If traditional systems are based on small populations striving to meet their 'living needs' (Orland), colonial farming in various territories – like Western agriculture in the age of 'Improvement' – was about profitable use of resources, often with positive disrespect for the original landscape (as in the several cases described in this volume of catastrophic erosion caused by ploughing at variance with the contours of the land). As Pacheco observes, 'In the early stage of frontier development, a logic of natural resource mining predominates'. At worst, this led to significant landscape degradation, for example in the 'profligate' system of sugar-cane farming in Brazil that relied on burning large areas of forest, cropping intensively for several years, then moving on to a new area of forest; and seemed premised on the notion that forest was a limitless resource. Pacheco's article shows a similar logic at play in the late twentieth century cattle-ranching boom in Pará, Brazil. Musemwa tells another story of rapacious resource extraction in Zimbabwe – with farmers and miners vying to 'to define [their] relationship to the environment by ensuring unfettered access to natural resources' with little thought of future land degradation. Cash crops (initially planted by colonisers, though Scales and Griggs trace the associated issues into the post-colonial era) are again and again associated with soil erosion and declining fertility – as explored by Moon, Oliveira and Winiwarter and Griggs relative to sugar cane in the Netherlands East Indies, Brazil and Queensland; and Musemwa and Scales for maize in Rhodesia and Madagascar. With excessive ploughing, soil erosion takes hold: a Commission into the problems associated with farming and mining in Southern Rhodesia found that, single-mindedly pursuing productivity farmers ploughed up natural landscape features that aided water run-off and thus prevented erosion, so that 'in a very short time, many acres of valuable pasture and woodland were converted into a donga-scarred waste'.[4]

---

3   See a paper for which, sadly, there was not space in this volume: Enric Tello, Natàlia Valldeperas, Anna Ollés, Joan Marull, Francesc Coll, Paul Warde and Paul Thomas Wilcox, 'Looking Backwards into a Mediterranean Edge Environment: Landscape Changes in El Congost Valley (Catalonia), 1850-2005'. *Environment and History* **20** (2014): 347–384. Mauro Agnoletti and Ian Rotherham also deploy this as a key concept in various works.

4   For the potentially catastrophic effects of cash-crop farming on vulnerable African soils, in this case in Malawi, see Wapulumuka Oliver Mulwafu, *Conservation Song: A History of Peasant-State Relations and the Environment in Malawi, 1860–2000* (Cambridge: The White Horse Press, 2011).

Introduction

As Moon explores, the demands of large-scale plantations had a deleterious effect on indigenous agriculture – sugarcane cultivation could survive a far greater drop in soil fertility than could the indigenous rice farming with which it was rotated in the Dutch East Indies. While it was commonplace for indigenous systems, particularly those based on slash-and-burn, to be decried by colonial planters, these were – as Oliveira and Winiwarter argue – usually well-balanced, using fire and rotation to access sufficient nutrients to ensure long-term fertility for a limited population. Both Clement and Scales suggest that the tendency to posit forest clearance as 'a vague chronicle of forest degradation' (Clement), linked to a 'Malthusian spiral of increased population and decreased land productivity' (Scales), ignores the often-sophisticated checks and balances in such endogenous systems. Larger-scale operations, whether conducted by colonial powers or as part of Western agri-business, have tended to be fuelled by external resources (slave labour, fertilisers, machinery) rather than those of the community inhabiting the land; and their products are marketed in areas remote from the site of production. Many early colonial plantations were 'profligate' systems: as resources were depleted, planters moved on to claim resources elsewhere, leaving fatally impoverished land behind. Oliveira and Winiwarter expose how, even while peddling a narrative of Brazil as a land of limitless plenty, Portuguese colonial agronomists were aware of the catastrophic effects of deforestation on soil fertility. They cite one manual that warns: '[The forests are finite ... Carthage and Troy did not turn into larger mounds of ash when pillaged than those you can see in the farms of Brazil, where they destroy these important and beautiful forests even today]'.

As Oliveira and Winiwarter make clear, agronomists in colonial Brazil were well aware that sugar and coffee cultivation could not be long sustained if they continued destroying forest apace; and had a degree of 'insight into the need of preservation of Brazil's biodiversity against the devastation wrought by colonial exploitation'. The same may be said of Dutch planters in the Netherlands East Indies (Moon offers a fascinating account of the contestedness of knowledge – 'who had the credibility to interpret' the situation and whose possible solutions should be trusted) and the authorities in Rhodesia (Musemwa) and Queensland (Griggs). The latter two cases, and that discussed by Soeterboek of scrub burning in the Australian Alps, illuminate how the views and practices of individual farmers do not necessarily accord with those of the authorities: Musemwa reports that, though farmers complained about the dire environmental effects of mining, they were mostly only 'goaded into taking up conservation measures' by fear that environmental destruction would translate into economic loss. Likewise Queensland sugar planters were slow to accept that soil was becoming less fertile, preferring to mask the decline with increased use of chemicals rather than adopting conservation measures – until it became economically expedient to do so. As Griggs notes, 'Acceptance that soil loss needed to be reduced required canegrowers to make two confessions: that a problem existed; and that this problem existed because of poor agricultural practices adopted currently or in the past', and this was a slow process, eventually hastened by a slump in sugar prices in the 1980s that made the reduced costs associated with such measures as trash blanketing and minimum tillage attractive. It was really only by-the-by that farmers taking up such methods were also able to self-present as environmentally concerned, combating soil erosion and cutting-back chemical use.

HELPING NATURE ALONG: MECHANISMS AND MYTHS OF PRODUCTIVITY

Richard Drayton, in his book *Nature's Government*, shows how commonplace it was in the early modern period to regard societies who did not cultivate the land as having few rights over it, a view that justified colonial enclosure and appropriation of lands that seemed insufficiently exploited by indigenous populations.[5] One of the epigraphs to *Nature's Government*, taken from John Winthrop's writings on 'The Natives of New England', circa 1660, encapsulates this view: 'They inclose no Land, neither have any settled habytation, nor any tame cattle to improve the land by, and soe have noe other but a Naturall right to these Countries'. There was a sense – both at home and in the colonies – that it was a human imperative to 'improve' land to maximise its productivity, and that in doing so – 'mix[ing] his labour with it', as John Locke put it – a man acquired rights of ownership.[6] Both the communality and the limited intervention of traditional systems, at home and abroad, were thus seen as dereliction of the duty to 'improve' and govern Nature. As populations increase and land becomes a limited resource, agriculture requires maximised efficiency and the need to 'improve' and assist Nature becomes more pressing and often reactive: soil fertility declines – add chemicals; pests threaten productivity – add chemicals.

In contrast to (but often alongside) the Enlightenment philosophy of Improvement – which was decidedly Georgic but also Protestant in its emphasis on labour – land that yields bountifully for minimum toil has been a human aspiration since Virgil's Arcadia, though the truth behind the myth (a rocky and barren landscape in the Arcadian case) has often been unpromising. The Brazilian green paradise described by Oliveira and Winiwarter was, like Eden, always-already threatened by human rapine; land that returns generously for minimal input of labour or other external factors is an enduring chimaera in human history – and one that was recurrently displaced onto the unfolding 'New Worlds' of the early-modern period. While the reality of farming has often been the thankless toil of wresting a living from marginal lands – as de-

---

5   Richard Drayton, *Nature's Government: Science, Imperial Britain, and the 'Improvement' of the World* (New Haven: Yale, 2000).

6   John Locke, *An Essay Concerning the True Original Extent and End of Civil Government* (1690).

scribed in Campbell's account of the old-time Ozarks farmers, or Frost's of the Australian farmers to whom rainforest was a 'demonised' obstacle – there has never been a shortage of optimistic souls hoping to get rich by ranching cattle in the Brazilian Pará, growing apples in New Zealand or maize in the forests of Madagascar. Roche describes how orchard land was conceptualised in New Zealand as providing the opportunity for families, 'by honest toil', to 'build a home and a business by harvesting nature's bounty'. He explores the 'values and attitudes invested in apple growing as a lifestyle', with the orchard posited as an mutually beneficial compact of humanity and nature, in an essay that both interrogates the schism between ideal and reality ('spraying was essential to the maintenance of the orchard as an "island" of nature') and chronicles the evolution of the New Zealand apple industry in a newly global market context where 'the local and the global ... are simultaneously and mutually constituted'.

This telescoping of the global and the local is now universal reality. Disquieting reports surface regularly in the media that – for example – precious irrigation water is being used in Kenya to grow flowers for Western supermarkets while local people struggle for subsistence. Conflicts over resource use are omnipresent and have a long history – as Moon shows in her description of efforts made even in colonial times to address points of conflict between local and global-market users. The modern watchword 'sustainability' is a highly manipulated concept, beset by nagging doubts that, for example, planting x saplings elsewhere is really valid compensation for the logging of old growth rainforest to fatten cattle. Madison shows that its pitfalls were anticipated by the Odum brothers in 1950s America, who perceived that the Green Revolution myth of ever-increasing productivity was just that, that the energy poured into the agricultural system in the form of petrochemical fuel and fertilisers could not simply be written out of the equation. It was simply not true that solar energy was being more and more efficiently used, 'with fertilisers and pesticides [allowing] any natural disasters and nutrient bottlenecks to be overcome' and such innovations as hybrid seeds and selectively-bred animals achieving the translation of solar energy into ever-greater calorific outputs. These developments relied on inputs that the Odums construed as often equating to more energy being put in to the system than the calorific output of food. They believed that there had been 'an interesting evolution in the history of agriculture toward declining net energy yields' – subsistence farming produced a sufficient energy yield, with minimal inputs, to sustain life; but technological advances had allowed the necessity of 'balancing the books' to be sidestepped, resulting in a surprising but 'strong correlation between auxiliary energy inputs (that is anything but solar energy) and decreasing net yields'. Since modern varieties of crop and farm animal are bred for maximum productivity under conditions of high energy input and would likely collapse without the 'power-rich management of man', the system becomes even more imbalanced, as 'profligate' in its way as the 'resource-mining' of frontier farming. As Suh observes, this situation is only magnified with the modern tendency to grow crops to feed livestock rather than for direct consumption by humans.

AUTHENTICITY, HARMONY AND RESTRAINT: POST-INDUSTRIAL DREAMS

In many of their observations the Odums prefigure the alternative agricultural ideas of the late twentieth century onwards, which have tended to value diversity, small scale and limited inputs; and their ideas – resolutely couched as scientific rather than ideological – were, as Madison observes, valuable backup for those favouring alternative agriculture for more nebulous, often 'pseudo-medical', reasons associated with 'personal redemption'. With the Odums' ideas about agro-ecology, the organic philosophy need 'no longer be based on vague misgivings about industrialised farming, but could now be justified as the avoidance of unstable expenditures of geological resources'. The Odums believed that agriculture ought to mimic 'a stable and self-maintaining mature ecosystem' and this conception also lies behind the permaculture systems described by Suh. The key concepts of permaculture are 'decentralisation; independence (reduced reliance on external sources); small rural communities (increased cooperation); harmony with nature; polyculture (integration of diverse crops and livestock); and restraint (internalisation of all external costs and increased reliance on renewable resources)'. These ideals are linked to low energy use, recycling and localism within an ecological framework that aims to make agricultural systems into 'human habitats' that echo the patterns and relationships off natural systems. Key to permaculture is the belief that the land should become better rather than poorer – instead of over-exploitation, rectified by chemicals, permaculture aims to keep nutrients within the system as traditional agricultural methods have tried to do (since, before the advent of agrichemicals, communities had to rely on 'permanent soil fertility'.)

Madison describes how the Odums' fear that America had 'forgotten how to farm without poisons' chimed not only with those promoting organic farming but with environmentalist thinkers such as Rachel Carson and Barry Commoner who popularised the view that agri-chemicals were 'dangerously at odds with nature and capable of endangering the broader populace'. Sheail takes up the story of increasing public misgivings about modern farming, exploring the efforts made by British government 'in adjudicating the impact and, more particularly, attributing responsibility for minimising any deleterious effects' and 'taking further hesitant steps in devising some kind of comprehensive statutory regulation', seeking an acceptable balance between the needs of farming, public health and wildlife. However, there were many who felt that legislation was not enough, that there was a need, environmentally and for the human soul, to get 'back to the land'. The urban escapees to the Ozarks described by Campbell were, of course, never going to change the world with their attempts to establish self-sufficient, communal farms that strove to replicate traditional ways of life. Human population continued to spiral, and productivity and efficiency remain the watchwords of agribusiness. However, the Ozarks back-to-the-landers were just one manifestation of phenomenon common in industrialised society – the desire to reclaim some sort of 'authenticity', a desire that is as much about self-actualisation as about environmentalism and sustainability.

While the Australian 'selectors' described by Frost and the ranchers studied by Pacheco laid their claim to parcels of land in the hope of economic gain, for the Ozarks settlers the goal was the spiritual health associated in the philosophically agrarian mindset with self-sufficiency: 'freedom, a piece of land that they can own outright to provide for themselves', to escape 'conspicuous consumption' and 'tap into a tradition of self-reliance'.[7] Their low-impact ways of farming might echo the traditional ways of the remaining Ozarks 'old-timers', but from ideology, not necessity. As Campbell explores many of the newcomers were 'misinformed and ill-prepared for self-sufficient living in a marginal landscape and they did not possess the traditional ecological knowledge necessary to survive on their own'; but it seems that that 'traditional ecological knowledge' and sense of environmental stewardship survived among the 'old-timers' and was somewhat revitalised by the influx of eager outsiders, however much the latter were mocked for their hippyish faith in solar heating, compost and the influence of the moon. While in the stories of colonialism, as told by Oliveira and Winiwarter, again and again 'indigenous knowledge was ignored' and 'the knowledge base' of outsiders was paramount, the modern 'colonists' arrived seeking to learn from the autocthonous population.

CHANGE AND ADAPTATION

Soeterboek hints at a similar transformation in approaches to indigenous eco-logical knowledge in his account of scrub burning by settler 'bushmen' in early to mid-twentieth-century Australia – this practice was clamped down on by the authorities as an environmental menace but was said by the herders at the time to be in the same spirit as the burning carried out by Aboriginal people, which is now understood as 'sophisticated environmental stewardship'. Indigenous ecological

---

7   It is interesting that the self is foregrounded in this philosophy – where traditional systems emphasised communality and harnessed the energy of a whole community to accomplish onerous tasks, back-to-the-landers seemed often to founder in attempts at communal living and to prize personal development above all. Suh suggests that, for permaculture to gain wider popularity, its emphasis on communality should be considerably watered-down. The achievements, potentialities and problems of modern communal systems on a larger scale – kibbutz and Chinese communist farming, for example – are, regrettably, outwith the scope of this volume.

knowledge, adapted into settler folk-tradition, against scientific understandings of the time, is now regarded as vital to understanding Australian fire-ecology.

The essays presented here recurringly foreground the impossibility of pinning down the values of farming and farmers. Soeterboek interprets Australian cattlemen's activities not as wantonly destructive but as perforce adaptive:

> They observed their environment and developed and adapted practical knowledge to help them to exploit it most effectively. In settler societies around the world, the earliest European invaders pushed beyond the bounds of 'civilisation' and were forced to come to terms with new and different landscapes from which they tried to live and profit.

As Moon notes, farming is a protean activity, in which the 'variability of local conditions' must be understood; and Griggs too highlights the contingent nature of farming in any given location: 'the transfer of models from other crops and physical environments is not always successful or possible'. Values, too, are mutable according to circumstances: Frost points out that Australian historical literature has found it convenient to posit nature-hating farmers who saw the forests as obstacles to productivity against urban aesthetes and conservationists who, given the luxury of not having to live in nature, could bleat about its protection. However, his paper shows that farmers were vocal in campaigns to designate National Parks, no doubt partly because of the economic potential of tourism ('diversification' was then, as now, a valuable survival strategy for farmers).

The authors in this volume repeatedly present the environmental positives of farming (biodiverse landscape mosaics, the preservation of old seed varieties in the Ozarks) as side-effects of farmers' need to survive, and so too are the negatives. We return here to Orland's starting proposition that farming equates to food production by means of human 'knowing and struggling with the natural environment'. The USFS caption to a photograph presented by Campbell of an abandoned Ozarks homestead signals the ultimate consequences of failure in that struggle: 'No, this home was not destroyed by a tornado. It was wrecked by poverty induced by an attempt to wrest from nature, lands never designed by her for ag[riculture]'. This domestic wreckage reflects in microcosm the local, national or global catastrophe whose spectre inspires farming's constant adaptiveness at every scale: it is imperative for humanity's survival that the link between 'landscape and nutrition' be unbroken.

The selected articles represent the views held by the authors at the time of original publication, and the state of scholarship at that time.

TRADITIONAL

Diverse Ecological, Economic and Socio-Cultural Values of a Traditional Common Natural Resource Management System in the Moroccan High Atlas: The Aït Ikiss *Tagdalts*

Pablo Dominguez, Alain Bourbouze, Sébastien Demay, Didier Genin and Nicolas Kosoy

1. INTRODUCTION

Local institutions linked to communal natural resource management have been studied for several decades and their contributions to 'sustainability' have been internationally recognised (Folke et al., 2007; Ostrom, 1990; Neves-Garca, 2004). However, few studies have analysed the existing traditional institutional *agdal* systems of communal natural resource management found in the Berber area of the Maghreb (Ilahiane, 1999; Venema 2002; Bourbouze, 1997; Cordier and Genin, 2008), even less so in a fundamentally multidimensional way (Auclair et al., 2007; Dominguez, 2010). The *agdal* can be defined as a system of seasonal prohibitions that limits access to one or more agro-sylvo-pastoral resources in order to allow them to recover from direct or indirect human pressure during their most critical period of growth. In fact, the *agdal* management system seeks above all to optimise the complementary use and productivity of the resources and to ensure their continuity. The prohibition of the removal of renewable resources during these critical periods allows the resource to grow exponentially and at the same time leaves it time for regeneration (in the case of the pastures, the flowering and establishment of new seeds), thus ensuring the continuity of the ecological cycles. But it also indirectly leads in many cases to other ecological benefits. One example of this is the maintenance of dense vegetal cover, a consequence of the biomass removal restrictions as shown in the case of forest *agdal* in the Aït Bouguemez (Hammi et al., 2007). Another is the appearance of higher rates of biodiversity conservation inside the *agdal* managed spaces than in comparable areas without or with less *agdal* management (Kerautret, 2005; Alaoui-Haroni, 2009; Dominguez and Hammi, 2010).

At economic level, various small scale studies have also made it evident that the contribution of collectively managed areas can have a positive impact on local economies (Ostrom, 1990; Neves-Garca, 2004). In the case of forest *agdals*, Genin et al. (in press) show that the leaves from the forest of *Querqus*

Pablo Dominguez et al.

ilex in the Ait Bouguemez valley represent an important complementary forage resource during winter snows when other pastoral resources (herbs, bushes, etc.) are covered by snow or in latency. In addition, in certain villages, the surpluses of pastoral *agdals* are exchanged for water use rights between neighbouring tribal fractions (Romagny et al., 2008). Pastoral *agdals* also contribute to agricultural enrichment through the recycling of nitrates and derivates that come from the highlands through animal manure (Demay, 2004).

With regard to social justice, the season and the duration of the prohibition of the *agdal*, the resources, and the spaces concerned by this prohibition, are decided solely by the tribal assembly (*jmaa*) on the basis of its own history, territorial heritage, political structure, knowledge and annual economic strategies, with the aim of ensuring the continuity of the local agro-pastoral system. This being so, as long as the tribal assembly (*jmaa*) continues to exist, the opening of the *agdal* will always have to wait until the date agreed by the majority (Ilahiane, H. 1999; Auclair et al., 2007). Hence the *agdal* guarantees a certain security of access to the group in especially critical moments and the continuity of resources for the benefit of the community as a whole. Nevertheless, this is not always so, mainly in the case of big tribal or intertribal *agdals*. Generally, in these cases, when agreement or respect of the rule are not reached by the community of users, the local governmental authorities play a referee role.

This ingenious system of agro-pastoral land rotation, where one space or resource is banned to users at a certain period while another is still available, is also intimately linked to a complex worldview. In fact the cultural and religious heritage underlying the *agdal* in Morocco constitutes a key element for reinforcing the sustainability of the current uses of natural resources and the social representation system which is strongly embedded in the *agdal* system (Mahdi and Dominguez, 2009; Simenel, 2008). For example, as we will describe later and as other authors have shown for many other High Atlas populations (Hammoudi, 1988; Rachik, 1992; Mahdi, 1999), religion strongly structures the pastoral life of which the *agdal* is a central part.

As we can see, failure to take into account the complexity of the *agdal* would be to distort any analysis of it. Hence, our main hypothesis in this article is that the *agdal* must be analysed multidimensionally if we are to understand its value fully. It is the aim of this paper to examine the many dimensions of the *agdal* system through a case study of the Ait Ikiss people in the High Atlas of Marrakech (central Morocco), who use different small *agdals* (locally called *tagdalts*) combined with a particular system of land rotation. We first explore the spatial functioning of this system, then estimate the contribution to livestock feeding of a small scale communal agro-pastoral territory managed under this system, and finally we assess the social and cultural importance of the *tagdalt* system as a whole.

The Aït Ikiss Tagdalts

2. CONTEXT OF THE STUDY.

The study area is the territory of the Ait Ikiss people, located at the heart of the High Atlas, less than 50 kilometres from the city of Marrakech. The climate is typically Mountain Mediterranean, with precipitation of about 500 mm/year, and monthly mean temperatures that range from 0 C° in December to 20 C° in July, and which strongly vary within the year and according to the geographical location. Like any other high mountain area, it is compartmentalised within a vegetation altitudinal gradient from about 1,300 to 3,000 m. This includes, from bottom to top, Mediterranean forest, scrub and brush, humid highland pasture, semi-dry steppe with strong presence of cushion shaped xerophytes, and at the mountain top, rough pasture and highly resistant Juniper.

This population is characterised to a greater or lesser extent by a traditional tribal organisation, previously described by British anthropologists as *segmentary structure* (Gellner, 1969). It is typified by fitting one social group (or *segment*) into another, from the smallest to the biggest, like a set of Russian dolls (Evans-Pritchard, 1970). For example, the Ait Ikiss continue to organise themselves in tribal fractions, sub-fractions, villages, clans and nuclear families. The population speaks *Tachelhit*, a South Moroccan Berber dialect. Most of the men speak Arabic as do some of the women, especially the youngest. It has been estimated (Bellaoui, 1989) that the relative contribution of the agropastoral sector to the local income of the Zat valley (the Ait Ikiss included) is approximately 75 per cent, which is usually combined with seasonal emigration or specialised local work such as masonry or smithy work or similar. Livestock consists mainly of cows, sheep and goats.

The Aït Ikiss group that we have studied comprises about 640 people, who occupy four different habitats: Azgour/Tifni, Ikiss, Warzazt and Yagour n'Ikiss (Figure 1). The Aït Ikiss group is patrilineal, as in other Berber societies. All decisions regarding a household's use of the agro-pastoral resources are taken by the male heads of each household, and in their absence, by the oldest adult male of the family. The Aït Ikiss, like the rest of the mountain *Mesioui*, are mainly defined in Morocco as non-orthodox Sunni Muslims and sustain beliefs and practices that are the result of a long cohabitation between pre-islamic religions and earlier Islam. They organise their activities and manage their communal territory through tribal assemblies called the *jmaa*. The seasonal *agdal* prohibitions imposed among their territories are decided by the *jmaa*, and nominated members serve as guardians (locally called *Ait Rbains*). When the *Ait Rbains* report a delinquent, graduated sanctions are usually established by the *jmaa*, according to the different types of offences. The extensive pastures of the Aït Ikiss and the dense humid prairies of the highlands mainly situated in the Yagour n'Ikis habitat (about 5 km²) are the most important basis for the existence

Pablo Dominguez et al.

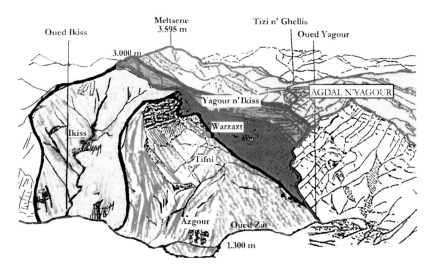

FIGURE 1. Territory of the Aït Ikiss and their four habitats (based on Dresch, 1939)

of the *tagdalt* system, but other spaces and resources are also important to this system as we will see later.

3. METHODOLOGY

Data collection

The qualitative data of this paper was collected through intermittent participative observation from 2004 to 2008, throughout all agro-pastoral seasons, adding up to one whole year of co-habitation with the Aït Ikiss people. More systematic open-ended interviews concerning the animal production structure and the uses of the Aït Ikiss territories were carried out with the assistance of Berber native speakers who helped with the translation and enquiry process during the months of June to September of 2004 and 2007. Questions were asked on topics such as the agro-pastoral uses and production of the territory, the main vegetal and fruit production per year, the way sheep, goats and cows were managed, number of adult animals, their selling prices at the local market, ownership of manufactured goods and each household's dependence on remittances.

The Aït Ikiss Tagdalts

Most interviews were carried out with the male household heads living in the village of Warzazt. In the case of absence of the male household head, the interview was carried out with the oldest man of the household present at the moment of our arrival at the house. Male household heads were mainly targeted due to the nature of the data (i.e., choosing the opening date of the *agdal*, the number of animals owned by a family, the broad agro-pastoral production decisions or the most important religious rituals). The systematic open-ended interviews were carried out with 83 of the 107 households that compose the Aït Ikiss, which represents about 77 per cent of its population and almost 100 per cent of the owners of livestock using the *tagdalt* system. The remaining 26 households were not interviewed because there was no male adult present at the time of the survey, mostly due to seasonal migration.

4. RESULTS

Results have a threefold focus: (i) the functioning of the Aït Ikiss agro-pastoral system, by mean of a spatial analysis; (ii) the contribution of the main Aït Ikiss pastoral territory to the nutrition of livestock; and (iii) the socio-cultural aspects of the system.

Space analysis of the tagdalt system of the Aït Ikiss

This area with heavy winter snow and harsh climate suffers from a fodder shortage for the herds during the winter and another at the end of summer in dry years. It is thus a challenge for these agro-pastoralists to manage the use of the territory in time and space in order to meet the nutritional needs of their livestock, and to ensure soil fertility in the cultivated areas by means of animal manure (Genin et al., in press). With this in mind, agro-pastoralists regulate access to the resources by means of various *agdals*. All these *agdals*, as a whole, are locally named *tagdalts* (small *agdals*) by comparison with the large tribal *agdals* which are managed by several villages. The agro-pastoral calendar in turn reflects the choices of the different herding restrictive periods which dictate the rhythm of displacements of animals and people, in relation to the availability of different natural resources throughout the year.

28 September—28 March
On September 28th and after two months of herding prohibition, the *agdal* of Ikiss opens. Until this date the prohibition concerned not only herding but also gathering walnuts as well as some other fruits. Starting from this date, all spaces are opened and there are multiple movements. In any case, it is nut gathering at the end of September which dictates the main movement of peoples (principal

Pablo Dominguez et al.

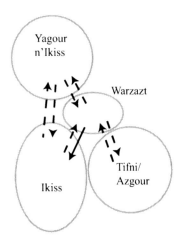

displacement is always indicated with a continuous arrow on our diagrams). At this time, most of the people leave for Ikiss because that is there where most of the walnut trees are located. The village of Ikiss is the location where most of the Aït Ikiss population spends the autumn and winter seasons. One of the reasons is the walnut gathering, but also because it is located at rather low altitude (1,700 m), and the weather is not as cold as at Warzazt (2,000 m) or Yagour n'Ikiss (2,200 m). Another other important reason for living in Ikiss is because it is the hometown of the entire group (hence their name, Aït Ikiss meaning 'those from Ikiss') from where, throughout the last century, some have moved to the other habitats. Thus Ikiss has more well-built houses and is more better-suited for spending the winter. Tifni is mainly an area for sheep-folds and Azgour, even if lower and with a better winter climate, has less agricultural lands and is an old sheep-fold transformed into a village only at the turn of the last century after the construction of the road at the bottom of the valley, and is less well equipped in infrastructure.

28 March—20 April (approximately)
On March 28th the whole of Yagour is put in *agdal* for three months (areas under the *agdal* prohibition are marked with discontinuous circles), in order to

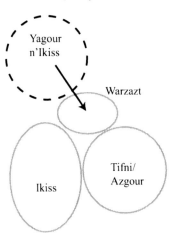

favour growth of graminaceous plants. In fact, the prohibition takes several weeks to become effective because there are always herders who stay a little longer and attempt local political manoeuvres to stay as long as they can. In practice, the prohibition is only imposed approximately from mid-April on every year. At this time, approximately fifty adult shepherds of the Aït Ikiss leave the Yagour. People who have been herding their sheep in the Yagour come down to Warzazt at this time. Caprines and bovines are kept all the year except in late spring and summer away from the colder regions, mostly in Ikiss and a very small minority in Azgour.

The Aït Ikiss Tagdalts

20 April (approximately)—20 May (approximately)

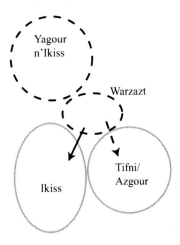

Warzazt is the second highest habitat of the Aït Ikiss, and the most extensively cultivated. According to the shepherds, the month of May is the most critical period for the growth of the pastoral plants at Warzazt. Thus, in mid-spring, the herding prohibition is also imposed on Warzazt, in order to allow the grass to grow back, particularly the strips between the cultivated fields. At this time of prohibition, the twenty families which have their main house in *Warzazt* (mainly sheep herders) are obliged to transfer their herds down to the Tifni sheep-folds at 1,900 m (on the way to Azgour), and especially to Ikiss at 1,700 m.

20 May (approximately)—10 July (approximately)

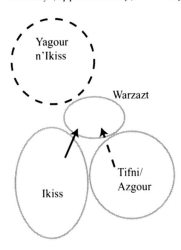

After one month of *agdal*, Warzazt opens again (around 20 May). This induces an opposite migration, of people going up from Ikiss or Tifni towards Warzazt, where about a month later (at the end of June) the barley harvest starts. At this period, virtually all the Aït Ikiss leave in transhumance towards the highlands. Also the two shops in Warzazt which had been closed since September reopen with the arrival of the population. In fact, the two shop owners (former local herders themselves) and the *fqih*/*imam*, move with the group from Ikiss to Warzazt. Just before, they sacrifice animals in the hope of a good stay in the high pastures.

10 July (approximately)—28 September

At the beginning of summer, the Yagour n'Ikiss is opened. Depending on the ecological conditions of the year, Yagour opens at around the beginning of July, according to the decision taken by the *jmaa*. The transhumant Aït Ikiss take their animals up the slopes leading to the richest summer pastures and prepare to enjoy the most abundant period for the community, which is seen by many children as a holiday period. The opening of the *agdal* of Yagour is an event of great local

Pablo Dominguez et al.

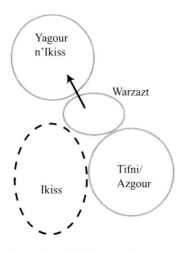

importance because it offers excellent grazing for the herds, and also an opportunity for people to meet the inhabitants of other lower villages after over 9 months of separation. Other ritual sacrifices of animals and festive events also take place at this time. In the areas most favourable for agriculture, the harvesting of cereals is carried out just a few days before the opening of the Yagour or in the following weeks. At the same time, the Aït Ikiss assembly authorises the collection of fodder for the winter season, where (only during the first three days after the opening of the Yagour) families are allowed to mow as much pasture as they can. A few weeks later, in mid-July, and following the same community logic, the Ikiss area is put in *agdal* too and herding remains prohibited once again. This prohibition particularly helps to protect fruit trees in the valley that are ripening at just this time, and also the lowland pastures that had previously been grazed, and that may be severely affected by the summer drought. Only every 15 days or so, the fruit tree *agdal* at Ikiss is lifted to prevent the rotting of certain fruits that had already matured on the trees before 28 September, when the whole *agdal* prohibition at Ikiss is officially lifted. In this way, the end of the cycle is reached, the whole system returns to the situation shown in the first diagram and most of the people come back to Ikiss for the gathering of the walnuts.

As we have seen, the dates and the access rules to the renewable resources of the Aït Ikiss are closely related to pastoral requirements, and also those of agriculture and to the system of production (availability of good pasture, mowing of fodder for the winter, fruit trees, cereals, horticulture, grass strips between the terraces, animal manure as agricultural fertiliser, etc.). We show below a calendar for the various *tagdalts* of the Aït Ikiss in relation with the agricultural calendar. This makes clearer the overlap between the *tagdalt* and the agricultural calendar.

The Aït Ikiss Tagdalts

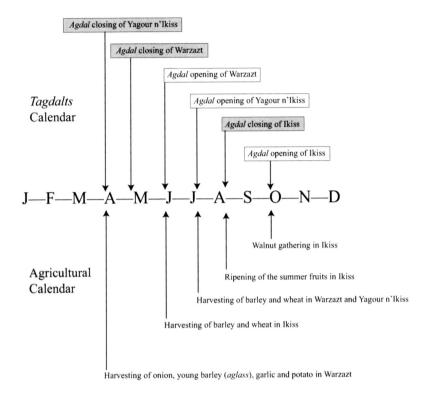

Agdal closing of Yagour n'Ikiss

Agdal closing of Warzazt

Agdal opening of Warzazt

Tagdalts
Calendar

Agdal opening of Yagour n'Ikiss

Agdal closing of Ikiss

Agdal opening of Ikiss

J—F—M—A—M—J—J—A—S—O—N—D

Walnut gathering in Ikiss

Agricultural
Calendar

Ripening of the summer fruits in Ikiss

Harvesting of barley and wheat in Warzazt and Yagour n'Ikiss

Harvesting of barley and wheat in Ikiss

Harvesting of onion, young barley (*aglass*), garlic and potato in Warzazt

Contribution of the Yagour territory to the Aït Ikiss livestock's nutritional requirements

We have attempted to assess the contribution of the Yagour n'Ikiss to the livestock's food requirements by quantifying the grazing days spent by different types of livestock within this territory. The reason why we chose to focus on the Yagour for this part of the analysis is because it is the most important herding territory for the group and therefore the most fully representative of their pastoral *tagdalt* system. We calculated the nutritional contribution of the Yagour n'Ikiss for the different types of livestock by counting the days passed on the Yagour by each type of animal, on the assumption that their grazing is not complemented with other feed resources during this period when the annual rainfall has been average. In fact, except in years of severe drought, it would appear that all animals obtain from the Yagour all their nutritional requirements (Demay, 2004).

Pablo Dominguez et al.

The number of head of livestock grazing on the Yagour n'Ikiss is very high (almost all the community's animals are there in summer) and corresponds to a little more than 3,000 animals including sheep (1,261), goats (1,794) and cows (138), corresponding to almost 3,600 Ovine Units (Jarrige, 1980). At the same time, we found that the food requirement of sheep appears to be strongly dependent upon the forage resources provided by Yagour since these represent 42 per cent of their annual food requirement. On average sheep spend 5 months per year feeding exclusively on the Yagour (5/12 months ≈ 42 per cent). This is an important contribution to the local economy since sheep represent the main cash income source. The price of a goat is less than half that of a sheep, and cows mainly contribute to the local economy only through consumption within the community, and a calf every two years. Contribution of the Yagour to the goats' annual diet is about 28 per cent, since they spend on average 3.3 months per year feeding on the *Yagour* (3.3/12 months ≈ 28 per cent), while cows obtain only 17 per cent of their annual food requirements from this territory due to their shorter stay in the highlands (on average 2 months per year). Overall, this analysis reveals the importance of one of the four pastoral lands of the Aït Ikiss, and explains quite well one of the main reasons why so much care and protection is devoted to the territory since it plays a crucial role with regard to feeding requirements at critical periods.

Cultural values associated with the agdal of Yagour

These spatial and quantitative analyses can only shed light on some aspects of the complexity of this management system; it is necessary to consider the symbolism of the *agdal* in order to achieve an in-depth understanding. The place of the saints in the tribal societies of Morocco and the North of Africa is well known (Montagne, 1930; Bel, 1938; Gellner, 1969; Mahdi, 1999). The shepherds traditionally put themselves and their herds under the protection of the local Muslim saints. The saint granted fertility and prosperity for cattle and the community. In compensation, shepherds honoured the saint's grave and those of his descendants with various gifts and sacrifices. For example, Gellner (1969) showed how, by the physical location of their mausoleums (at the borders between tribes), many saints and especially their descendants played a role of referee in conflicts, acted as guarantor for various types of agreement and assured safe passage to the high pastures at the end of the *agdal* prohibition each year.

The *tagdalt* system of the Aït Ikiss follows the same pattern, and is traditionally placed under the patronage of certain saints (Dominguez, 2010). We will mainly refer here to the Saint *Sidi Boujmaa* and his descendants, who had the great *agdal* of Yagour under their patronage, and still have a certain symbolic importance in this matter. This Saint is buried in a village of the same name. The village is located at one of the symbolic gateways to the Yagour, within the territory of the tribe of the Mesioua, and its importance derives from its traditional crucial relationship with the *agdal* of Yagour. In fact, local tradition asserts that all the different saints of the Yagour came to an agreement to give Sidi Boujmaa the key to the *agdal*. Therefore, no one could enter the *agdal* before his descendants announced it at the weekly market of the *Arbaa Tighdouine*. This was the case until about the 1970s, when unusually high demographic growth, economic transformation, possible climate change and cultural transformations, among other factors, forced a change. The decline of traditional religious observances triggered at that time is due to strong social pressure that resulted in their being seen as archaic and out of phase with true Islam. This perception may be related to the introduction of compulsory state schooling, generational cultural change, the rise of a more orthodox form of Islam that is intolerant of the veneration of saints, television, the market economy and other agents of cultural globalisation such as NGOs and local government. Nevertheless, the saints are still seen today by most of the local population as the symbolic mainstay and guarantor of the order that prevails in the *agdal* and the organisation of the transhumance. Access to the *agdal*, with its dates of opening and closing, is thus closely related to the person of the Saint, who is portrayed as a sacrosanct character who confers dignity and authority on the rules of *agdal*. This same tradition specifies that it is the *zaouia* (the brotherhood) of Sidi Boujmaa which keeps the papers certifying the rules of access to the *agdal*. This same institutional and symbolical organisation appears often with other saints, when moving to or from other areas before and after the various *agdal* prohibitions. As indicated above, rituals such as *maarouf* (animal sacrifices in honour of the saints), are performed throughout the pastoral cycle and in particular at key moments such as the opening of various *agdals*.

After the 'prohibition period' of the *agdal* consecrated by the saints, in which fear and malediction are imposed on the *agdal* area, with the arrival of the *agdal* opening time, there is a new sacred 'permissive period'. The opening of the *agdal* and the stay of the shepherds in the mountain pasturelands is accompanied by important festivities and recreation and high ritual activity. Also at the opening of new areas for agriculture or the inauguration of water channelling, and especially when the shepherds return to their summer farms with the best pastures, major or minor rites are celebrated to seek reconciliation with the genie that had occupied these areas during their absence. For example, it is reported that some herders slaughter small ruminants when they arrive at the *agdal* in honour of Allah and the saints. Others cook couscous and divide it collectively. Others offer pieces of entrails to the toads symbolising the devils of the places. All of this is done in the hope of peaceful resettling in the pastoral sites. Despite an evident decline in of these practices, the Aït Ikiss population continues to honour its saints through gifts offered near the graves or mausoleums in the form of grains, blood sacrifices and so on. In compensation, from the

Pablo Dominguez et al.

Saint they hope to obtain prosperity, fertility of their cattle and the protection of the group and its territory.

Concerning the sacredness of the territory, like the sacred legislation that governs the *tagdalt* system, the area managed by the *agdal* is also sanctified. This sacred status and *horm* (prohibition) generally decreed by the saints is reinforced by a sacred topography of the *agdal* managed territories. Popular narratives reinforce its sacred character, and refer to other magic or religious elements, in the hope of offering proof and attempting to intimidate the unbelievers. There are sayings such as: 'a man dressed in white on his white horse appears every year when the *agdal* prohibition is introduced at the Yagour, to safeguard the pastures from the dishonest by punishing the offenders in different ways'. Or 'the 365 saints turning around the Yagour to ensure the supervision of the *agdal* with their horses and their camels'. Some of the different saints located within the Yagour, as many as the days of the year, have tombs more or less clearly marked within and in the vicinity of the Yagour, but others do not have any place of sepulchre, just natural features or piles of stones that serve as a reference. These Saints are supposed to be responsible for overseeing the good functioning and the respect of the *tagdalts* of the Aït Ikiss.

In the Aït Ikiss territories, many areas are subject to strong local religious beliefs, which are not without ecological consequences. For example, some sacred *agdals* located near the mausoleums of the saints or the cemeteries of the different villages, are provided with abundant vegetation, and are permanently protected from consumption by livestock and from harvesting, by the sole power of the beliefs and social prohibitions that protect these sacred places. They constitute useful seed banks and plant dispersion areas. Other elements also are also involved in this sacred geography of the territory. For example, there are more or less sacred or magic water springs, which guarantee health and the *Baraka* (benediction). Or spaces populated by genies, associated most often with humid areas or areas far away human habitats. These genies are not necessarily described as malevolent beings, but they inspire strong fear, and they are especially present during the various prohibitions of the *tagdalts*.

5. DISCUSSION

Being universal features of the *agdal*, we have emphasised the three aspects that comprise the added value of the *agdal* management system: (i) conservation of the bio-physical environment; (ii) the performance of the current local economy; and (iii) the maintenance of current social cohesion and cultural coherence. The conservation of the bio-physical environment is favoured by non-intensive exploitation of rangelands that is characterised by a relatively wide dispersal of cattle in the various grazing areas throughout the year, as we have seen. The only exception is at the opening of the great *agdal* of Yagour, in the first weeks of summer, when most of the sheep, cows and goats are concentrated within the same area of about 40 hectares with a particularly rich sward. In any case, the intensive herding at this time is well supported by the vegetation that is at its most abundant and sumptuous, with grasses reaching almost one metre in height and a density that completely conceals the ground. The generally extensive approach to ecosystem management culminates precisely in the *tagdalt* system which allows to each type of area an annual resting period. This herding prohibition makes it possible for the vegetation to accomplish its phenological cycle, allows the establishment of young seeds and thus favours the continuity of the ecosystems and provision of the associated ecosystem services (Aloui and Alifriqui, 2009). Hence, the vegetation cover and vegetal diversity found in the Aït Ikiss *tagdalts* are higher than those found in similar areas outside *agdals* in 'open access' or under degraded *agdal* management (Dominguez and Hammi, 2010). Because of this, the *tagdalts* also seem to be a helpful governance regime to combat soil erosion. Finally, the system of different *agdals*, closed and open at different times, results in an *ecological mosaic* effect throughout the whole territory of the Aït Ikiss due to a specialised and differentiated use of the four areas, which is considered as a favourable feature in landscape ecology and a characteristic of the historical 'design' of Mediterranean landscapes (Blondel, 2006).

With regard to the performance of the local economy, we found that the *agdal* management system favours the sustainability of the pastoral economy since it ensures the continuity of the current ecosystems, as explained above. In fact, the current system of ecological and economic equilibrium would simply collapse and change in character if the *tagdalts* disappeared, since without a resting period the different areas would lose part of their *carrying capacity*. But the economic benefits derived from the *agdal* are not only constrained by the ecological sustainability of the system that permits the continuity of the economic system. For instance, at a higher organisational level, all Aït Ikiss benefit from the *agdal* system that guarantees access of rangelands to the different shepherds, and prevents conflicts through the regulation of competition between the different users. Zootechnical and ecological arguments are of limited importance when the social organisational character of the *agdal* is taken into account, and this is a key part of the economic system, if not one of the most fundamental.

We also found that the main pastoral territory, the *Yagour*, can contribute up to 42 per cent, 27 per cent and 17 per cent of the diet of the sheep, goats and cows, respectively. In this context, it is important to highlight the economic weight in the agro-pastoral gross income that derives from this especially rich pastoral territory, due to its importance for animal nutrition. Less relevant in terms of forage production but similar in their pattern of land exploitation, the other sectors of the Aït Ikiss together provide forage and income to the agro-pastoral system in a similar way. Such an important fodder contribution of the

Pablo Dominguez et al.

four *agdal* managed areas is especially significant as it must be noted that the agro-pastoral economy represents about 70 per cent of the total economy of the community (Dominguez, 2010) and thus greatly determines the current landscape and natural resource management system in the region.

Moreover, the fodder contribution accumulated after the herding prohibitions in all the territories subject to the *agdal* arrives in late spring, summer or at the end of summer, when the other pastures are dry, and when fodder demands are higher as young animals are still suckling or mothers are already pregnant with the second annual offspring. This gives an added economic value to the fodder that has been accumulated during the different *agdal* prohibition periods. In any case, monetary accounting is difficult (if not impossible) with such a multifaceted system as the *agdal*, which involves diverse ecological, economic and cultural aspects, in a society that capitalises weakly on its own production. And even if we had wished to focus solely on the monetary aspects of the system, the findings would have presented an underestimation of the *agdal* managed territories' contribution to the Aït Ikiss' overall household economy. In fact, the actual figures would have been higher than those found in this article (i.e. higher feeding rates in summer than in winter, production of milk, wool, manure, honey, and water, are not easily taken into account). However, the fact that the real contribution of the pastoral territories would be higher than that derived just from the nutritional percentages that we have calculated in Section 4 only reinforces the fact that the *agdal* system is a fundamental element of the performance of the current local economy.

Nevertheless, as we have seen, it would be highly reductionist to stop the analysis at this point, since the *tagdalt* system has other functions that are less directly materialistic or quantitative. It involves other social and symbolic functions that are not quantifiable. First of all, it allows an equitable right of access to pasture for all the community and guarantees a certain degree of social justice among herders. Even if it is always the richest cattle owners that benefit most from the communal pastoral resources, they will never be allowed to herd during the critical prohibition periods. While the majority still maintains the rule of the *agdal*, all (rich and poor) will be equally capable of meeting their respective fodder needs at the opening of the different *tagdalts*. At the same time, the *tagdalts* also play a role in identity building and the feeling of belonging to a certain group that confers rights and duties over those territories. On a more cultural level, the *agdal* is also one of the main vehicles of a complex belief system that we have briefly described and that gives protection and meaning to the local social organisation and livelihoods. As we have seen, transhumance towards the *agdal*-managed territories is the occasion of an intensive deployment of social and cultural activity, accompanied by an atmosphere of religious celebrations. All of this takes place in a highly multidirectional and interconnected way, and contributes to consolidating the collective discipline that ensures the success of this management model. This is a type of social organisation where the sacred holds an important place and gives meaning to material activities. The *tagdalts* for their part entail meanings and rites that periodically bring people together, contribute effectively to social cohesion and guarantee the continuity of the cultural order.

6. CONCLUSION

This study has examined the multiple dimensions of the *tagdalt* system of the Aït Ikiss with the aim of achieving a better understanding of an ingenious agro-pastoral system which is intimately linked to a complex worldview. This longstanding endogenous resource management system appears to fulfil almost all the accepted institutional principles that are considered to be important for sustainable resource management (Ostrom, 1990; Armitage, 2005):

1. A clear definition of right owners allowed to exploit resources.
2. An identification of the types of resources involved, and well-defined boundaries between the areas differentially managed.
3. Participation of all local stakeholders in defining rules, by the mean of local assemblies.
4. A graduated scale of sanctions for contraveners.
5. Mechanisms for conflict resolution.
6. Existing self-monitoring systems whereby resource users are accountable for their own actions.

This system, as a result of a long history of Environment–Society interactions illustrates concretely the indigenous knowledge and the traditional ecological knowledge framework developed by Blaikie et al. (1997), and Peloquin and Berkes (2009), among others, as well as giving weight to the quadriptyc proposed by Berkes (2008) of knowledge-practices-institution-worldview for undertaking complexity found within traditional natural resource management systems. The various sets of practices, rituals and institutional arrangements have to be undertaken all together in order to assess their capacity to value scarce natural resources and to promote forms of adaptation to changing conditions or needs (Berkes et al., 2003, Olsson and Folke, 2001).

As we have seen, the *agdal* can be easily included within the Maussian 'total social fact' framework (Mauss, 1968). The highland territories are of great multidimensional importance for the Berber agro-pastoralists of the High Atlas, and they are to a large extent managed by various *agdal* systems. The joint symbolic weight of the local sacredness that accompanies the *agdal*, as we have seen above, associated with the economic benefits of these practices, appear to be

Pablo Dominguez et al.

fundamental for the sustainability of the highland pastures, and vice-versa. The development of an active economy, the coordination of the movement of a large number of herds in a reasonably just and peaceful way, and the maintenance of a cultural system seem to have co-evolved all together.

If we look at many other areas in Morocco, it is apparent that there have been many *agdals* that have broken down over the last 30 years, for example in the Middle Atlas. The explanations for this are multiple but are usually associated with a weakening of the vitality of customary practices. This has in turn led to such over-grazing, that the systems of animal rearing have been totally altered and in some extreme cases have also led to new restrictions for the herders. The animal production area has usually not decreased, but has changed its nature: construction of higher animal folds, modern fodder production or purchase, mechanisation, new transport, full integration into market economy, etc. (Bourbouze, 1999). Even if such a transformation could be a viable alternative for maintaining or improving the levels of agro-pastoral productivity, it would not be consistent with collective socio-cultural coherence and ecological sustainability that the traditional *agdal* systems seem to offer. In fact, such transformations and intensification of the ways of production in the mountains of the Maghreb generally lead to biodiversity loss, erosion of pastoral ecosystems, rupture of fundamental elements of the traditional system of social justice and loss of many patrimonial or cultural values (Auclair et al., 2007). Hence, traditional resource and land management systems can be of great interest for a renewed approach to development (Mosse, 2005). What we may perhaps conclude from this study, is that a hybrid quantitative and qualitative approach is necessary to assess the full dimensions of the *agdal* systems. There are different possible lines of research. First, to analyse accurately and objectively how the reinforcement of the saints' and *agdal's* symbolic traditions could contribute effectively to the maintenance of this pastoral management system that has proved its sustainability over the centuries. On a general note, we think that the advantages of a traditional cosmological and agro-pastoral production system that has provided sufficient, sustainable and successive levels of productive yields over time should be emphasised, in a context where this Berber religious tradition is continuously under pressure, as explained above.

Secondly, from a more materialistic point of view, it could be interesting to seek more profitable time-space scale readjustment of *agdals* in order to improve their natural resource production, since the scale has not always been chosen through an integrated eco-territorial analysis, but frequently on the basis of centuries old ethno-territorial and historical traditions that now could be revised in many cases. A third future line of research could be to explore the possibilities of the *agdal* in new approaches to sustainable agricultural development. Although these points should be explored from a different perspective to that described in this article, we think that there is a potential for establishing a policy of *local agdal*

The Aït Ikiss Tagdalts

product certification by directing the production of the *agdals* towards products adapted to the local ethno-ecological conditions (local ways of production) and the external market (red meat, cheese or honey). This type of product certification could be based on the specific territory-resource-culture characteristics of the *agdals*. It is also possible to explore other formulas of using the *agdals* for the improvement of the agricultural productivity of these Berber agro-pastoral systems. Finally, one further line of research that seems plausible to us would be that the *agdal* system could help guarantee the conservation of the landscape and the current image of the territory, a major source of attraction for the growing tourism industry. Thus, after showing the services rendered by the *agdal* as a qualified and especially legitimate local institution in natural resource management, it could perhaps now be of interest to focus on the services it could offer with regard to the protection of certain touristic resources too. The high pastoral *agdals* play a role in sculpting and maintaining the local landscapes with images of abundant grass, open spaces, the solitude of uninhabited mountains and efficient agricultural production by the fact of being prohibited for year-round habitation (Dominguez, 2007). In other words, the *agdal* system is a uniquely Moroccan form of landscape shaping, and it would therefore be interesting to study in greater depth whether the *agdal* communitarian resource management system could be promoted as a profoundly Moroccan approach to the farming of territories, that is economically sustainable and ecologically enriching, as well as relatively just and cosmologically coherent.

ACKNOWLEDGEMENTS

Research was funded by Programme AGDAL (« *biodiversité et gestion communautaire de l'accès aux ressources sylvopastorales* » / Institut Français de la Biodiversité – Institut de Recherche pour le Développement: financement n° 2886), the 'Formation à la recherche' scholarship from the Agence Universitaire de la Francophonie, the 'Field Work' scholarship from the UNESCO Fellowship Program and the 'BECAS MAE-AECID' from the Spanish Agency of International Cooperation and Development. The authors also thank the group AHCISP from the Universitat Autònoma de Barcelona who gave constant support to our work, the organisers of the Ethno-ecology workshop at the IXth biannual congress of the International Society for Ecological Economy (New Dehli, India), Ricardo Godoy and Victoria Reyes-García, at which a previous version of this paper was presented. We also wish to thank Mjid Mourad, Taoufik El-Khalili and Simohamed Aït Bella for their excellent work as translators and interviewers, all the members of the local NGO *Association des amis du Zat* and especially its president Ahmed Bellaoui, for having granted us so many facilities, both human and infrastructural, to enable us to approach the local communities. We finally thank very warmly all our informants who where so patient and cooperative.

Pablo Dominguez et al.

BIBLIOGRAPHY

Alaoui-Haroni, S. 2009. *Les pelouses humides dans le haut Atlas : Biodiversité végétale, dynamique spatiale et pratiques de gestion coutumière.* Ph.D Dissertation, Univ. Cadi Ayyad, Laboratoire d'Écologie végétale, Marrakech.

Alaoui, S. and M. Alifriqui. 2009. 'Recent dynamics of the wet pastures at Oukaimeden plateau (High Atlas Mountains, Morocco)'. *Biodiversity and Conservation* **18**(1): 167–189.

Armitage, D. 2005. 'Adaptative capacity and community-based natural resource management'. *Environmental Management* **35**(6): 703–715.

Auclair, L., A. Bourbouze, P. Dominguez and D. Genin. 2007. *Les agdals du Haut Atlas. Biodiversité et gestion communautaire des ressources forestières et pastorales.* (CD-ROM document) Final report of the AGDAL program. Institut de Recherche pour le Dévelopement / Institut Français de la Biodiversité, Marseille, 196 p.

Bel, A. 1938. *La religion musulmane en berbérie. Esquisse d'histoire & de sociologie religieuse.* Paris : Ed. Librairie Orientaliste Geuthner.

Bellaoui, A. 1989. *Les pays de l'Adrar-n-Dern. Etude géographique du Haut Atlas de Marrakech.* Ph.D dissertation, Université de Tours, Département de Géographie, Tours, 500 p.

Berkes, F. 2008. *Sacred Ecology. Traditional Ecological Knowledge and Management Systems,* 2nd edn. London: Routledge.

Berkes, F., J. Colding and C. Folke. 2003. *Navigating Social-Ecological Systems: Building Resilience for Complexity and Change.* Cambridge: Cambridge University Press.

Blaikie P., K. Brown and M. Stocking. 1997. 'Knowledge in action: Local Knowledge as a development resource and barriers to its incorporation in natural resource research and development'. *Agricultural Systems* **55**: 217–237.

Blondel J., 2006. 'The "design" of Mediterranean landscapes: a millennial story of Human and ecological systems during the Historic period'. *Human Ecology* **34**: 713–730.

Bourbouze A., 1997. 'Des agdal et des mouflons'. *Courrier de l'environnement* **30**: 1–13.

Bourbouze, A. 1999. 'Gestion de la mobilité et résistance des organisations pastorales des éleveurs du Haut Atlas marocain face aux transformations du contexte pastoral maghrébin', in M. Niamir-Fuller (coord.), *Managing mobility in african rangeland: the legitimization of transhumance* (London: IT publications: FAO and Beijer intern. instit. Ecolog. Economics), pp. 236–265.

Cordier J.B., and D. Genin. 2008. 'Pratiques paysannes d'exploitation des arbres et paysages forestiers du Haut Atlas marocain'. *Revue Forestière Française* **60**(5): 571–588.

Demay, S. 2004. *Diagnostic agraire dans le Haut Atlas marocain. Territoire des Ait Ikiss.* Master dissertation, INA, Paris-Grignon.

Dominguez, P. 2007. 'Transformación de instituciones religiosas tradicionales en el Alto Atlas de Marrakech (Marruecos) y su impacto en los ecosistemas sub-alpinos. Caso del sistema pastoral del agdal'. *Perifèria* **7**: 26, http://antropologia.uab.es/Periferia/Articles/ecologia_yagur_pdominguez.pdf.

Dominguez, P. and S. Hammi. 2010. 'L'agdal du Yagour, écologie et pastoralisme', in K. Fernández (coord.), *Proceedings of the conference Ecología y Pastoralismo.* Donostia: Ed. Koldo Mitxelena, **2**: 34–56.

Dominguez P. 2010. 'Approche multidisciplinaire d'un système traditionnel de gestion des ressources naturelles communautaires: L'agdal pastoral du Yagour (Haut Atlas marocain', PhD dissertation, École des Hautes Études en Sciences Sociales / Université Autonome de Madrid.

Dresch, J. 1939. 'Caractères généraux de la vie pastorale dans le massif du grand Atlas'. Proc. 4ème congrès de la Fédération des soc. Savantes de l'Afrique du Nord, t. II. Alger, 493–497.

Evans-Pritchard, E. 1970. African political systems International. London: Oxford University Press, 302 p.

The Aït Ikiss Tagdalts

Folke, C., L. Pritchard, F. Berkes, J. Colding and U. Svedin. 2007. 'The problem of fit between ecosystems and institutions: Ten years later'. *Ecology and Society* **12**(1): 30, http://www.ecologyandsociety.org/vol12/iss1/art30/

Gellner, E. 1969. *Saints of the Atlas*. London: Weidenfeld and Nicholson.

Genin D., Kerautret L., Hammi S., and Alifriqui M. In press. 'Biodiversité et pratiques d'agdal: un élément de l'environnement à l'épreuve de ses fonctions d'utilité pour les sociétés rurales du Haut Atlas'. In L. Auclair and M. Alifriqui (eds.), *Agdal. Patrimoine socio-écologique de l'Atlas marocain*. Rabat, Morocco: IRCAM Editions.

Hammi, S., M. Al Ifriqui, V. Simonneaux and L. Auclair. 2007. 'Évolution des recouvrements forestiers et de l'occupation des sols entre 1964 et 2002 dans la haute vallée des Ait Bouguemez (Haut Atlas Central, Maroc)'. *Sécheresse* **18**(4): 271–277.

Hammoudi, A. 1988. *La Victime et ses masques : essai sur le sacrifice et la mascarade au Maghreb*. Paris: Ed. du Seuil.

Ilahiane, H. 1999. 'The Berber *agdal* Institution: indigenous range management in the Atlas Mountains'. *Ethnology* **38**(1): 21–45.

Jarrige R. (dir.) 1980. *Alimentation des ruminants*. Paris: Ed. INRA.

Kerautret, L. 2005. *Entre Agdal et Moucharika*. Master dissertation, Université de Provence, Laboratoire Population-Environnement-Développpment, Marseille.

Mahdi, M. 1999. *Pasteur de l'Atlas: Production Pastorale, Droit et Rituel*. Casablanca: Fondation Conrad Adenauer.

Mahdi, M. and P. Dominguez. 2009. *Regard anthropologique sur transhumance et modernité au Maroc*. AGER, no. 8: 45–73.

Mauss, M., 1968. *La fonction sociale du sacré*. Paris: Ed. de Minuit.

Montagne, R. 1930. *Les berbères et le Makhzen dans le sud du Maroc*. Paris: Ed. Félix Alcan.

Mosse, D., 2005. *Cultivating Development. An Ethnography of Aid Policy and Practice*. London: Pluto Press.

Neves-Garca, K. 2004. 'Revisiting the tragedy of the commons: Ecological dilemmas of whale watching in the Azores'. *Human Organization* **63**: 289–300.

Olsson, P., and C. Folke. 2001. 'Local ecological knowledge and institutional dynamics for ecosystem management: a study of Lake Racken watershed, Sweden'. *Ecosystems* **4**: 85–104.

Ostrom, E. 1990. *Governing the Commons. The Evolution of Institutions for Collective Action*. Cambridge: Cambridge University Press.

Peloquin C., and F. Berkes. 2009. 'Local knowledge, subsistence harvests, and social-ecological complexity in James Bay'. *Human Ecology* **37**: 533–545.

Rachik, Hassan. 1992. *Le sultan des autres, rituel et politique dans le Haut Atlas*. Casablanca: Ed. Afrique Orient.

Romagny, B., L. Auclai and A. Elgueroua. 2008. 'La gestion des ressources naturelles dans la vallée des Aït Bouguemez (Haut Atlas): la montagne marocaine à la recherche d'innovations institutionnelles'. *Mondes en développement* **63**(1): 63–80.

Simenel R. 2008. ' L'origine est aux frontières : espace, histoire et société dans une terre d'exil du Sud Marocain', PhD dissertation, Université de Paris X.

Venema B., 2002. 'The vitality of local political institutions in the Middle Atlas, Morocco'. *Ethnology* **41**: 103–117.

Alpine Milk:
Dairy Farming as a Pre-modern Strategy of Land Use

Barbara Orland

1. INTRODUCTION: ALPINE ENVIRONMENT AND AGRICULTURE

The link between the landscape and nutrition has ceased to exist in the mindset of modern industrialised societies. There are only a few products left, such as selected wines, where geographical origin plays a role. Most foodstuffs have seen a drastic reduction in the amount of land necessary for their production. The separation of animal and plant production has fostered animal production, independent of the area, and even some vegetables and herbs can now grow on an artificial culture medium. Regionally and locally variable crop farming has been replaced by agribusiness which is characterised by a highly advanced division of labour and which involves the production, processing, transportation, storage and marketing of foodstuffs.

Today agriculture can orientate its production much more to the demand efforts than to landscapes and soil. This remove from the soil enabled agricultural specialisation and the concentration on single products that are further processed and marketed in very different ways. The markets are largely open, and thanks to conservation technologies and transportation systems even highly sensitive products like fresh milk can be transported hundreds and more miles. As a consequence a regional, site-adjusted agricultural promotion policy has become rather nonsensical.

In stark contrast to this, most European Union members have their own regional agricultural promotion programmes. Until the end of the 1970s, they were justified mainly for social and economic policy reasons. In the past two decades, however, environmental policy has increasingly taken their place as the main justification.[1] Due to overproduction, product-based subsidies are being continuously reduced and incentives are being provided so that farmers will give up, or at least reduce, agricultural production (e.g. through set-aside programmes). At particular locations, however, farm-based direct payments are increasingly being granted.

The Alpine mountain regions are one of those locations where farming is highly subsidised in order to protect the flora and fauna and to preserve regionally typical landscapes. At first due to the purposes of the tourism industry, care of man-made landscapes has become an important buzzword and the farmers are discovered as indispensable gardeners. Even as early as 1975 a journalist came up with a perfect description of the problem, saying: 'First the cows went, and then the tourists – who's left to milk!'[2] As for the manufacture of food, mountain farming is hopelessly unprofitable. The massive input of labour by the mountain farmer, who is at a disadvantage in using mechanical implements, is not in the least bit profitable from an agronomic perspective. As 'environmental producers'[3] and standard-bearers of the man-made landscape, though, farmers are sorely missed in the Alps these days.[4]

So far, the influence of agriculture on the environment has mostly been described in terms of over-use as results of technological change. Applications of chemical fertilisers, pesticides, fungicides or herbicides, concentration of animals and monoculture, these and other technologies of industrialised farming raised broad public concern.[5] In the Alps, however, geographers were among the first to point out that ecological problems are the result of not only technologically determined over-use but instead of the under-use of nature. According to the contents of several Swiss reports to the UNESCO programme *Men and Biosphere*[6], weed infestation, abandoned roads and crumbling buildings had caused the landscape to be structurally simplified and impoverished. Places for flora and fauna to live, which were only created by agrarian land use in the first place, disappeared with farmers. Swiss social scientist and economic geographer Paul Messerli therefore concluded that the best method of maintaining the land and protecting vegetation was through continuous, site-adapted agricultural land use.[7]

Such well-meaning suggestions, however, cannot deny the fact that farmers never shaped their landscape according to aesthetic or environmental protection considerations. The Alpine landscape actually developed as a 'measure of what was possible' (Paul Messerli), which earlier generations developed in order to meet their living needs. From an historical perspective, this means that Alpine farmers and their animals not only affected their environment by their first appearance, but also continued to influence the perceptions of each new generation of people. As Dale Porter put it in his study on the *Embankment of the Thames*, the purposes for which agricultural field patterns originally were designed, and the contexts in which they functioned, can change or even disappear: 'As objects, however, they tend to persist more or less in their original form'.[8]

From this perspective, environmental history has to take into account that landscape is both cause and result of agricultural development, and it has to distinguish between the land created by agricultural work and the popular imagery of an area that tends to freeze the existing landscape as natural scenery. This distinction also has consequences for the assessment of technology. It is not always technology that changes the environmental attitudes and practices of land use. Rather than the simple cause of environmental problems, technology itself can be the outcome of environmental conditions. In the words of Dale Porter, production methods and technology are much more mediators 'between cultural values, social groups, and institutions on the one hand, and the natural environment on the other.'[9]

The same can be said for the products of agricultural work. Pre-modern dairy products should not primarily be perceived as the aim and outcome of agricultural production; the meaning of milk goes far beyond the idea that farmers support a region's or nation's demand for food. Rather the story of Alpine milk gives an impression of the ways food production was developed as a means of knowing and struggling with the natural environment. Whereas today the locality of products may be seen as externally imposed imagery, celebrating the virtues of a landscape, pre-industrial butter and cheese making reflected much more the connection between local land resources, farmer's skills, tools and practices. Richard White's plea to take man's shaping of nature through work into greater account in environmental history for me seems to be evident.[10]

2. METHODOLOGICAL ASPECTS

In this sense, I shall introduce the history of the Alpine dairy farming as a case study which illustrates the changing relations between natural resources, economic interests and cultural values that form the production methods people choose. Although I discuss the interplay between land use and agricultural production methods, this paper is not intended to be a landscape history and, less than that, a piece of historical geography. My main focus is the history of milk production as an interconnection of regional living conditions, skills and practices. Drawing upon literature and source material mainly from Switzerland, and occasionally from South Germany and Austria[11], the first part of this paper will describe in an ideal-typical model how dairying was integrated in mountainous farming practices. By looking explicitly at the Swiss case, the article, in a next step, shows that already in the sixteenth to eighteenth centuries the mountain dairy system was not only part of a peasant's subsistence economy but also served the requirements of a nationwide and even international butter and cheese trading network.

Spatial perceptions of agriculture and site-specific production methods adapted to the local environment were neither static nor inevitable. Although dairy was an accepted mountain farming system, the character of the landscape did not inherently force people to engage in this type of farming. Even in the Alps the variety of what we call Alpine mountain farming was so great that it is nearly impossible to present it in brief form for the whole Alpine region.[12] From this point of view, dairy farming can at best be described as a situated knowledge and activity, a flexible production method within limits set by the environment: a result of labour division between different geographical areas. As such, it was an integral part of Alpine farming systems and was influenced by community rules that regulated access to, and use of, mountainous meadows.

However, the connection of place and production should not be seen as a self-evident environmental protection practice. If some environmentalists argue 'that work on the land creates a connection to place that will protect nature itself'[13], such a view cannot be confirmed by this paper. Over the centuries, the notion of over-use and under-use of disposable land resources was constantly mentioned, and methods and regulations to balance the use of Alpine pastures can be read as strategies of conflict management. Furthermore, the commercialisation of pre-industrial dairy products was a source of conflicts that functioned between the limits of milk production and different economic interests on the market. As will be shown, some of today's 'traditional' dairy products shape these interrelations between nature, technology and the market.

In order to understand the character of milk production as a site-specific method, the following remarks, in the main, will concentrate on the period between the sixteenth and eighteenth centuries. The final third of the paper, however, gives a brief summary of the nineteenth century and the developments that precipitated the crisis of the Alpine dairy farming system that finally led to the notion of under-use. Comparing former conflicts surrounding the management of shared resources and today's problems with the under-use of the mountain regions, we can demonstrate differences in the evaluation of the landscape: until the turn of the nineteenth century, agriculture, as a site-specific production system, was hardly called into question. If this changed in the nineteenth century, it was because of the transformation of a rural perception of the landscape to one of nationwide agricultural territories with open markets. As will be argued in this paper, one of the most significant indicators of change that shaped modern milk production was the idea of dairy as an area-independent branch of agricultural production. Across Europe and without any regard for soil conditions or vegetation zones, agricultural regions which had been used for grain production were being converted to intensive dairy farming, and to a new system of milk processing and trade, no longer determined by regional farming practices. Thus, dairy farming acquired an entirely new goal: whereas traditional dairying had served agricultural strategies of land use, the modern milk industry yoked itself to the service of the market.

Most histories of dairying date important stages and activities associated with the industrialisation process of dairying back to the 1870s. Since that time, new breeding techniques and feeding regimes; the introduction of mechanical machines like the milk centrifuge; experiments with milking machines and pasteurisation technology; the fast introduction of fresh milk supply, followed by new 'factories' for diversified milk products, radically changed dairy manufacturing.[14] What is often neglected, however, is that the dynamics that caused the rise of an industrialised and highly engineered milk production were based upon the re-evaluation of agricultural strategies of land use.

To make this evident, the goal of the last part of the paper is to explore some of the challenges to old perceptions of the landscape that happened within the so-called 'organic phase'[15] around the turn of the nineteenth century. A closer look at these stages of agricultural modernisation before 1870 and the consequences of these developments for mountain dairy farming will be described, especially the fact that cheese production moved from the mountains to the valley, plunging mountain dairy farming, which had hitherto been extremely adaptable to changes in the economy, into a crisis. It will be shown how local agricultural reformers reacted: how they tried to overcome the crisis and stabilise the 'traditional' farming system under the conditions of an emerging international milk industry.

3. THE 'NATURAL' ENVIRONMENT OF DAIRY FARMING

Whereas narratives of an industrialised dairy farming system usually neglect landscape and soil but start with the cow as productive force, our ancestors judged dairy farming as a strategy of land use. Not only in the Alps but also in other regions of Europe the saying 'Dairy farming is land use'[16] was taken for granted. With this sentence people reflected the long-standing rule that animal production for human food has been linked to soil. To balance the equilibrium between grain production and animal husbandry was one of the fundamental rules of all pre-modern farming systems.[17] Thus, the production of extraordinary dairy products in great quantities was the expression of an adequate fodder supply. An old farmers' rule holds that 'cows give milk through their mouths!'.[18]

Unlike today, there existed no nationwide markets for animal feed. The available fodder was instead dependent on the way the land was being used, i.e., the relationship among all types of land management such as arable farming, permanent meadows, forests, pastures, orchards, vineyards, etc. In general, farmers did not use the arable crop in order to produce fodder, only periodically the fallow land supplied fodder for livestock. Today, specialisation and market access have resolved this issue. Milk can be produced on a year-round basis with a feeding system of silage, hay, and grain. In earlier times, farmers needed to adjust more to the variety of land areas which had developed through the symbiosis of rock formations, elevation, climate and vegetation. All mountainous areas were characterised as potential *dairy zones* due to their steep cliffs, rocky terrain and difficult climatic conditions. In addition, coastal wetlands, river meadows, highland and lowland moors, which could only be made arable through large-scale cultivation work, seemed like obvious locations for animal husbandry and dairy farming.[19]

Up to the nineteenth century, and despite the fact that people drew agricultural boundaries in different ways, it was a strong belief that most plants and animals had their natural habitat, outside of which they did not exist. In regions with deciduous forest, one agrarian reformer wrote in 1853, the farmer would be well advised to breed goats, rather than cattle or sheep; in oak and beech tree forests he should prefer pig breeding. Dry and spacious areas should be allocated to sheep, fertile gardens and orchards to bees, lush meadows to cows.[20]

Against this background, the idea of dairy zones as a synonym for spacious and lush meadows, good air and healthy herbage evolved in conjunction with rural experiences, farming practices and increasingly scientific expertise (in the Alps since the sixteenth century).[21] Over the centuries, it became a core concept of knowledge, more and more autonomous and self-evident, and hardly traceable back to its origins. In any case, the idea of characteristic dairy zones survived regional farming traditions and – as will be shown later – guided even the work of agricultural reformers during the industrialisation of agriculture and dairying.

In the Alps, the defining natural feature of dairy zones were the Alpine pastures, those mountainous areas at various elevations which, due to the high humidity and cold weather that induced a short vegetation period, only lend themselves to animal husbandry. These permanent Alpine pastures gave the mountain range its name. However, the German words Alm (Bavarian) and Alp (Alemannic), which predate the Roman era, also denote the cultivated areas in the mountain range, which, through clearing and grazing, constantly changed in size and form over centuries.[22]

The evaluation of these Alpine pastures was highly dependent on the differences of elevation in connection with climatic variations. An old farmers' proverb holds that the grass is always better the higher one goes, and at the top it is so good that even farmers might like to eat it.[23] In fact, with increasing elevation plant growth diminishes and with it the yields, but since the intensity of sunshine increases, Alpine plants process greater amounts of energy; their protein and fat contents are higher. Animals react in a similar manner. Because of the demands of Alpine living on their bodies, animals are slower to fatten than during the same length of time in the valley, and milk output falls by a considerable amount in the Alps. However, milk is creamier when manufactured at higher elevations. Still today, it is a scientifically proven fact that Alpine milk contains between 15 percent and 30 percent more fat than in the valley.[24]

Moreover, Alpine products were considered to be tastier and healthier because of herbs found only there, containing high percentages of ethereal oils.[25] Butter produced in the Alps in the summer could be valued by its yellow hue and improved spreadability – an advantage that was not unimportant in trade. In addition, the high percentage of ethereal oils was said to improve the salivation and digestion of cattle; and it was claimed that the mountainous climate was generally much healthier for cattle and gave them considerably greater resistance to disease.[26]

In addition, Alpine pastures should benefit from the cultivation. Grazing animals, with their dung, improved the natural soil quality and changed the plant cover by favouring certain plants while eschewing others. Old-growth grass, tall-herb communities and high-growing grass disappeared with animal husbandry and especially cattle farming. They were replaced by plants that were said to have a favourable effect on milk. All yellow-coloured plants were held responsible for the deep yellow of the butter. This is how the buttercup acquired

Barbara Orland

its name. And dandelions are still called *Milchkraut* (milkwort) in many Alpine villages. A lot of botanical names have their roots in these traditional folk beliefs.[27]

The local flavour of the high mountainous meadows showed up again in their products, which is why the cheeses were named after the Alpine pastures where they were produced: *Urnerbödeler* is cheese made in *Urnerboden*. *Albulataler*, *Entlebucher*, *Appenzeller*, *Saanen* or *Emmentaler* cheese: whole regions were identified with certain types of cheese, or, like in the *Domodossola* region, names such as *Val Formazza* showed their special relationship with dairy farming. As these descriptions attest, the links between landscape, animal and product represented a strain of thought with several components. The notion of dairy practices encompassed perceptions about the influence of the terrain on plants, animals and their products, or more generally, the ties between different parts of nature.

Consequently, it is not surprising that when the first naturalists and doctors directed their attention to the Alps, they too soon discovered these links. The products of the lowlands could not match the quality of Alpine products, said Hippolyt Guarinonius, doctor and humanist, in the sixteenth century.[28] In 1541 the famous doctor Konrad Gessner wrote his *Libellus de lacte et operibus lactaria*, describing the fertility of mountainous nature in the quality of its products.[29] To the educated classes in Europe the fame of mountain products was soon such a cliché that travellers were surprised not to see cows 'walking up to their bellies in grass'. Ludwig Wallrath Medicus, a German author who in 1795 wrote a report on Alpine farming, was certainly disappointed to find 'just short and rather low grass'.[30]

4. DAIRYING AS INTEGRAL PART OF MOUNTAIN AND VALLEY FARMING

In contrast to farming systems in other regions, the Alpine farmers faced specific problems resulting from the mountainous area. The diversity of landscapes was dominated by the combination of altitudinal belt, varying terrains and difficult climatic conditions. The first and most important effect of these environmental conditions has been restrictions to grain production. Swiss historian of economics Christian Pfister estimated for the canton Bern that in the eighteenth century valley regions could use three-quarters of the farmland for grain production; lower highlands less than 40 percent, upper highlands about 15 percent; and mountainous regions only 4 percent.[31]

Moreover, the environmental impact of land use in the Alps could change within very small areas; this becomes even clearer if we compare the basic conditions in only two Swiss cantons. While contemporaries in the canton Bern distinguished three agricultural zones – the so-called 'cornland' dominated by three-field rotation, the Emmental (a hybrid of permanent meadows and farmland for grain and potatoes production) and the 'upperland' (mountainous areas with grass – especially swampland in the valleys), in other parts of Switzerland, e.g. *Grisons* (located in the East), permanent grassland dominated the strategies of land use in so far as the canton, on average, was located at an higher altitude. Estimates from around 1900 placed available pasture resources at about 50 percent of all productive soil; at the same time in the canton *Valais* about 35 percent of all productive soil was defined as pasture.[32]

But even if the natural environment forced many farmers to concentrate on livestock production and animal husbandry, this did not result in the same kinds of livestock production. With regard to forage requirements and land resources, farmers, in general, can annually support only a restricted number of animals, whereby the daily plot allowance is variable with the sort of animals and their differing needs in nutrition, frequency of returning to pastures, plant growth and climate conditions. Like other land management systems concentrating on animal husbandry, Alpine farming required decisions to be taken regarding the composition of the herd, whether animals were to be fed through the winter or purchased in the spring or – with regard to cattle and dairy farming – whether more cows for milking or breeding purposes were desired.[33]

According to these preconditions, not only did the amount of small livestock and horses differ considerably within various cantons during the seventeenth and the eighteenth centuries, but, as Swiss historian Jon Mathieu pointed out, even the cattle farming based upon decisive distinctions highly influenced the possibilities of dairy farming.[34] Three specialisations can be distinguished: in regions in which mainly livestock farming (milk cows, bulls or draft oxen)[35] was conducted, to a great extent milk was needed to suckle young cattle, with little left over for butter and non-fatty cheeses. However, if the focus was on butter and particularly marketable fatty cheese, then, conversely, farmers decided against extensive cattle farming. A third and important specialisation was the production of fodder that was sold to cattle breeders, cattle traders and dairymen who tried to maintain their herds through the winter. As early as the late Middle Ages this kind of agricultural specialisation could be found in the core dairy farming areas in Switzerland.[36] It meant that farmers largely did not own cattle but instead rented out and leased farmstead meadows and lowland pastures or sold grain and hay fodder (particularly in the winter). In fact, the regional and local division of labour in livestock production was so well developed by the eighteenth century that there was an active exchange of grain for cattle and milk products between mountainous regions and lowlands.[37]

Whatever the relationship between different regions and various parts of the farm has been, it is impossible to describe the historical variety of farming systems that developed under these circumstances and that could change even within a small municipality. Thus, the following section constructs an ideal model of several characteristics of Alpine cattle farming as they can be synthesised from a huge amount of literature on this topic.[38] At first, all types of production were governed by the dictum that the use of the limited land resources had to be coordinated depending on the season. The unifying feature of mountain farming compared with lowland farming, which was largely concentrated on arable farming, was a graduated farm organisation. That meant that mountain farms consisted of permanent settlements, and farms in the valley or lower altitudinal belt and one or several settlements at various elevations which were used only part of the time.[39]

The mountain farming schedule envisaged homestead grazing in the spring, i.e., until mid-May or early June. Then the farm family took its cattle and migrated for approximately one month to the spring meadow, which was cut later in the year. In mid-June cattle were herded up to the highest Alpine meadow elevations. In the *Vorarlberg* region (Austria) cattle were supervised in community Alpine meadows by hired shepherds while the families returned to the spring meadow or the valley to keep producing hay.[40] From mid-September to early October the cattle were once again put out to pasture at the lower altitudinal belts before returning home to graze. The hay made in the early summer on the spring meadow was stored there and brought to the valley in the winter months according to requirements. In early November the family went back and only around mid- to-late December did the animals return to the stalls for the next five months.

The number of stages depended both on the mountains' elevations and how soon the cliffs jutted out once the forest, meadow and evergreen tree zone had been passed.[41] Another feature was that the Alpine meadows were divided by farming stages. The lowest and the most fertile Alpine meadows, which were most important in economic terms, were mostly hay- making and cow pastures. When milk cows grazed there, then butter and cheese were produced. Above them were the pastures for non-milk cattle production. The roughest, driest and steepest slopes were the exclusive domain of sheep.

Moreover, location dictated the degree of utilisation. With increasing altitude, vegetation started to grow later in the year, the first snow came earlier, and the number of grazing days went down. Some high Alpine pastures could only be used for a few days, and others up to six months. Thus, the division of summer and winter resources was a complicated business for which it was difficult to define general rules. The effective use of one and the same Alpine pasture, too, sometimes required that cows and heifers had to be moved from one area to another. Over centuries, for example, the cows of the village *Törbel* in the canton Valais were grazed on the *Moosalp* following a regular schedule, with either one or two weeks in five different sections and four weeks at the homebase. 'The cows were grazed in one spot after the morning milking, then taken higher up in hot weather to escape insects, returned to pasture in the late afternoon, milked again around 5 P.M., given further good grazing in the early evening, and turned into their enclosure at nightfall.'[42] In one case, owing to the steepness of the landscape, grazing was avoided from a certain elevation upwards, and in another case hay was only harvested at the lowest altitudinal belts.[43] On many Alpine pastures, even as early as the late Middle Ages rented or purchased cattle were brought to pasture and in the autumn either driven back to where they had come from, sold or slaughtered.

The type of Alpine use also determined what equipment needed to be taken from one grazing stage to the next along with the cattle. For a long time there were only barns, stalls and implements on cow pastures. A mountain hut, near a stream, was a sure sign of dairy farming that was conducted only on those meadows, which could be used for several months. Since the seventeenth century it became more usual to set up barns for shepherds as well as for Alpine animals other than cattle; above the tree line they were made of stone, and in the forests, of massive logs.[44] Their architecture hardly changed in the period thereafter. A living area for the Alpine herdsmen or dairymaids, a cattle shed for the animals and a feeding station were all the comfort that they afforded. The centre of the hut was occupied by an open fire, over which the large copper cauldron was fixed to a rotatable, wooden crane (a tree trunk with a branch jutting out at nearly a right angle). In the autumn it was taken down into the valley and in the spring it was carried back up. The rest of the equipment was rather modest; milk was processed using a set of pails, bowls, cloths, wooden stirring instruments, storage and transporting vessels, and a butter keg.

5. BALANCING OVER- AND UNDER-USE OF MOUNTAINOUS RE-SOURCES

Since the higher elevations were not suited to year-round settlement, the question of who owned or regulated access to Alpine meadows was decisive from an economic and an ecological standpoint. A cooperative solution was needed for questions like who could take what cattle to what Alpine meadow, who was supposed to participate in the vast work of taking care of the landscape, and to whom the buildings and the products of a typical Alpine summer (cheese, butter, dung) belonged.[45]

These issues became all the more pressing when the manorial system was abolished, population grew noticeably, and city markets flourished, all of which thrust mountain meadows since the sixteenth century into a position of outstanding economic importance. About the same time, the use of Alpine pastures for sheep and goats was replaced by cattle as the demand for both cows and cheese grew from the more densely populated plains. As a consequence, many Alpine regions in Switzerland became more integrated with economic development in north Italy and in the Swiss lowlands. No longer was clearing back forest and creating new glades in the mountains a risk-free proposition. Already in the sixteenth century, several northern Alpine areas were judged by contemporaries as dangerous frontiers, and many farmers were forced to convert their farmstead meadows in the valley to hay meadows.[46]

Barbara Orland

Whatever ownership forms managed to prevail in the period thereafter,[47] in most cases people strove to keep ownership and rights of usage separate. That afforded flexibility in the use of meadow resources and the products of an Alpine summer, yet at the same time required exact cooperation. Nearly everywhere only village or commune residents were eligible to join the cooperative. Citizens of other communities were barred from joining but could let their cattle graze on an Alpine meadow for a hefty fee. Even on private Alpine pastures, the only legal form where ownership and rights of usage were identical, foreign cattle were a frequent sight. If an Alpine pasture was in the community's collective ownership, this, in turn, did not automatically mean the milk was processed in cooperative fashion. On very large and highly productive community Alpine meadows, the facilities and implements were often privately owned and milk was independently processed. There were consequently many mixed forms between strict individual cheese dairies and cooperative operations with hired shepherds, farm labourers and dairymen. It often happened that the owners of the Alpine pastures herded their cattle jointly but sold the milk to a dairyman who bought it at a fixed price and processed it at his own risk. Or, the owners, farm families and village neighbours, exchanged milk, after the milk of the entire herd was, in turns, processed by every member of the Alpine cooperative on his/her own account.[48]

The most decisive reason for the separation of ownership and rights of usage was the problem of finding that fine line between the overgrazing and selective grazing of feeding grounds. Overgrazing, in particular, required from an early date resource-conserving measures that were regulated in a specific legal document – the Alpine ordinances.[49] This codified legislation, which appeared as early as the fifteenth century and more often along with the intensification of Alpine farming from the sixteenth centuries onward, sought to strike a balance between limited natural resources and growing economic interests, similar to agreements on the time of sowing and harvests of the fields, irrigation rights, forests or use of commons.

The Alpine ordinances came about due to specific disputes and changes in ownership and demarcation, but also often to increase harvests.[50] In one way or another the issue almost always boiled down to preventing over-use of the Alps. Cattle overpopulation on an Alpine pasture could cause ecological problems: their vegetation cover destroyed by too many animals, and unable to recover quickly. Conversely, if there were too few animals, they were selective in grazing, which meant that some fodder plants gradually disappeared and that the pasture turned to weed. In a long-term perspective, this, too, would decrease the available fodder and the biodiversity of the meadow.

Since a basic rule was that eligible herders could only drive their cattle to the meadow if they had kept them over the winter without purchasing additional feed, often cattle had to be rented if a pasture faced the threat of under-grazing. Apart from this, there were very specific rules to prevent the short-term overgrazing of the meadows. Maximum limits on use (maximum head of cattle, expressed as cow rights = the right to bring one cow to an Alpine pasture in the summer[51]) and strict use time periods (neither too early nor too late, etc.) were intended to prevent this problem.[52]

The proper number of animals, grazing in a specific time period, was the important factor. However, the same holds true for the beginning and end of the grazing period. Joint herding was the key. What has today been degraded to a folklore event, in fact served to ensure previously that none of the eligible herders was first to show up with his cattle at the fresh pasture. Compliance with the set numbers and times was strictly enforced.[53] To sum up, the Alpine pastures were not only used as the most productive economic zone but at the same time were also recognised as the most ecologically sensitive zone and used with special care. Through ownership and possession the use and care were provided. But also a certain amount of social control was organised to give the cultivated land additional stability. This included a great deal of maintenance and repair work of barns, paths and water resources, improving eroded areas, and clearing away debris from avalanches on Alpine pastures in spring. Such work also aimed to prevent obstruction (blockage and damming caused by trees, rocks, etc.). This maintenance and repair work always required a great deal of labour that had to be resolved as a collective task. Thus, not only a certain political framework was necessary, sometimes even feudal lords (the counts of Tyrol, for example) supported farmer's cooperatives because they had realised that short-term exploitation of their land would undermine their own sovereign authority as well as their power.

6. BUTTER AND CHEESE – PRODUCTS OF THE SUMMER MEADOW

Rules met practical needs in land use and were tailored to characteristics of the cooperative's environment. Careful control of access, fair division of maintenance and repair work and, particularly, the proper number and assortment of animals shaped the centuries old form of Alpine farming systems. Within this system of flexible grazing and range management, the production of butter and cheese also helped to strengthen the utilisation of mountainous areas. It enabled flexibility within the limits set by the natural environment, the strategies of land use and the social structure of the communities. Here, too, the need of every community's families for food and communal property existed side by side. In the next section I shall address in more detail the materiality of the dairy practices, to shed more light on the inherent logic of the work with milk under the conditions of a specific environment.

In principle, butter and cheese can be produced year-round, and, of course, one can assume that the Alpine people did so in pre-modern times. Nevertheless, butter and cheese were represented as products of the summer meadows and irrespective of the extent to which dairy farming was done in the summer on the Alpine pasture, there existed no commercial cheeseries (dairies) in the valleys before the beginning of the nineteenth century.

Since the end of the eighteenth century, several authors have argued that the output derived from the Alpine pastures was negligible compared with the quantity of milk produced on the meadows located on lower elevations, in the valleys or near by the farms. Estimates provided by different methods of calculation produced a range of about 30 percent for the amount of milk produced and processed on the Alpine pastures in the summer months.[54] Why then, one wonders, have Alpine pastures been the dairy zone par excellence? As the previous sections made quite clear, the intensive use of scarce resources encouraged diversification that utilised available pastures in different ways. But I would argue that it was not only a specific kind of labour economics that defined the value of mountainous meadows. Butter- and cheese-making on the Alpine pastures should also be explained by cultural values, e.g. the evaluation of the landscape and the quality of the products derived from it.

As previously argued, until the nineteenth century it was a widely accepted belief that the best dairy products derived from the mountains. The notion of butter and cheese as summer meadow products encompassed observations about topography, climate, plants and animals as well as visions of nature and health. However, what is important to mention is that these beliefs coincided with some technical problems in the manufacturing of dairy products in premodern times.

First, one should bear in mind that if dairying does not work with a constant flow of milk like today, people have to make decisions. If the amount of available milk is restricted, then, on the one hand, one can concentrate on butter and produce only a low-fat cheese from the skimmed milk. In this case, the milk stood one or more days in a flat wooden vessel until cream separated and floated to the surface. The cream was then skimmed off and churned into butter in a large keg. This skim milk (buttermilk), with the addition of sour whey, was then used to make a cheese. This old method of using milk was called either butter making or acid curd cheese making.[55]

On the other hand, if one decides to produce cheeses for a longer shelf life then one needs, in general, more milk and substantial parts of the milk fat.[56] This becomes all the more clear if one looks at the ratio of milk to raw yield of butter and cheese. For 1 kilogram of fatty cheese around 10 to 12 litres of milk are needed (the reader should bear in mind that 1 loaf of Emmental cheese now weighs around 100 kilogram).[57] There were different amounts of milk depending on the season. During the winter months, a considerable decrease of milk production was natural. Already in November, cows in calf gave less milk or even nothing, and from January to March almost all milk was needed to suckle calves (if the priority objective was cattle breeding even longer). Moreover, in the last few weeks before giving birth to a calf, cows do not give milk, and even throughout the rest of the year their output varies. If one takes nineteenth-century figures as a yardstick,[58] good cows provided 8 to 10 litres in the best time of the year, in June, yet in autumn their daily output often shrivelled down to 2 litres of milk a day. Since farmers saw to it that calves were born at the end of the winter, wherever possible, and that they were milked in the first few weeks (since the amount of milk provided by a cow rises with the weight of a calf), the Alpine summer coincided with the period of high milk yields.

To be able to manufacture 10 kilograms of cheese even in the best months – May to July – the milk of 10 to 15 cows was still necessary. Because most valley meadows were commons, Alpine villages and communities had limits on their cattle herds; having 6 to 8 cows in the stall was already a sign of prosperity.[59] Only joint and often cooperative operation of cheese manufacturing on large Alpine pastures was able to bring together in one place enough cows to make it profitable to produce cheese that would keep. In the valley, one family might manage two or three cows without the help of farm labourers; on the Alpine meadows three or four labourers could take care of 50, 60 or even more cows together. According to data about an Alpine pasture of the Valais canton, raised at the end of the nineteenth century, two herders and their assistants supervised the grazing of about 210 cows and some 60 heifers and calves. The three dairies located on this pasture handled some 70 cows, each was staffed by a cheese-maker, a female milker, and a weigher. These full-time workers, altogether about 15 persons, relieved over 100 households of the daily chores of feeding and milking the livestock.[60] The work of this hired staff allowed the farm families meanwhile to concentrate on the demanding tasks of haymaking and grain harvesting etc. And even the owner of just one cow could purchase butter and cheese products from such places.[61]

7. DAIRY PRODUCTS FOR SELF-SUPPLY AND THE MARKET

The storage time of dairy products and the expert knowledge of herders and cheese-makers finally brings us to the market relations of early modern Alpine dairy farming. In general, the butter and cheese making in the mountains did not only serve the needs of the local population. Since the fifteenth century, the export of Alpine cheese did its part to spread the wondrous reputation of the Alpine milk. Unlike the rest of Europe, in the Alps many feudal landowners or monasteries began early on to share their 'fief meadows' with the rural populace without levying taxes in all cases. The population in the Swiss mountain cantons were the first to enjoy more freedom and independence than other farmers. Here, commercial cheese making was being practised as early as the fifteenth/ sixteenth centuries.[62]

What is important here, however, is that the storage time of dairy products increased their sale chances, but not always their prices. Salted butter may be made from either sweet or soured cream. To ensure preservation, however, it had to be salted, more or less. Butter merchants usually tasted the products and mostly graded them after the salt was added. Thus, it was an important factor whether they found the butter mildly or heavily salted. While quality standards in the butter trade depended on the salt, cheese prices were up to the hardness of the sales offer. The harder a cheese was, the longer it could be stored and transported, the higher was its price.

In some parts of Switzerland, e.g. the canton of Glarus, butter and cheese (the so-called *Schabziger*) were exported; in other cantons, e.g. the Bernese Oberland, butter was the main commodity traded up to and into the sixteenth century.[63] The mountain farms in this region that later became famous for their fatty cheeses supplied domestic demand and also conducted some trade with various cities on the Upper Rhine (Strasbourg, for one). Since 1481, however, city councils of the canton Bern repeatedly banned butter exports in order to guarantee the domestic supply of butter.[64] Paradoxically they achieved the exact opposite effect. During the sixteenth century many cheese farms, starting in the *Fribourg* alps, began to intensify the production of fatty cheese and to make such vast improvements in the technology that the product was easier to transport and therefore easier to export.[65]

From now on national and international trading exerted influence on the Alpine dairy practices. Prior to this, cheese was only exported at irregular intervals, usually accompanying cattle along the north-south transit routes. Only the Schabziger cheese from the canton of Glarus achieved fame beyond its regional borders; it was an acid curd cheese made of skim milk, which obtains its characteristic green colour from a certain herb (Trigonella caerlea).[66] Once the cheese farms, hit hard by the Butter Regulations, realised that fatty cheese was more lucrative,[67] an increasing number of them went over to fatty cheese production. Thus, the spread of fat cheese production enhanced the cheese trade. Between the sixteenth and eighteenth centuries the important Swiss cheese regions were the Greyerzerland (canton of Fribourg) and the Emmental, with its still renowned cheeses; the Bernese Oberland, with its famous *Sbrinz* or *Spalenkäse*; and the canton Appenzell. Alongside these existed several specialities like *Bellelay*, *Schwyzer*, Entlebucher or *Ursner* cheese and the already mentioned Glarner Schabziger.

The shift in the method of production had some operational implications. Now, all elements of milk were used to make fatty cheese: i.e., the cream was not skimmed off, nor was it churned into butter. The use of rennet had been known for a long time but it was only the increasing trade that had such a decisive influence on the technology of making fatty cheese. The improvement spread by dairymen throughout the northern Alpine region since the end of the sixteenth century involved the hardness of the cheese. By subjecting the cheese mass to greater heat, cutting it into smaller pieces in the cauldron and storing the loaves in the pantry for a longer time, the cheese became considerably harder and thus more durable. In contemporary terminology, the cheese was 'burned'.[68]

The fact that the loaves of cheese became smaller through this method, though, was not in the merchants' interest. Duties on the cheese were determined by quantity and not by weight. Therefore, larger loaves of cheese were easier to sell than smaller loaves. The large Alpine pastures, having large cattle populations and thus being able to produce large quantities of milk, were especially suited to meeting this customer demand. The latter was only limited by the volume of the cauldrons, and they soon could hold over 100 litres. This explains the success of certain types of cheese, such as Gruyère or Emmentaler. As summer cheeses they sold well abroad; the soft cheeses made in the autumn remained in Switzerland.[69] To meet international demand, other regions (e.g. Entlebuch) began during the eighteenth century to declare their cheese as Emmentaler and to sell it under that name.

At the same time this reorientation led to the standardisation of cheese-making methods across the country. The method originally brought to the Emmental Alps by Fribourg dairymen spread throughout the northern Alpine region.[70] Here, the milk was heated to at least 25–28°C, generally to 30–36°C, and tested on a forearm. When the desired temperature was reached, the pot was removed from the fire and filled with rennet from calves' stomachs.[71] The curd was then, in the cauldron, cut in the smallest pieces possible with a shaven pine twig, since the hardness of the cheese is in inverse proportion to the size of the pieces of curd. Time was of the essence since the mass would otherwise have become too sour. Once the whey was removed, the leftover quark was put into round wooden cutters with openings on the sides which allowed the excess cheese liquid to seep out. Under pressure from stones, the cheese then sat still for some time before being taken to a cellar for salting and maturing. After three to four months, the cheese, stored in kegs, could be transported down from the mountains by mule, and then shipped throughout Europe.

Between the sixteenth and eighteenth centuries many who could not subsist through farming entered the fatty cheese business. Labourers, milkers, cheese farmers trained by Alpine cooperatives, and the sons of farmers who did not inherit enough land to live on, reacted to the rising demand of cities for milk products.[72] In the canton of Bern, these so-called *kuher* jointly leased cow herds or bought pregnant cows and went from farm to farm with their animals through the farmstead meadows and Alpine meadows of their region in order to buy stall space and hay from farmers. With the success of this system, it became even more attractive for them to engage in dairy products trade.

This situation led to a unique contract between the farmers in the valleys and the cow herders. When the *kuhers* came down from the Alps in autumn, needing a home for themselves and their cattle, they were sheltered at the farms. The farmer fed and lodged everyone, human and animals, until his stores of hay for the cows and provisions of food for the humans were exhausted. The *kuher* then moved to the next farmer on contract, who again lodged him for a few weeks. This rotation went on for all the winter months, until it was time to ascend to the Alps again.

There was a great advantage for both parties. The *kuher* had shelter in winter, receiving food and everything he needed for his family and his animals. The farmer got the dairy products and also the dung he needed urgently for his fields. Soon wealthy city dwellers saw how profitable it was to invest in Alpine meadows. Patricians from Bern, who owned the most profitable alpine meadows, leased their meadows, complete with buildings and cheese vessels, to this newly developed guild of cow herders over the summer; the latter, as a class of self-employed entrepreneurs, gradually purchased large herds of cows and became specialists in cheese manufacturing and trade.

Business with fatty Emmental or Gruyère cheese, which were increasingly being marketed under the collective trade name 'Swiss Cheese', went so well that in the eighteenth century, that other parts of Switzerland, like the Zurich region or Grisons, entered the dairy business. Regions like the Glarnerland or Appenzellerland, which had focused mainly on cattle trade until then, also began to specialise in dairy production and processing.[73] At the time many noted that a cattle epidemic would bring greater misfortune to Switzerland than a human epidemic.[74]

8. AGRICULTURAL MODERNISATION AND THE UPSWING OF VALLEY CHEESE FACTORIES

The Alpine farming system thereby proved to be economically quite adaptable and made farmers in well-connected regions and along the old transit routes rather prosperous. Between the sixteenth and the early nineteenth centuries the Alpine pastures were a highly productive economic zone for Alpine farmers in regard to livestock production and dairying; as the eighteenth century came to a close, this system began to disintegrate.

For a long time the literature mentioned the beginning of the Industrial Revolution as an explanation of this process, with not only the classical industries (textiles, mining and precision engineering) but also tourism, energy production and the expansion of transportation being cited.[75] This view, however, made farmers look like backward *'hicks'* who were lagging behind commercial and industrial growth. More recently, though, scholars have largely debunked all these clichés of farmers as Luddites.[76] In the Alps, too, agricultural modernisation was by no means an exogenous process; in fact, innovations with effects lasting well into the future were launched using local knowledge and experience.[77] That dairy farming lost its territorial roots and thus gave rise to the danger of potential under-use of the mountains was not the result of industrialisation and the economic marginalisation of agriculture.

In actual fact, the post-eighteenth-century agricultural reforms had the undoubtedly unintended side effect of pushing aside Alpine farming, and of driving a nail in the coffin of butter and cheese production.[78] Around 1760, the new physiocratic school of economic thought deemed land to be the key to economic progress. Consequently, its proponents demanded that agriculture be modernised. The idea was not to eliminate mountainous agriculture but to accomplish a fundamental land reform with a view to using more effectively the natural nitrogen cycle.

All proposals and measures for agricultural reform were based on four innovations: 1) abolishing the old style of meadow farming in all arable farming regions (i.e., no longer allowing land to lie fallow and parcelling the communal pasture); 2) planting feed plants, such as clover and sainfoin; 3) summer stall feeding to make it easier to collect cattle dung and liquid manure; and 4) requiring that fields and meadows be fertilised more intensively with the increased quantities of dung.

In the past, lowland farmers – without questioning the complex system of mountain and valley farming in its entirety – had been able to share in the profitability of Alpine farming through leasing, renting out land or selling hay. However, as agricultural reform began to take effect, the vested interests began to change. Lowland farmers extended their cultivation of feed in the fields, left their cattle in stalls, and began to cultivate their meadows and fields more intensively. In some regions, the protoindustrial years of crisis (1770–1773) supported these ideas. In these years of fame, local authorities of the canton Glarus, for instance, started to prohibit the nationwide and international cattle sale (the centuries old and famous *Welschhandel*) and encouraged farmers all over the country to improve the milk production. Fodder scarcity and inferior feed quality, the harsh consequence of these measures, were accepted. Moreover, they motivated local authorities to accelerate land reforms.[79]

The process of change lasted decades; nevertheless, more and more farmers in the lower-lying grain-growing regions hoped to increase their incomes through artificial meadows or artificial feed and expanded their cattle husbandry.[80] Regions where, despite all the natural advantages of abundant vegetation, farmers had stubbornly adhered to the practice of planting grain, likewise reacted. For instance, the Allgäu region in the area at the foot of the Alps, where, around the turn of the eighteenth/nineteenth centuries, 'farmers, having received foreign ideas, thought of the natural possibilities of [...] making the "yellow" and "blue"Allgäu (flax growing) into the blossoming "green" Allgäu'.[81] This reorientation was so radical that today, in the Bavarian part of the Allgäu, the farm landscape is said to have been completely 'greened'.

However, when the feeding situation in the lowlands improved, the next logical step was to bring dairying down into the valley. Swiss author Jeremias Gotthelf, in his 1850 novel *Die Käserei in der Vehfreude*, described this with stunning acuity:

Barbara Orland

'clover, sainfoin, and alfalfa came into the land, and stall feeding became possible, the forests were opened-up, meadows made arable and potatoes planted 'en masse', not just as a sort of dessert. Where cattle were in stalls, there was dung, large and small dung, and it was used copiously and sensibly. As more dung was available, arable land grew, as did the herds of cattle and specifically the cows that could be used [...] As the number of cows grew, so did the milk, for everything is interlinked, and one grows from another in an uncommon manner, and often in such a fine line that man does not even see the thread, a much finer thread than between cows and milk.'[82]

Yet Gotthelf also noted that farmers at first remained sceptical. To be sure, in an initial movement between 1790 and 1810 many of them introduced fodder cultivation, stall-feeding and the manufacture of fertiliser, and also participated in a reform of community ownership of land. Nevertheless, there was little support for bringing cheese making out of the mountains. In fact, fatty cheese production, in connection with stall-feeding, became the subject of decades of heated debate, especially among economic schools of thought:[83] too deeply rooted was the conception that high-quality fatty cheese could only be made out of Alpine milk and that the good reputation of Swiss cheese had its roots in mountain farming.

In fact, the first commercial cheeseries, established in the valleys around 1820, were not established by farmers but by agricultural reformers[84] or dairymen with business acumen, who could sense that Alpine pasture farming was headed downhill, or cheese traders.[85] The latter were primarily concerned with shortening their transportation routes when they attempted to convince valley farmers of the virtues of dairy production. In addition to cheese factories, they created warehouses where they could store unripe cheese purchased from the Alpine meadows and allow it to age. Gradually non-farmers began to acquire expertise in cheese making and to refine it as they pleased.

These cheeseries, however, presented farmers with a new type of competition. Soil quality, fodder, care of animals and Alpine pasture farming were no longer relevant. Cheese traders naturally were no longer interested in those issues. Their sole concern was cheese as a raw material, which – as people were slowly beginning to realise – could easily be produced in large quantities and sufficient quality in low-lying areas. However, a sensible method of organising the procurement of the quantities of milk needed to produce fatty cheese was needed, since no single valley farmer had herds the size of those brought together in the Alpine meadows. Joining forces to form cooperatives seemed to be the answer, and not just to this problem. United, cooperative farmers could resist the prices forced upon them by cheese traders and – if they could reach an agreement – run cheese making operations themselves.

Despite initial objections and resistance among farmers and cheese traders, the cooperative model spread quickly from its origins in western Switzerland.[86] The constant improvements in the transportation infrastructure from the 1840s and the reduction (in some domestic transportation routes, total abolishment) of customs duties served to convince the last doubters that cheese could also be manufactured in the valleys. In 1847 an agricultural census in the canton of Bern counted 217 valley cheeseries; a decade later this number had already swollen to 355.[87] Only mountain cheese production failed to benefit from the general upswing; its share of Alpine cheese production steadily declined.[88]

9. DOUBTS CONCERNING PROGRESS

Because the production conditions in the valley were by no means more favourable than on the mountains, and technological improvements in cheese making were not implemented immediately, farmers – in the beginning – had good reasons to be sceptical. Into the 1860s the equipment in valley cheese factories was no different from that of their counterparts in the mountains. Only around this time did it become standard practice to use walled-off cheese cauldrons with outer covers, an iron grate, ventilation and a direct link to the chimney instead of open fires.[89] And it was only in 1877 that Wilhelm Lefeldt (1813–1913) invented the milk centrifuge, giving cheese-makers a useful way of skimming cream off the milk more quickly and efficiently.

Predictably, the upswing and profitability of dairy farming were accompanied by efforts to extract the maximum profit from meadows and milk cows. To natural dung were added all sorts of artificial fertilisers, and natural fodder was joined by many varieties of artificial feed. The cheese factories soon realised the cost. Despite improved facilities and careful use of production techniques, Swiss cheese factories began to see a noticeable increase in defective cheese output in the 1870s (i.e., it was improperly fermented, did not keep as long, or the dough crumbled, etc.) and exporters noted that the general quality decreased. Operators of cheese factories were agreed that the cause of this decline in quality was to be found in the intensification of feeding, the use of artificially preserved liquid manure and all other types of artificial fertiliser. In an ironic twist to history, the end of the nineteenth century saw calls for a return to natural feed and meadow grazing.

Nevertheless, the new system was not fundamentally called into question. Only mountains and mountain farming were considered to be impediments to innovation. Interest in the economic potential of the higher elevations died; it is telling that Bernese citizens sold their Alpine properties one after another around the mid-nineteenth century. The Emmental cow herders' foray into the world of free enterprise came to a quick end. By the 1860s the former capital of cheese making in the Bernese Oberland had been reduced to a cattle-supplying area for the lowlands.[90]

By this time politicians, civil servants and the educated classes (especially the clergy and doctors) felt the increasing need to save Alpine farming. During the heyday of Alpine farming complaints had often been made about overgrazing of Alpine meadows. With the advent of valley cheese factories the arguments were turned around: now accusations were being made of apathy and neglect of Alpine meadows. The fact that the highest Alpine pastures were gradually being abandoned altogether and that, at more accessible Alpine pasture locations, cattle were being left alone without herders, was considered to be a sign of indifference on the part of farmers.[91] Critics noted that buildings were falling apart, roads were being buried by landslides, and fences were no longer being repaired. Moreover, piling fertiliser next to the huts at the milking stands instead of spreading it over Alpine meadows seemed to be a scandalous waste of valuable fertiliser.[92]

Mountain farmers were increasingly being criticised for being backwards, averse to innovation, and hidebound. However, backwardness was not being seen as a sign of ecological friendliness or as a result of increasing competition; on the contrary, Alpine reformers saw such backwardness as a sacrilege to an inherently bountiful natural environment. They claimed that humans caused much greater damage than landslides and other natural disasters. It was not only the 'unmitigated thoughtlessness with which the high Alpine forests were, and sometimes still are, being laid to waste and destroyed at the hands of humans'[93] which encountered harsh criticism. What was worse, they said, no lush meadows were being planted in place of the high Alpine forests: only barren landscape and devastation remained.

Riding a wave of national sentiment, the mostly bourgeois, urban reformers in Switzerland felt compelled to take matters into their own hands. On 25 January 1863, some 30 men from all parts of the country convened to form the Swiss Alpine Farming Association.[94] Their job was to unite activities to improve the Alpine landscape. In the founding document, '[a]ctivity in the pure Alpine air and a diet of cheese, butter, milk and meat' was praised as a source of health, physical strength, and the spirit of freedom of the Swiss people. The shepherd, who embodied these Swiss virtues, could not be allowed to fall into oblivion, nor could his regions be allowed to suffer under the indifference of modern man. It was said that there were *'immense treasures to be had'* in the mountains.[95]

More and more, the Alpine dairy tradition was transformed into national milk symbols that could be used as marketing instruments in the transition of the *old* Swiss dairying practices into an industrialised milk business under the conditions of an emerging international trade. As a new kind of crisis management it did not only help to integrate the different local identities within Swiss agriculture (lowland and mountain farmers; region and nation). The creation of a typical Swiss product was also aimed to improve the competitive position of Swiss products, 'because we have to confess, that since the 1870s it (the Swiss dairy industry, BO) in its quality evolution did not withstand the competition of foreign countries.'[96]

One of the brightest thinkers among the founders of the new association, the country pastor Rudolf Schatzmann (1822–1886) who hailed from the Bernese Oberland,[97] wrote in 1872: 'Milk is the fundamental material of our nation, to build and support its ability to work'.[98] With those ideas in mind, reformers of the Alpine dairy farming responded in two ways. On the one hand, the aim was to catch up to the economic boom in the valley by improving Alpine farming. People like Schatzmann thought that science and technology would provide the tools to that end. They used lectures, essays, courses, exhibitions and inspections of cheese factories to preach tirelessly the gospel of the latest achievements in the science and technology of dairy farming to the mountain inhabitants. Thinking that in order to save Alpine farming dairy farming needed to be modernised, Schatzmann's greatest triumph was the opening of the first Swiss milk testing station in Thun on 1 September 1872. This was later joined by another station in Lausanne.[99]

On the other hand, promoters of the Alpine dairy farming system used their activities to persuade the urban population about the healthy and tasty products of the mountain meadows. In his *Alpwirthschaftliche Volksschriften* in 1873, Schatzmann praised 'the aromatic Alpine herbs and the fresh, healthy Alpine air' as the important conditions 'for healthy, strong and beautiful cattle and tasty butter and cheese.'[100] Schatzmann, like his companions, made no secret of the fact that every innovation in the valley made the soil there more attractive, thus exposing ever more painfully the disadvantages of higher elevations. Nevertheless, he never got tired of reminding his fellow citizens that Alpine nature was the pride of the Swiss national economy. And if the whole nation would comprehend this relevance, he predicted, this would usher in a golden future of mountain dairy farming.

10. COMPETITION THROUGHOUT EUROPE AFTER 1870

These predictions would turn out to be a fallacy. By this time the writing was on the wall: Alpine milk producers, whether in the mountains or in the valleys, would soon face unimagined competition. Across Europe and without any regard for soil conditions or vegetation zones, agricultural regions, which had been used for grain production were being converted to intensive dairy farming beginning in the 1870s. In all those places where one used to say that 'cattle guarantee true pure yield only in those open and fertile pasture areas not capable of arable farming which are used for cattle husbandry, but not in general in usual arable farms where cattle are and must be only held out of sheer necessity'[101], now farmers were beginning to grow interested in cattle and its products as a goal of production.

Between 1870 and the beginning of World War I, cow's milk became a mass consumer good which was increasingly being sold as fresh produce, especially

Barbara Orland

as a beverage. Within just a few decades, dairy production in a country such as the German Empire achieved a production value on a par with that of coal mining.[102] Even Emmental cheeses, despite being traded as a Swiss specialty, were soon no longer the exclusive preserve of Switzerland. Starting in 1914 fake Emmental cheese was produced in Germany, France, Italy and even in Finland and Russia.[103]

The spread of dairy farming throughout Europe represents an agricultural evolution of gargantuan proportions, and its story cannot be told in this paper.[104] Let me say this much: the fact that livestock husbandry, dismissed by many central European farmers as useless and burdensome, suddenly seemed profitable, was foreshadowed in the debate on agricultural reform following the end of the eighteenth century but only came into being once developments in international grain markets created the harsh conditions for this change. Europe's last major famine was in 1846/47; in the years that followed, the liberalisation of the grain trade and the effects on distribution of an increasingly wider and more closely linked network of railroad lines were able to offset an under-supply and not just at the local level. National and international markets were able to establish themselves for the long haul, which was soon followed by price wars. Agriculture based traditionally on grain production became unprofitable and needed to be replaced with other sources of income. The structural change in agriculture was accelerated by the fact that numerous farmsteads were sold or changed from full-time to part-time farms and by the migration of wage-earning farm labourers to the cities.

This change of heart among farmers, which did not take place in an uninterrupted fashion, was also accelerated by the processes of urbanisation and industrialisation. All across Europe, the growth of cities exposed the limits of the traditional system of food supply. The hunger for land of industry and residential areas pushed agriculture increasingly from central locations to the fringes; and trade in foods, which were transported across large distances, naturally became more and more important. Whereas relatively durable foodstuffs such as grain were quite adaptable to the new sales requirements, the market for perishable fresh produce reacted with price increases. In pre-industrial Europe, the dairy industry, for stock-building purposes, had largely meant butter and cheese, but in the last thirty years of the nineteenth century fresh milk (as a beverage and an ingredient in fresh dairy products) was characterised as a peoples' foodstuff in the cities. That made it rather lucrative. Converted into Reichsmark, between 1855 and 1865 milk cost between 9 and 10 pfennigs per litre when purchased from a farmer in the city of Berlin. By 1875 the price had gone up to some 14 to 15 pfennigs, and in the 1880s it held firm at around 18 to 20 pfennigs, of which the producer pocketed 12 to 15 pfennigs.[105]

The farmers based in the immediate vicinity of large consumer centres were understandably the first to react to the observation that the price of fresh milk was rising faster than that of other agricultural products. Many of them sold their farmland as construction sites, often involuntarily, and only practised pure dairy farming on their remaining property. In extreme cases, these farms had no independent agriculture anymore but instead purchased nearly all the necessary fodder on the fodder markets, which were growing fast due to imports. Without their own fodder base, they could only purchase highly pregnant or freshly lactating cows, milk them for a lactation period, and then sell them for slaughter.[106]

Such highly specialised 'milking-only' farms were not the only consequence of the increasing urban demand for milk. Milk production, which began to fragment, just before the turn of the century, into individual markets for fresh milk, butter, cheese and industrially manufactured durable dairy products, became more and more enticing to all farmers. Even the more remotely located farmyards were able to share in the milk boom. Their milk was distinctly cheaper than that of the milking-only farms, especially once hygienic measures and improved transportation networks helped to resolve the tiresome transportation problem. Conditions became increasingly favourable for making cow's milk a multipurpose raw material.

11. IMPLICATIONS

The implications of this development for mountain farmers were obvious. Alpine mountain farming, traditionally based on livestock husbandry, could not easily share in the profits from the growing consumption of milk and dairy products. To this day the transportation of fresh milk from the mountains to the vicinity of consumption centres has remained a technological problem. And cheese factories, in light of the sudden torrent of milk products, only had a chance to compete in expanding markets if they had enough ways and means of distributing their products to city markets.

Although some Swiss regions were particularly well suited to this type of distribution, international competition still made its presence felt almost immediately. From the mid-1880s the number of new cheese factories began to stagnate. Even in the highly productive canton of Bern, there were no more cheese factories in 1923 than in the 1880s (625), although the quantity of processed milk had risen from 1.2 million litres in 1884 to 2.07 million in the 1910/1911 farming year. The region, though a participant in the upswing, had lost its site-related advantages.

The fact that the intense competitive pressure had a less severe impact on the Alpine meadows can once again be attributed to the adaptability of mountain farming. Over the course of the nineteenth century the number of abandoned Alpine farms gradually decreased. It was especially because of the change in the makeup of the Alpine livestock that the loss of Alpine meadow area was held in check. Cows were increasingly disappearing; animals not requiring human herders (pups and heels) remained. The tradition of butter and cheese making

Barbara Orland

in the summer decreased; the use of the Alpine pasture livestock breeding persisted. An awareness of tradition, economic necessity, the availability of labour in sufficient quantities and tourism in the Alps were further reasons why many Alpine cooperative farmers and individual farm operators insisted on sending cattle up to the Alpine meadows in the summer.

Only after World War II did the Alpine population become less willing to farm the land under increasingly difficult farming conditions; and as increasing numbers of young people left the mountain villages, the system collapsed. At that time, however, townspeople attached value to the aesthetic and ecological functions of mountain farming. In view of the increasing environmental problems in major metropolitan areas, the desire to flee the cities for a pure environment during increasingly long holiday periods increased and formed new forms of land use. These new tendencies also influenced Alpine agricultural policy. Thanks to regional promotion measures, mountain farming survived. Since the end of the 1970s even more cows have been brought to the Alps to graze. In some places the milk is delivered to the valley in appropriately equipped vehicles, by cable car, or even by helicopter.[107]

From an agronomic perspective such measures seem dubious. At the same time when the mountainous milk production was stimulated, on a European level milk quotas were enforced in order to stop overproduction of milk. Since 1984 European farmers are allowed to produce a stipulated maximum amount of milk. But what seems plausible from an agronomic perspective seems necessarily desirable from an environmental policy point of view. Because mountainous food production is promoted in order to create a landscape which is valued as a 'natural environment,' today's dairy farming has become a kind of gardening technology.

NOTES

1. For the history of the European agricultural policy in general, see: Ingersent 1999; Priebe 1985. For the ecological change of mind in agricultural policy, see Heißenhuber 1994. I would like to acknowledge the great help of Michael Dear, who has translated an early version of this paper.

2. Quoted in Niederer 1996, 372.

3. Greif, Franz and Klaus Wagner, Österreichische Berglandwirtschaft - Stütze der Marktproduktion oder ökologische Restlandwirtschaft, in Anwander Phan-huy 1997, 39. See also Dax 1998, 153–9.

4. The Alpine panorama marketed by the tourism industry includes not only rough and inhospitable mountain ranges but also the rural man-made farming landscape with its specific fauna, buildings and implements. Unlike in North America, where the concept of 'wilderness' plays a key role in environmental policy debate, historically developed cultural landscapes determine what constitutes nature in Europe. This consensus in society that addresses farmers' reproductive achievements in nature led to the tasks of *'Erhaltung der natürlichen Lebensgrundlagen, Pflege der Kulturlandschaft, dezentrale Besiedlung des Raumes'* ('Maintaining natural foundations of life, taking care of the cultural landscape, and the decentralized settlement of the area') being

incorporated into the Swiss Federal Constitution in a referendum in 1996. As a consequence, direct agriculture-related payments have risen dramatically, from 100 million Swiss francs in 1993 to 745 million francs in 1998. Whereas in valley farming some 15 % of gross agricultural earnings come from direct subsidies, in the highest mountain farming zones this figure is a whopping 50%. See *Mehr Natur in der Landwirtschaft - eine Zwischenbilanz* 1999.

5. Over-use, pollution and the waste of resources became not only the main *leitmotifs* of environmental policy but also of environmental history. See the report of Stine/Tarr 1998.

6. This was tendered by UNESCO in 1971 to promote the planning and testing of new forms of land use. Along with 100 other countries Switzerland and the other Alpine countries, in the mid-seventies, developed the mountain project *Der menschliche Einfluss auf die Gebirgsöko-systeme*. During the 1980s this was followed by 10 further similar projects (2 French, 4 Swiss, 3 Austrian and 1 West German project). A summary is to be found in Messerli 1989.

7. Ibid, 65.

8. Porter 1998, 219.

9. Ibid, 8. As Donald Worster suggested, the way we use the land reflects our understanding of nature. See Worster 1993. William Cronon pointed out that our understanding of the relationship of humans and nature is itself dynamic. See William Cronon, Introduction: In search of nature, in Cronon 1995, 23–56. In this sense, research of the last decade is increasingly focusing on the interrelatedness of land, labour and technology, especially with regard to the construction of urban environments. Very influential has been Cronon 1992. Conceptions of property and products, practices of land use developed as expressions of specific living conditions, political systems and cultural values are already explored in his book Cronon 1983. See also White 1991; Steinberg 1991. Compiling a history of land use is a large undertaking, done by many scholars working in a variety of fields. Differences in the scope of various disciplines are described by Matt Osborn, Sowing the Field of British Environmental History, http://www2.h-net.msu.edu/~environ/historiography/british.htm#9.

10. I refer to his essay 'Are You an Environmentalist or Do You Work for a Living?': Work and Nature, in Cronon 1995, 171–85.

11. The significance of this farming system to the Alps is reflected in a wealth of literature. There are a large number of studies – dated and recent, regional and national, social scientific, cultural-geographical, historical and legal in nature – which have addressed this type of farming. One of the few studies to deal with Alpine farming as a land-use strategy from an ecologic perspective is the work of US cultural anthropologist Robert M. Netting. See Netting 1981. Above and beyond the literature quoted in the further course of this paper, I would like to refer the reader to the literature overview in Mathieu 1992, 233–4.

12. It was French geographer Emmanuel de Martonne who in 1926 made the now classic attempt to bring some order to the variety of types which had developed encompassing not only various sovereign territories but also within them owing to differences in land ownership, legal relationships and nature-related methods of use. To greatly simplify the issue, he distinguished between six regions seeming to follow along an invisible north-south line originating in the earliest periods of human settlement. All are combined farming types, between mountains and valleys, between grain and cattle husbandry/dairying/forestry, the cultivation of vegetables, fruit and wine, and chestnut orchards. See Bätzing 1991, 150.

13. White, Richard 1995, 'Are You an Environmentalist or Do You Work for a Living?': Work and Nature, in Cronon 1995, 171.

14. Melanie Du Puis tells the story of the US dairy industry. In her path-breaking work she explains how milk was constructed as 'nature's perfect food' and how this ideal shaped patterns of consumption as well as the implementation of a new production system. Du Puis 2002.

15. Following the model drafted by Paul Bairoch the first 'organic' phase started in the late 18th century and ended during the major depression that hit the European agriculture from 1875 to 1890. The second phase is described as the mechanic phase, lasting until the Great Depression.

Barbara Orland

The third phase is classified as being in the period of the welfare state and thus the post-World War Two period. See Pfister 1995, 176.

16. Schneider 1916, 7.

17. Still a good reflection on the complex interplay between grain and animal production offers Riemann 1953.

18. Rüger 1851, 7.

19. Besides the coastal countries of Netherlands (the Belgian nickname for the Dutch is *kaaskoppen*), Denmark, Ireland, Sweden etc., it was particularly the northern Alpine countries, at first Switzerland, which had a pronounced and often highly developed dairy farming system. In France, besides the Alps, especially the Massif Centrale, Les Vosges, Les Pyrenees and Corsica have been the *old* cheese regions. In the more arable-farming-oriented regions, the milk of cows, goats and sheep was a by-product of animal husbandry (but a well-liked one), available in pitifully insufficient quantity and only on a seasonal basis.

20. Stamm 1853, 372.

21. I took the phrase from Sally McMurry who describes how dairying families in Oneida County, NY, thought this way when they started an intensive milk production in the middle of the nineteenth century. McMurry 1995, especially. Even today, Italian farmers in a small Alpine village think in terms of dairy zones. See Grasseni 2001.

22. In French the high-elevation meadows above the tree line are called *montagne, alpage, alpe*; in Italian, *alpeggio;* in Slovenian, *planina*. In the following I will use 'Alpine pastures' for the grazing areas and 'Alps' for the mountains. For a definition of the German-language terminology see Bätzing 1997, 15. At the fringes of the Alps and in the eastern Alps the meadows were and are relatively small, whereas in the high mountains of the Central Alps they can today be as large as 25 sq km with elevations varying by up to 1,000 metres. Alpine farming was done in all Alpine countries and, in evolved forms, also in other European mountain ranges such as the Pyrenees, Jura, Vosges, Apennine and Carpathians.

23. Quoted in Bätzing 1991, 29.

24. See Mathieu, Jon, *Bedeutung des Alpwesens in der frühen Neuzeit*, in Carlen and Imboden 1994, 97.

25. During the nineteenth century studies were repeatedly run to prove these relationships. In most cases the assumptions were confirmed. See Anderegg 1894: 504–9.

26. Cattle traders called this increased resistance to disease the dowry of Alpine summer herding. It was assumed that cows raised on an Alpine meadow would never lose this constitution. According to today's knowledge, the life-span of an Alpine cow is roughly six months longer. See Bätzing 1991, 29.

27. See Frehner 1919.

28. See Stremlow 1998, 53.

29. Gessner 1541/1996.

30. Medicus 1795, III/IV, 23–4.

31. Pfister 1995, 158-160.

32. Mathieu 1992, 235.

33. Mathieu refers to a statistic of the year 1866 that counted, on average, six calves to ten cows in Grisons and only two calves to ten cows in Valais. In this specific case, the data do not point to a greater milk production in Valais. In relation to cattle owners and local population, both cantons possessed quite the same number of cows. Mathieu 1992, 235.

34. Ibid, 234/5.

35. Regional differences in Swiss cattle raising between the seventeenth and nineteenth centuries are reported in Anderegg, vol. 2, 1898, 363–75. Farmers and cheese makers in Appenzell left the raising of cattle to cattle farmers in the Bregenz woods, Montafon and the Valois valley, and they bought young stock every year at the cattle markets in the Vorarlberg, in Bludenz,

Schruns, Dornbirn, Feldkirch, and Schwarzenberg (today Austria). See Weishaupt 1998, 17. See also Orland 2003.

36. Examples of feeding contracts from the seventeenth and eighteenth centuries can be found in Ramseyer 1991, 46–9.

37. For more precise information see Braun 1984, 58–69.

38. There are precise descriptions of this system for several centuries and for various regions. For the following section I will refer to: Hösli 1948; Mathieu 1992; Netting 1981; Frödin 1940/1941; Penz 1978; Aegerter 1983; Vogler 1987/1804; Berchtel 1990; Groier 1990; Nägeli-Oertle 1986; Carlen/Imboden 1994; Kärntner Landwirthschafts-Gesellschaft 1873.

39. At times the farms were so far apart that one had to march for several days to get from one to another. The community of Törbel in Valais, for example, used the Oberaaralp near by Ober Hasli, in the canton of Berne, which was still a three-day trek. See Netting 1981, 26.

40. See Groier 1990.

41. Depending on the natural circumstances a distinction was made between two or several grazing stages. In addition, the Alps were divided into lower alps (< 1,300 m), middle alps (1,300 – 1,700m) and high alps (> 1,700 m), though the borderlines, especially to the lower-lying spring meadows, were fuzzy. On the definition of the German terminology see Berchtel 1990, 47/8.

42. Stebler, F.G., Die Vispertaler Sonnenberge, in: Jahrbuch des Schweizer Alpenclub, 56 (1922), quoted in Netting 1981, 64.

43. The spring meadows, farming areas lying below the Alpine meadows, were an in-between stage between the valley and the high Alpine meadows and were most likely to have stalls, hay huts or abodes. In the Italian Alps and Ticino, there are to this day spring meadow communities, small villages with chapels, and in some cases even schoolhouses.

44. In general, the lower the grazing altitude, the more complete and more comfortable the buildings were. Especially in the spring meadow, which was grazed in the spring and again in the fall/winter, there were rather comfortable buildings where the farm family and all its accoutrements moved to in the spring. For one thing, these abodes shortened the distance between the valley farm and the Alpine meadow farm, and for another, they made the hay harvest, which was scheduled for the summer, much easier since hay harvesting was concentrated mainly on the spring meadows.

45. Besides the above literature, see also Stolz 1949/1985; Burmeister, Karl-Heinz, *Rechtsver-hältnisse an den Alpen (Mit besonderer Berücksichtigung von Vorarlberg)* in Carlen 1994, 17–36.

46. Hösli reports Alpine pastures in the canton of Glarus which, out of their owners' greed, were made useless as early as the sixteenth century through over-use because avalanches and wild streams destroyed the remainder of the forests and piled ashes and rubble onto the meadows. Hösli 1948, 313.

47. Four types of ownership, with many mixed and transitional forms, developed since then and can be found even today: 1. commons, which belonged to all the farmers of a village or community, 2. cooperative Alpine meadows in which farmers from several places in a valley or region joined forces to use an Alpine meadow together, 3. private Alpine meadows which were owned by a family, and 4. Alpine meadows which were farmed on behalf of clerical or secular landowners. Following the French Revolution, in the north and west numerous meadows were sold as national property and came into the possession of individuals. In the Eastern Alps (Slovenia), by contrast, many farmers remained serfs of the large landowners until and into the nineteenth century, yet had the rights to chop wood and take animals to pasture on the Alpine meadows. See Stolz 1985, 147–270.

48. A more detailed discussion of the differences between individual and cooperative milk production and processing is offered by Mathieu 1992, 234–51. In order to find out how much cheese and butter (including lean cheese) every member of the cooperative was to receive at the end of the season, the milk of his cows had to be measured. The yield of butter and cheese to be expected at the end of the Alpine farming season depended on this measure of the sample

milk production. The measurement days were a rather tense matter, for everyone tried to get as much out of it as possible: cheese farmers tried to milk as little as possible during trial milking, while the farmers were speculating on their cows yielding as much milk as possible. Even before the measuring day supervisors were sent up to the Alpine meadow to ensure that milk was not drawn the day beforehand or that the Alpine meadow personnel didn't drive the animals around and tire them out, which would reduce their output of milk.

49. Many of those ordinances have remained unchanged over centuries and, like the *'Grindelwalder Taleinungsbrief* dating back to 1404, are still in force to this day. What is interesting about the Grindelwald case is that this union of Alpine pasture owners took the development of the tourism into its own hands. See Aegerter 1983; Nägeli-Oertle 1986.

50. Burmeister, Karl-Heinz, *Rechtsverhältnisse an den Alpen (Mit besonderer Berücksichtigung von Vorarlberg)* in Carlen 1994, 19.

51. The number of animals allowed on to the meadow was set depending on the type of soil of the meadow, its fertility (say, in terms of special grasses that grow there), and its general state (e.g. rocky terrain, swamp). Cow rights were converted depending on the amount of fodder the animals needed. 1 cow right = 2 dry cows, 4 calves, 6 sheep, 12 goats or 1 foal. One horse was worth 2 cow rights. See ibid.

52. If that was not enough, then in some cases the foreign animal was summarily sent home, the grazing period of domestic animals shortened, only 'valuable' cows allowed but not goats, etc. See Anderegg, vol. 2, 1898, 650–3.

53. Generally someone from among the Alpine meadow owners was elected in charge of enforcing these rules and also decreed that all members of the cooperative were to participate in the building and improvement of paths, sidewalks, fences and huts, as well as the cleaning, care and fertilization of the meadows and pastures.

54. Data from Mathieu 1992, 236.

55. See Gutzwiller 1923, 21.

56. This has been reflected to this day in the character of the types of cheese. Tomme, which came about in the north of Savoy, is to this day a small cheese which originally developed from individual meadow farming where there was not enough dairy production to make larger cheeses. For the history of individual cheese types see Nantet 1994.

57. The amount of butter produced depended on the skimming method. Only in the 1870 were centrifuges developed, which allowed 3.8 kg of butter to be made out of 100 kg of milk. Before then, butter output was much lower. See Anderegg, 1894, 528.

58. See *Die Milchwirthschaft im bayerischen Allgäu* 1895, 6. In it, it says that milk output decreased by the day starting in mid-August. Some dairymen said that this autumnal decrease in milk output was attributable to the blooms of various plants that only grow during that time of year. See also Anderegg 1894, 493.

59. From the Allgäu it is reported that the communes allowed day labourers to own one or two cows 'out of mercy and charity.' Flad 1989, 13.

60. Data quoted in Netting 1981, 65.

61. In some places the production of rennet cheeses was shifted to the high Alpine meadows, whereas butter and lean cheese were made on the spring meadows on the way up to the higher elevations in the spring and on the way down again in the fall.

62. There was certainly plenty of justification for the Swiss Alpine population gaining a reputation for industriousness and Switzerland being termed a pre-capitalist, agro-bourgeois society. According to Biucchi 1985, 53. To the Swiss cattle and cheese trade in this period, see Braun 1984; Grass 1988; Ruffieux/Bodmer 1972.

63. More exact data for the canton of Berne is provided by Schatzmann 1861, 8–19.

64. For the centuries-old conflicts surrounding the supply of butter in Switzerland see Anderegg 1898, vol. 2., 534/5; Gutzwiller 1923, 44–6.

65. According to Gutzwiller the two areas where hard rennet cheese had the longest tradition were Greyerz (Gruyère) and Schwyz. In 1548 the middle and upper Bernese Oberland manufactured hard cheeses. The new method of manufacturing cheeses was taken further north into the Simmental and Emmental valleys by French Swiss tenant farmers from Alpine meadows. At the same time it was these tenant farmers who in the Emmental region began to export hard cheeses. Gutzwiller 1923, 22.

66. In the Middle Ages Glarus conducted extensive cattle trading. The vast majority of milk was used to raise calves, with the rest being used to make butter or Schabziger cheese. As early as 1464 the Landsgemeinde (cantonal assembly) in Glarus decided to affix a seal on their cheese to protect it from imitation. See Anderegg 1898, vol. 2, 534. In the fifteenth and sixteenth centuries trade in Schabziger cheese is said to have become considerable, and there is evidence that in the seventeenth century this cheese, along with other products from that region, was loaded onto the region's own ships and transported to Rotterdam, the Netherlands, and from there on to England, Russia and even as far away as India. See 50 Jahre Schweizerische Milchwirtschaft 1937, 11.

67. According to a study conducted in 1788 revenue from the manufacture of fatty cheese, in respect to the market prices of that period, exceeded that of butter and lean cheese making by nearly one-third using the same amount of milk. See Anderegg 1898, vol. 2, 504.

68. See Ramseyer 1991, 57.

69. This exacerbated the butter shortage; therefore, some dairymen made a sort of pre-curd butter. Pre-curd was the layer of fat floating on top of the milk being heated in the cheese-making cauldron. However, this butter was of inferior quality and therefore hard to sell. Ibid.

70. In Hungary cheese making using rennet became a part of the processing of cow´s milk, from at least the early seventeenth century through the influences of foreign dairy specialists employed on particular seignorial farms. See Eszter Kisban, 'Milky ways on milky–days: the use of milk products in Hungarian foodways', in Lysaght 1994, 15.

71. Various substances can be used to coagulate the milk quickly, causing a firm curd to develop. The most frequently used substance is the digestive juice contained in the stomachs of calves, kid goats and other ruminants. In the southern Alpine regions, where young stock was too valuable to be slaughtered for their rennet stomachs, the rennet of deer was used, as were thistle blossoms and fig tree juice. The rennet is added to heated milk, the trick being to find the perfect balance between time, temperature and amount of rennet. Milk then separated into solid components (curd in the case of rennet cheese and quark in the case of acid curd cheese) and its fluid component, whey.

72. Ramseyer 1991 studies this development in depth.

73. In the German areas at the foot of the Alps (Allgäu and Swabia) the political and economic implications of a non-unified Germany prevented a similar upswing from taking place. Dairy farming was only practiced as part of self-sufficiency farming. The chaos of the Napoleonic wars, which hit this area particularly hard, as well as the parcelling of the already-postage-stamp-sized farms which intensified following the end of the eighteenth century all did their part to put the region in economic limbo. The Austrian Alpine regions of Tyrol, Steiermark, Bohemia, Moravia, Galicia and Bukovina fared no better. See Lindner 1955, 79–81.

74. Quoted in Hauser 1961, 135.

75. See e.g. Franz 1994; Schmidt 1990.

76. Today economic historians give agriculture more credit for the economic upswing than economic theorists did for a long time. See Mathias/Davis 1990; Hudson 1992; Buchheim 1994, 49–54.

77. Pfister 2002, 7–14.

78. For this whole section see Pfister 1995, 175–202.

79. See Hösli 1948, 45–52.

Barbara Orland

80. Data on the size and composition of cattle herds demonstrated the new priorities. In the canton of Berne, around 1760 the grain-growing areas still had a large number of draught animals, particularly oxen, whereas in the mountainous regions the cows were in the majority, commensurate with their significance for cheese making. Transitional areas, also called field grass areas, were also already heavily oriented toward dairying. After 1790 (until 1911) the cow population began to outgrow that of horses, oxen and sheep in all parts of the country by an incredible, yet varying extent. Pfister 1995, 189.

81. Lindner 1955, 20.

82. Gotthelf 1850/1984, 235.

83. See Gutzwiller 1923, 88.

84. See Guggisberg, vol. 2, 1953, 123/4.

85. Many of the cheese dealerships had been established by former dairymen or cow herders. In the 1810-1820 decade many of them emigrated to the Bavarian Allgäu region and there tried to manufacture the same quality of heavy and fatty cheese from their home region. A local cheese trader from Allgäu, Karl Hirnbein (1807–1871), imported another type of soft cheese from Belgium: *Limburg* cheese. See Flad 1989, 26–32.

86. The farmers of a village established an association to build a cheesery. The capital was usually obtained by issuing stocks; it bore interest and was paid back in increments. Less frequently, the capital was obtained and repaid by only one entrepreneur. Once the interest had been repaid, it was used as a reserve fund for new buildings, repairs, etc. In addition to the building company, there existed an operating company which comprised all milk deliverers, who did not have to be identical with the members of the building company. The operating company provided cheese makers and laborers for an operating period, usually only for the summer. Only toward the end of the nineteenth century did year-round operations become standard practice. For more details see Anderegg 1894, 87–9. According to Anderegg envy, jealousy and mistrust among farmers led to several cheese cooperatives being given up soon after having been established. The milk was sold instead.

87. This period of growth slowly came to an end in the 1870s. Exact figures are in Pfister 1995, 198.

88. It is estimated that in 1870 in all of Switzerland there were about as many valley cheeseries (2,600) as Alpine dairy farms (2,800). See Anderegg 1894, 87.

89. In the 1880s a type of cart-borne fire was developed in which, unlike open fires and closed fires, the heat could be discharged using a portable cart, called the fire cart; when not in use, this cart could be moved away from the cauldron and to a water heater. Attempts had been made since the 1860s to use steam heating methods but this innovation found little support because the steam often found its way into the milk, and customers claimed to be able to detect an aftertaste in the cheese. The first working steam cheese factories were not built until after the turn of the century. In 1913 there were five cheese factories with open fires and hanging cauldrons in the canton of Thurgau, 132 cheese factories with cart-borne fires, and 40 steam-operated cheese factories. See Gutzwiller 1923, 98–100.

90. To avoid bankruptcy, some of these cheese specialists – as an irony of history – now sold the hay in the valley. Or they intensified livestock breeding to satisfy the demand for cows in the valleys. See Ramseyer 1991.

91. They claimed that these cattle grazed grassy cliffs barren with impunity, and that in some cases the grass was even uprooted, causing the grass cover to thin out and exposing it to trampling. They also criticised that fact that once the cattle were finished, sheep and goats could take their turn at grazing, causing further damage to the grass cover, and that on the lower spring meadows horses could graze with impunity. See, for instance, Schild 1852; Wilhelm 1868; Trientl 1870.

92. von Tschudi 1865, 299.

93. Ibid, 298.

94. See Schild 1865.

95. Ibid., 14.
96. As one dairy expert put it in retrospect: Widmer, A., Aus der Vorgeschichte, in: 50 Jahre Schweizerische Milchwirtschaft 1937, p. 23.
97. For his personal history see Wahlen 1979.
98. Schatzmann, Rudolf, Die Milchfrage vor der gemeinnützigen Gesellschaft des Kantons Bern, Aarau 1872, quoted in: Kollreuther 2001, 22. Since 1859 Schatzmann edited the *Alpwirth-schaftliche Monatsblätter*, published annually altogether six times until 1866 by publisher J.J. Christen in Aarau.
99. These were modeled on Alpine testing stations already in place in several locations in the German and Austrian Alps. See Die Alpenversuchsstationen 1867.
100. Schatzmann 1873, 3.
101. Found in Theoretisch-praktisches Handbuch der größeren Viehzucht from 1810, quoted in Abel 1986, 66.
102. In 1934, 237 billion litres of milk were manufactured, having a value of 2.3 billion Reichsmark. The value of coal produced that year was 1.94 billion RM, and that of raw iron production 0.66 Milliarden Reichsmark. See Reif/Pomp 1996, 77.
103. Schneider 1916, 43.
104. See Orland forthcoming.
105. See Martiny 1891, 10.
106. See Reif/Pomp 1996.
107. See Nägeli-Oertle 1986, Brechtel 1990, Groier 1990.

BIBLIOGRAPHY

50. Jahre Schweizerische Milchwirtschaft. Festschrift unter gefälliger Mitwirkung einer Anzahl von Fachleuten 1937. ed. Schweizerischer Milchwirtschaftlicher Verein. Schaffhausen: Kühn & Comp.

Abel, Wilhelm 1986. *Massenarmut und Hungerkrisen im vorindustriellen Deutschland.* 3rd ed. Göttingen: Vandenhoek & Ruprecht.

Aegerter, Rolf 1983. *Grindelwald: Beiträge zur Geschichte der Besiedlung und Landwirtschaft vom Mittelalter bis ins 19. Jahrhundert.* Bern: Diss. Uni Bern.

Anderegg 1894. *Allgemeine Geschichte der Milchwirtschaft.* Zürich: Füssli.

Anderegg, Felix 1897/1898. *Illustriertes Lehrbuch für die gesamte schweizerische Alpwirtschaft.* 3 vol. Bern: Steiger.

Anwander Phan-huy, Sybil and Hans Karl Wytrzens (eds) 1997. *EU-Agrarpolitik und Berggebiete. Beiträge der gemeinsamen Tagung der Österreichischen Gesellschaft für Agrarökonomie und der Schweizer Gesellschaft für Agrarwirtschaft und Agrarsoziologie, 26. und 27. September 1996 in Innsbruck.* Kiel: Schweizerische Gesellschaft für Agrarwirtschaft und Agrarsoziologie.

Bätzing, Werner 1991. *Die Alpen. Entstehung und Gefährdung einer europäischen Kulturlandschaft.* München: Beck.

Bätzing, Werner 1997. *Kleines Alpen-Lexikon. Umwelt-Wirtschaft-Kultur.* München: Beck.

Berchtel, Rudolf 1990. *Alpwirtschaft im Bregenzerwald.* (Innsbrucker Geographische Studien, vol. 18) Innsbruck: Universität Innsbruck.

Biucchi, Basilio M. 1985. Die Industrielle Revolution in der Schweiz 1700–1850. In Carlo M. Cipolla and Knut Borchart (eds.), *Europäische Wirtschaftsgeschichte, vol. 4, Die Entwicklung der industriellen Gesellschaften.* Stuttgart, New York: Uni TB.

Braun, Rudolf 1984. *Das ausgehende Ancien Régime in der Schweiz. Aufriß einer Sozial- und Wirtschaftsgeschichte des 18. Jahrhunderts.* Göttingen, Zürich: Vandenhoek & Ruprecht.

Barbara Orland

Buchheim, Christoph 1994. *Industrielle Revolutionen. Langfristige Wirtschaftsentwicklung in Großbritannien, Europa und Übersee.* München: Deutscher Taschenbuch Verlag.

Carlen, Louis and Gabriel Imboden (eds) 1994. *Alpe - Alm. Zur Kulturgeschichte des Alpwesens in der Neuzeit.* Brig: Rotten-Verlag.

Cronon William (ed.) 1995. *Uncommon Ground. Rethinking the Human Place in Nature.* New York, London: W.W. Norton & Company.

Cronon, William 1983. *Changes in the Land: Indians, Colonists, and the Ecology of New England.* New York: Farrar, Straus and Giroux.

Cronon, William 1992. *Nature's Metropolis: Chicago and the Great West.* New York, London: W.W. Norton & Company.

Dax, Thomas 1998. Die Probleme der Berggebiete in Europa. Eine vergessene Dimension in der EU-Regionalpolitik. In Agrar Bündnis (ed.) *Landwirtschaft 98. Der Kritische Agrarbericht.* Kassel: Bauernblatt-Verlag.

Die Alpenversuchsstationen im landwirthschaftlichen Bezirke Westallgäu 1867. In *Landwirtschaftliche Blätter für Schwaben und Neuburg VI*: 129–33, 137–40.

Die Milchwirthschaft im bayerischen Allgäu 1895. Erinnerungsgabe für die Besucher der Allgäuer Kollektivausstellung gelegentlich der deutschen Molkereiausstellung. Lübeck: o.O.

Du Puis, Melanie 2002. *Nature's Perfect Food. How Milk became America's Drink.* New York/London: New York University Press.

Flad, Max 1953. *Die agrarwirtschaftliche Entwicklung des württembergischen Allgäus seit 1840.* Hohenheim: Diss. Landwirtschaftliche Hochschule.

Flad, Max 1989. *Milch, Butter und Käse. Ein Beitrag zur Geschichte der Milchwirtschaft in Württemberg.* Stuttgart: Theiss

Franz, Herbert (ed.) 1994. *Gefährdung und Schutz der Alpen.* Wien: Verlag der österreichischen Akademie der Wissenschaften.

Frehner, Otto 1919. *Die schweizerdeutsche Älplersprache, Alpwirtschaftliche Terminologie der deutschen Schweiz, Die Molkerei.* Frauenfeld: Huber.

Frödin, John 1940/41. *Zentraleuropas Alpwirtschaft.* 2 vols. Oslo: Ascheoug.

Gessner, Konrad 1541/1996. *Das Büchlein von der Milch und den Milchprodukten.* translated into German by Siegfried Kratzsch. Mönchengladbach: Schlosser.

Gotthelf, Jeremias 1850/1984. *Die Käserei in der Vehfreude: Eine Geschichte aus der Schweiz* (reprint of the edition of 1850) Zürich: Rentsch.

Grass, Nikolaus 1988. Vieh- und Käseexport aus der Schweiz in angrenzende Alpenländer, besonders im 16. und 17. Jh.. In *Wirtschaft des alpinen Raumes im 17. Jh..* hg. von Louis Carlen and Gabriel Imboden. Brig: Rotten-Verlag.

Grasseni, Christina 2001. Developing skill, developing vision. Manchester: Ph.D.thesis.

Groier, Michael 1990. *Die 3-Stufenwirtschaft in Vorarlberg. Entwicklung - Bedeutung – Perspektiven.* Wien: Bundesanstalt für Bauernfragen.

Guggisberg, Kurt 1953. *Philipp Emanuel von Fellenberg und sein Erziehungsstaat.* 2 vols. Bern: Lang.

Gutzwiller, Karl 1923. *Die Milchverarbeitung in der Schweiz und der Handel mit Milcherzeugnissen.* Schaffhausen: Dr. Kühn.

Hauser, Albert 1961. *Schweizerische Wirtschafts- und Sozialgeschichte.* Erlenbach, Zurich, Stuttgart: Rentsch.

Heißenhuber, Alois, Jens Katzek, Florian Meusel, Helmut Ring 1994. *Landwirtschaft und Umwelt, (Umweltschutz, Bd. 9).* Bonn: Economica.

Hösli, Jost 1948. *Glarner Land- und Alpwirtschaft in Vergangenheit und Gegenwart.* Glarus: Tschudi & Co..

Hudson, Pat 1993. *The Industrial Revolution.* London: Arnold.

Ingersent, Ken A. 1999. *Agricultural Policy in Western Europe and the United States.* Cheltenham: Edward Elgar.

Kärtner Landwirthschafts-Gesellschaft (ed.) 1873. *Die Alpenwirthschaft in Kärnten*. Klagenfurt: Ferdinand v. Kleinmayr.

Kollreuther, Isabel 2001. *Milchgeschichten. Bedeutungen der Milch in der Schweiz zwischen 1870 und 1930*. Basel: Lic. Phil. Uni Basel.

Lindner, Karl (ed.) 1955. *Geschichte der Allgäuer Milchwirtschaft. 100 Jahre Allgäuer Milch im Dienste der Ernährung*. Kempten/Allgäu: Milchwirtschaftlicher Verein.

Lysaght, Patricia (ed.) 1994. *Milk and milk products from medieval to modern times: Proceedings of the Ninth International Conference on Ethnological Food Research*. Edinburgh: Canongate Academic.

Martiny, Benno 1891. *Die Versorgung Berlins mit Vorzugs-Milch. An Hand der Geschichte erzählt*. Bremen: Heinsius.

Mathias, Peter and John A. Davis (ed.) 1990. *The First Industrial Revolutions*. Oxford: Blackwell Publishers.

Mathieu, Jon 1992. *Eine Agrargeschichte der inneren Alpen, Graubünden, Tessin, Wallis 1500–1800*. Zürich: Chronos.

Mathieu, Jon 1996. Agrarintensivierung bei beschränktem Umweltpotential: der Alpenraum vom 16. bis 19. Jahrhundert. In *Zeitschrift für Agrargeschichte und Agrarsoziologie* 44: 137–161.

McMurry, Sally 1995. *Transforming Rural Life. Dairying Families and Agricultural Change, 1820–1885*. Baltimore: Johns Hopkins University Press.

Medicus, Ludwig Wallrath 1795. *Bemerkungen über die Alpen-Wirtschaft auf einer Reise durch die Schweiz*. Leipzig: Gräff.

Mehr Natur in der Landwirtschaft – eine Zwischenbilanz, article in the *Neue Zürcher Zeitung* newspaper, September 22, 1999.

Messerli, Paul 1989. *Mensch und Natur im alpinen Lebensraum. Risiken, Chancen, Perspektiven. Zentrale Erkenntnisse aus dem schweizerischen MAB-Programm*. Bern, Stuttgart, Wien: Haupt.

Nägeli-Oertle, Rudolf 1986. *Die Berglandwirtschaft und Alpwirtschaft in Grindelwald (Schlussbericht zum Schweizerischen MAB-Programm Nr. 21)*. Bern: Bundesamt für Umweltschutz.

Nantet, Bernard et al., *Käse. Die 200 besten Sorten der Welt*. Köln: Dumont.

Netting, Robert M. 1981. *Balancing on an Alp. Ecological change & continuity in a Swiss mountain community*. Cambridge, London, New York, Melbourne: Press Syndicate of the University of Cambridge.

Niederer, Arnold 1996. *Alpine Alltagskultur zwischen Beharrung und Wandel. Ausgewählte Arbeiten aus den Jahren 1956 bis 1991*. ed. Klaus Anderegg and Werner Bätzing, 2nd. ed. Bern, Stuttgart, Wien: Haupt.

Orland, Barbara 2003. Turbo Cows. Producing a competitive Animal in 19th and early 20th century. In Schrepfer, Susan R. and Phil Scranton, eds, *Industrializing Organisms: Introducing Evolutionary History*. New York: Routledge.

Orland, Barbara forthcoming. Milky Ways. Dairy, Landscape and Nation Building until 1930. In Sarasua, Carmen and Peter Scholliers, eds, *Agriculture, Food and Technology*, Oxford: Berg Publishers.

Osborn, Matt 2002. *Sowing the Field of British Environmental History*, http://www2.h-net.msu.edu/~environ/historiography/british.htm#9.

Penz, Hugo 1978. *Die Almwirtschaft in Österreich. Wirtschafts- und sozialgeographische Studien*. Kallmünz/Regensburg: Lassleben.

Pfister, Christian 1995. *Im Strom der Modernisierung. Bevölkerung, Wirtschaft und Umwelt im Kanton Bern 1700–1914*. Bern, Stuttgart, Wien: Haupt.

Pfister, Ulrich (ed.) 2002. *Regional Development and Commercial Infrastructure in the Alps, Fifteenth to Eighteenth Centuries*. Basel: Schwabe & Co.

Porter, Dale H. 1998. *The Thames Embankment, Environment, Technology, and Society in Victorian London*. Akron/Ohio: University of Akron Press.

Barbara Orland

Priebe, Hermann 1985. *Die subventionierte Unvernunft: Landwirtschaft und Naturhaushalt*. Berlin: Siedler.

Ramseyer, Rudolf J. 1991. *Das altbernische Küherwesen*. 2nd ed. Bern, Stuttgart: Haupt.

Reif, Heinz and Rainer Pomp 1996. Milchproduktion und Milchvermarktung im Ruhrgebiet 1870–1930 In *Jahrbuch für Wirtschaftsgeschichte* 1: 77–108.

Riemann, Friedrich-Karl 1953. *Ackerbau und Viehhaltung im vorindustriellen Deutschland*. Kitzingen-Main: Holzner Verlag.

Ruffieux, Roland and Walter Bodmer 1972. *Histoire du gruyère en Gruyère du XVIe au XXe siècle*. Fribourg : Ed. Universitaires.

Rüger, D. 1851. *Die neue chemisch-praktische Milch-, Butter- und Viehwirtschaft*. vol. 1 Löbau: Dummler.

Schatzmann, Rudolf 1861. *Schweizerische Alpenwirthschaft*. Aarau: J. J. Christen.

Schatzmann, Rudolf 1873. *Alpwirthschaftliche Volksschriften*. vol. 1, Aarau: J. J. Christen.

Schild, Josef 1852: *Die Zunahme der Land- und Abnahme der Alpen-Wirthschaft der Schweiz*. Zürich: E. Kiesling.

Schild, Josef 1865. Bericht über die Aufgabe des schweizerischen alpwirthschaftlichen Vereins und dessen bisherige Arbeiten In Rudolf Schatzmann (ed.), *Schweizerische Alpenwirthschaft*. No. 6: 3–17.

Schmidt, Aurel 1990. *Die Alpen – schleichende Zerstörung eines Mythos*. Zürich: Benziger.

Schneider, Ida 1916. *Die schweizerische Milchwirtschaft mit besonderer Berücksichtigung der Emmentaler-Käserei*. Zürich/Leipzig: Rascher.

Stamm, Ferdinand 1853. *Die Landwirthschafts-Kunst in allen Theilen des Feldbaues und der Viehzucht. Nach den bewährten Lehren der Wissenschaft, der Erfahrung und den neuen Entdeckungen in der Natur, gründlich, faßlich und ermuthigend erläutert*. Prag.

Steinberg, Theodore 1991. *Nature Incorporated: Industrialization and the Waters of New England*. Cambridge et al.: Cambridge University Press.

Stine, Jeffrey K. and Joel A. Tarr 1998. At the Intersection of Histories: Technology and the Environment. In *Technology and Culture*. 39: 601–40.

Stolz, Otto 1949/1985. *Rechtsgeschichte des Bauernstandes und der Landwirtschaft in Tirol und Vorarlberg*. Reprint from 1949. Hildesheim, Zürich, New York: Olms.

Stremlow, Mathias 1998. *Die Alpen aus der Untersicht. Von der Verheissung der nahen Fremde zur Sportarena. Kontinuität und Wandel von Alpenbildern seit 1700*. Bern, Stuttgart, Wien: Haupt.

Trientl, Adolf P. 1870. *Die Verbesserung der Alpenwirthschaft*. Wien: Gerold.

Vogler, Werner (ed.) 1987/1804. *Werdenberg um 1800: Johann Rudolf Steinmüllers Beschreibung der werdenbergischen Land- und Alpwirtschaft*. Reprint from 1804. Buchs: Buchs Druck.

von Tschudi, Friedrich 1865. *Landwirthschaftliches Lesebuch für die Schweizerische Jugend*. Frauenfeld: Huber 1865.

Wahlen, Hermann 1979. *Rudolf Schatzmann 1822–1886. Ein Bahnbrecher der schweizerischen Land-, Alp- und Milchwirtschaft und ihres Bildungswesens*. Münsingen: Buchverlag Fischer Druck AG.

Weishaupt, Mathias 1998. 'Viehveredelung' und 'Rassenzucht'. Die Anfänge der appenzellischen Viehschauen im 19. Jahrhundert. In Mäddel Fuchs (ed.), *Appenzeller Viehschauen*. St. Gallen: Typotron AG.

White, Richard 1991. *Land Use, Environment, and Social Change, The Shaping of Island County*, 2nd edn. Washington: University of Washington Press.

Wilhelm, Gustav 1868. *Die Hebung der Alpenwirthschaft. Ein Mahnwort an die Alpenwirthe Oesterreichs nach den in der Schweiz gemachten Wahrnehmungen. Im Auftrage des hohen k.k. Ackerbauministeriums*. Wien: Staatsdruckerei.

Worster, Donald 1993. *The Wealth of Nature: Environmental History and the Ecological Imagination*. New York: Oxford University Press.

Spanish Wood Pasture:
Origin and Durability of an Historical Wooded Landscape in Mediterranean Europe

Vincent Clément

INTRODUCTION

In Mediterranean Europe, human action on forests is generally perceived in a negative way. Rural communities are thought to have dramatically depleted the forests. Because of this well-established dogma, analysis of the interaction between humans and the forest frequently amounts to a vague chronicle of forest degradation. However, this assumption is sometimes contradicted in reality. Without denying the existence of important deforested expanses in the Mediterranean area, the impact of rural communities on forest evolution has not always been a bad thing.

In fact, in many cases rural communities have contributed to the shaping of valuable forest landscapes. Spanish wood pastures, called *dehesas* (*montados* in Portugal), are a vivid example of this. Dehesas are among the most original forest landscapes in Mediterranean Europe. They are composed of two strata: a more or less sparse blanket of trees (mostly evergreen oaks) and, under the trees, a herbaceous stratum often used for grazing. This cultural landscape occurs in the western Mediterranean area (Sardinia, Morocco, Spain, Portugal) as well as in the east (Crete, Cyprus, Greece). Dehesa landscapes are not restricted to the Mediterranean countries of the Ancient World: extended wood pastures cover large surfaces in central Chile and also in North America, especially in Michigan and Texas. But they are not as well represented as in the Iberian Peninsula, where dehesas stretch out over 3,000,000 hectares, of which more than 90 per cent are located in South-west Spain (Figure 1). They represent the biggest wood pasture cover in Europe.[1]

Environmental history is vital to our understanding of this cultural landscape. The first aspect of this article is to look for the origins of the dehesa landscape. In the particular context of South-west Spain, the border logic and the medieval *Reconquista* process are elements that undoubtedly played a decisive part in its genesis. But in this area of ancient agrarian civilisation, it seems necessary to raise the question of the possible existence of dehesa landscapes before medieval times, as well as the question of the place of the dehesa in the well-known triptych *ager*, *saltus* and *silva*. The second aspect concerns the spreading of

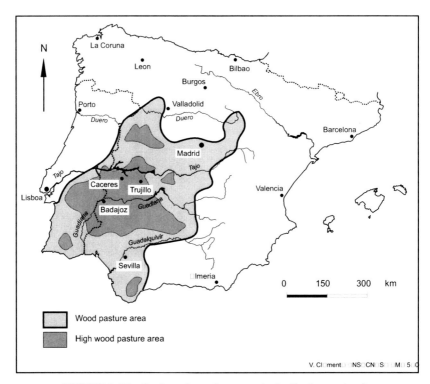

FIGURE 1. Distribution of wood pastures in the Iberian peninsula

this landscape from the Middle Ages onwards. Dehesas are usually associated with the large properties owned by military orders who took an active part in the Christian Reconquest. Is the spreading of the dehesa landscape really connected with the large properties which belonged to military orders, or is it the outcome of the rise of the transhumance from the thirteenth century onwards? The third question is how can the surprising durability of the dehesa landscape be accounted for? What are the historical factors that explain why the dehesa landscape has been preserved over the centuries and is still widespread today in South-west Spain?

I. DEHESAS AS AN ANTIQUE LANDSCAPE

Our contemporary reading of rural landscapes in the Mediterranean area is largely dependent on the famous Roman triptych *ager*, *saltus* and *silva*. What is the place of the dehesa in this apparently simple rural picture, in which every form of land use seems to fit neatly into one of those three types of landscapes? Do we have to consider the dehesa as a completely atypical landscape? Or should we challenge the landscape triptych, which seems suspiciously simple?

In the classic conception, *ager* corresponds to cultivated areas, *saltus* to grazing lands and *silva* to forest stretches.[2] Nevertheless, these three categories, which are used for describing the Mediterranean rural landscape, are in fact legal concepts inherited from Roman legislation. They were used for distinguishing different kinds of properties. In the Roman Empire, *ager* and *saltus* were provincial properties, while *silva* belonged to the public domain of Rome. *Ager* was a kind of individual property corresponding to the cultivation plots which were assigned to the Roman settlers after a centuriation process.[3] The individual plots as a whole composed what the Romans called *ager cultus*. The plots that were not allotted to any particular settler were broadly named *saltus*. The *saltus* areas were intended for common use.[4]

Ager, *saltus* and *silva* did not imply that the land was intended only for either agricultural, pastoral or forest use, or that they were well-determined landscapes. *Ager cultus* was certainly intended first and foremost for agriculture, but herds were not excluded from it.[5] The Romans set up the two-field system in which lands were cultivated only every two years. Every year, half of the cultivated lands were left fallow and herds were led onto them to graze. The fallow lands became temporarily a *saltus* appendix.[6] Herds' intrusions into *ager cultus* were not merely anecdotal. According to the *Lex de modo agrorum*, mentioned by Cato in 167 B.C., the Roman legislators wanted to regulate such a practice in order to protect the cultivated plots better because herds invaded them too often.[7] *Saltus* was not a scrub landscape without trees as is often claimed; on the contrary, the *saltus* concept was applied to all the wooded areas.[8] As opposed to *silva*, *saltus* was a forest close to the cities and villages and used as wood pasture. It could include some temporary cultivation plots.[9] *Silva*, on the other hand, was a deep forest, remote from cities and villages, where tall trees were occasionally cut down to supply the Empire's needs for building or navy timber. But even then, the difference between *saltus* and *silva* was not very strict. Sometimes, *silva* took the name of '*silvae et pascua publica*', which leads to suppose a use of these forested areas for breeding.[10] Some *silva* parts, called '*pratum*', were reserved as pasture for animals belonging to the Roman army (horses, mules, donkeys). Finally, extended *silva* blankets that were less accessible were out of the control of the Roman legions. The rebel Iberians found shelter there and moved around in the forest with their livestock.

So the idea of an antique rural landscape, which was rigorously divided into *ager*, *saltus* and *silva* is largely a myth.[11] The different types of rural space in Roman times more or less included agricultural, pastoral and forest activities. The borders between fields, pastures and forests were often vague.[12] Latin agronomists recommended this agrarian pattern. Columella, born in Cadiz

Vincent Clément

(*Gadès*, Spain), defended the *coltura promiscua* system in which the farmer could associate on the same plot crops, fodder trees for the feeding of herds (in particular oaks, elms and ash trees) and a vineyard.[13] Cato insisted on the necessity of preserving forests for firewood (*silva caedua*) and forests with acorns (*silva glandaria*) for pig feeding. He established a hierarchy between nine productions: vineyard, irrigated plots, willow trees, olive trees, meadows, cereal crops, coppice, orchards and forests with acorns.[14] Such a classification implicitly proves the considerable contribution of trees and woods to the rural economy. It also underlines that in Roman times many forests were integrated into the rural space, in particular for stock breeding. At times, they were eventually transformed into wood pastures.[15] For example, Roman Umbria (Italy) was famous for its good acorns and for its pig breeding. Oaks played a major role in pig feeding, so the farmers absolutely had to maintain trees on their lands, especially oaks called *quercie camporili*. Desplanques' point of view is that the oaks preserved in Umbria formed a particular landscape similar to the Iberian wood pastures.[16] Barker considers that most of the Italian forests were affected, more or less, by this landscape dynamic, because of the development of livestock breeding.[17]

In Roman South-west Spain, the rise of the transhumance system had contributed to the shaping of dehesas.[18] But it is difficult to assess the real incidence of such a process. Surprisingly, the landscape dynamic of the forest produced by the increase in pastoral activity is little understood. Traditional historiography laid more emphasis on Roman agriculture, showing little interest in the relationships between the breeding system and forest evolution. This is because agricultural activity left a lot of built structures (for example, Roman villas and small hydraulic dams), whereas remains of pastoral activity are obviously less numerous and more difficult to find.[19] But it is quite clear that the Romans were not the inventors of the dehesa. The analysis of pollen and wood charcoal carried out by Stevenson and Harrison set the origin of the dehesa landscape at about 4,000 years B.C. in the Huelva province.[20] Other researchers have confirmed those results. In the Medellín region, pollen studies realised by Almagro Gorbea highlight the fact that the dehesa landscape dates back to 2,500 years B.C., long before the Roman occupation.[21] Davidson and Chapman, both archaeologists, have also demonstrated that the transhumance system and the formation of dehesas were perfectly integrated into Palaeolithic and Mesolithic cultures of South-west Spain.[22] For Edmonson, the dehesa landscape's appearance in prehistoric times is connected with developments in livestock breeding.[23]

In conclusion, the landscape of wood pastures has its roots set in prehistoric times, before the Roman occupation. In Roman times, such a kind of landscape was also present in the countryside. But, because of insufficient research concerning its spread in antiquity, it is difficult to have a definitive idea of its relative importance in Spain. Nevertheless, the dehesas were certainly not as extensive as the croplands, vineyards and orchards along the Guadiana and the Guadalquivir valleys (called *Anas* and *Betis* in Roman times) where the Roman settlers were chiefly concentrated. After the fall of the Roman Empire, invasions of the Early Middle Ages (Barbaric invasions, Arabic conquest) widely contributed to depopulation of this area, in particular in what is today the Extremadura region, where dehesas are currently so widespread. The dehesas of Antiquity mostly disappeared under the combined effect of human depopulation and forest recovery. In the twelfth century, Extremadura was still a thinly populated march.[24] So the spreading process of the current dehesa in South-west Spain is certainly a genuine product of the Reconquest.

II. DEHESAS AS AN INHERITANCE OF RECONQUEST

Dehesas are undoubtedly one of the most visible consequences of the Christian Reconquest in the rural landscape.[25] But we must never forget that during the Medieval period the dehesa was above all a legal concept before being a kind of landscape. The term was used to refer to an enclosed forest.[26] In the Fuero Juzgo, the Visigoths' Code of Law enacted in 654, the dehesa appears in its ancient form of *pratum defensum*, which stemmed from Roman legislation.[27] According to the Corominas dictionary, the word *defesa* was only mentioned for the first time in 924. The dehesa concept occurs in this form in the Fuero of Sepúlveda of 1076.[28] This particular status of forest can also be found in other countries of Mediterranean Europe. In Lombardy (Italy), a *gazium* (plural, *gazzio*) was an enclosed forest, which was reserved for hunting and grazing.[29] In Languedoc (France), the words *devesa*, *devès* or *devèze* also indicated enclosed forests.[30] Nevertheless, the spreading of the enclosed forests in Spain has no equivalent in Mediterranean Europe. After the Christian Reconquest the greater part of Extremadura was progressively covered by dehesas. The dynamic began in the late thirteenth century. But as demonstrated by Martín and Oliva the process accelerated considerably during the course of the fourteenth and the fifteenth centuries.[31] Why was there such an expansion in enclosed forests after the Reconquest? Did the dehesas grow either with the development of the large properties of the military orders as is usually accepted, or rather because of the rise of transhumance?

Before analysing this point, it is necessary to have a clear idea of what the concept of large property was in the Medieval period. The large properties of South-west Spain had their origin in the process of territorial appropriation after the Christian Reconquest. In the south Tagus River, the Reconquest progressed in a significant way only after the storming of Alcántara in 1213. Southern towns of Extremadura were conquered some decades later: Cáceres in 1229, Badajoz in 1230, Trujillo in 1232 and Medellín in 1234. The repopulation of the

Vincent Clément

conquered territories had a strategic aim.[32] In order to strengthen the Christian Reconquest, it was necessary to attract new settlers into this forested and weakly populated march. The appropriation of the territory took place according to two modalities. The military orders (orders of Alcántara, Calatrava and Santiago) played a very active role in the advance of the Christians. They were composed of monk-soldiers. To reward them, the king gave them large properties. Around 1350, the orders of Alcántara and Santiago owned almost two million hectares between them.[33] As in the north Tagus River,[34] the king also granted large territories called '*comunidades de villa y tierra*' to the communities of colonists. They were usually situated around fortified towns like Cáceres, Trujillo and Badajoz, though this was not systematic. In the case of Mérida, the king gave the land to the military order of Santiago.[35] These modalities of territorial appropriation thus created a dual situation: on the one hand, there were the large properties of the military orders and, on the other, what we can qualify as the large properties of communities, for the territories of the *comunidades de villa y tierra* had often tens thousands of hectares.

The idea of large property was rather different from our current conceptions. The owners, whether from the military order or from the community, did not have an exclusive use of their territories. They had to organise the military defence of their territory, and they were also allowed to levy taxes and to exercise justice. But there was no rigorous separation between seigniorial, community or individual ownership. In the territory of Mérida, which stretched over 35,000 hectares, the Order of Santiago could only keep a third of this estate for its own use. The remaining two-thirds had to be given up to the colonists. So the new settlers used the greater part of the territory of Mérida, mostly in a collective way. But they could also possess individual plots or small dehesas in full property.[36] In the territories of the communities, as in the case of Cáceres, the king or the council of the community also allowed some settlers to possess private plots or dehesas.

In the large properties of the military orders, breeding of livestock was an effective way to exploit the territories conquered against the Moors. Moreover in Extremadura the omnipresence of the forest was an ideal condition for the development of the transhumance.[37] In the eleventh century, most of the convents in the northern Duero area (Cardeña, Oña and Sahagún) possessed large flocks of sheep, and they already practised transhumance over short distances.[38] But transhumance increased thanks to the progress of the Reconquest towards the south, because the military orders could develop a system of transhumance over long distances. The lands conquered in the territory of *al-Andalus* favoured the herds' moving over more than a thousand kilometres. Thus, functional links were established between the summer pastures in the north (*agostaderos*) and the winter pastures (*invernaderos*) in the south of the country.[39]

From the late thirteenth century onwards, the transhumance system increased in an unprecedented manner because of two historic events: in 1230, the union of the realms of León and Castile, and in 1273, the creation of the *Honrado Consejo de la Mesta*, a powerful association of transhumant breeders. King Alfonso X (1252–1284) granted vast privileges to the members of the association.[40] The breeders of the Mesta were allowed to take their herds in all the forests of the kingdom.[41] They had a monopoly on the use of every highway and byway (*cañadas*, *cordeles* and *veredas*) of the transhumance routes. The *Comendadores* of the military orders were among the Mesta's biggest breeders. So, at the end of the Middle Ages, the territories under the authority of the military orders, all of the Mesta's members, had become a succession of enclosed forests. In the fifteenth century the Order of Santiago already possessed at least 11 dehesas in the territory of Mérida alone.[42] In all of the territories of Calatrava, there were at least 114 dehesas belonging to this military order.[43] Transhumance routes connected all of the dehesas of the military orders. The breeders of the Mesta led up to the enclosed forests their numerous sheep, which were around 2,700,000 in 1467.[44]

What was the situation in the large estates of the communities? The enclosed forests also increased in those territories between the fourteenth and the fifteenth centuries.[45] For instance, at the beginning of the fifteenth century the territory of Badajoz already had at least 25 dehesas.[46] The reasons for such an increase were the opposite of those applicable to the military orders. In fact, the dehesas of the communities grew in reaction to the development of the transhumance herds.[47] The settlers wanted protection against the sheep of the Mesta, which too often invaded the forests of the peasant communities. The uses of the communities' dehesas were quite different from those of the military orders, as shown in the old Ordinances of Badajoz (1500).[48] The magistrates of the community compiled the document in order to regulate all the common practices in the territory of Badajoz. The document is composed of 52 articles, of which the first 15 concern the dehesas of Badajoz, and underline their importance for the community. The enclosed forests were named *dehesa boyal*, as in other communities.[49] They were reserved for the oxen, which were the most valuable animals for the farmers (article 3).[50] Sheep, goats and pigs were excluded from the dehesas (article 1). Guards, called *boyeros*, had to watch over the dehesas (article 5); they had to stay and live all the time in the dehesa (article 13).[51] The agreement of the ploughmen of Badajoz was necessary for designating the boyeros (article 15).[52] In return, the community undertook to respect the cañada which crossed their territory (article 25).[53]

Nevertheless, these Ordinances did not put an end to the conflicts between the breeders of the Mesta and the community. The clauses of the new Ordinances of Badajoz (1767) were almost the same clauses which had been used to resolve the conflicts inherited from the Late Middle Ages.[54] The memory of the conflicts

Vincent Clément

between Mesta breeders and peasant communities persisted from age to age, because they were often violent. One of the most famous examples concerns the village of Fuente Ovejuna. In 1792, the geographer of the king, Tomas López, recalled the incident: in 1476, the villagers put to death the *Comendador mayor* of Calatrava, Fernándo Gómez de Guzmán, because of his abuses concerning the dehesa boyal.[55] The event inspired Lope de Vega (*Fuente Ovejuna*, 1612).

So the growth of the dehesas was chiefly due to the development of transhumance, rather than to the existence of the large properties belonging to the military orders. In the territories of the military orders, the forest had to be enclosed to ensure that the breeders had enough pasture to feed their sheep. The turning point came in the second half of the thirteenth century, when transhumance breeding experienced an unprecedented rise. The enclosed forests continued to spread afterwards, mostly over the fourteenth and the fifteenth centuries. At the same time, surprisingly, transhumance also caused the development of dehesas in the territories of the communities, so that they might protect their own forests against the sheep of the Mesta. But did the dynamic of enclosure necessarily imply the development of the wood pasture landscape?

III. DEHESAS AS WOOD PASTURE LANDSCAPES

Researchers have neglected the question of the origin of wood pasture landscapes in the dehesas. Llorente Pinto recently underlined the lack of detailed research on the subject,[56] as did Grove and Rackham: both scientists blamed Julius Klein, author of a masterly thesis on the Mesta, for failing to back up his statements on the landscape of the dehesa with historical sources.[57] That is why two contradictory dogmas have resulted from this vacuum. For some academics, the appearance of the wood pastures goes back up to the Reconquest, while others assert that the landscape of wood pasture dates no farther back than the eighteenth century.

The origin of the dehesa as a wood pasture landscape is indeed a difficult matter. Written medieval sources often mention the existence of dehesas but without giving any description of the landscape. Nevertheless, there are some sources which have been barely exploited and which supply invaluable information: the boundary surveys. Such demarcations were numerous in the fourteenth and fifteenth centuries. They were drawn up for the lawsuits establishing the owners' rights on particular dehesas. Boundary surveys are legal documents with great reliability.[58] They were made under the authority of magistrates who represented the different parties involved. The magistrates followed the limits of the dehesas on the ground and confirmed or added boundary marks, in order to fix the borders of the controversial rustic goods. Later the magistrates drew up a report describing exactly the place and nature of every boundary mark. Trees were often used as boundary markers.

The example of the territory of Cáceres is quite interesting because it is situated in the heart of the large dehesa expanses of South-west Spain (Figure 2). Several boundary surveys were made in this territory in the fourteenth and fifteenth centuries. The three demarcations described here are respectively situated at the east, south and west parts of the territory of Cáceres. They concern different kinds of owners, either collective or private ones.

In the first document, dated 20 February 1300, the magistrates of Cáceres, responding to a royal demand, carried out a boundary survey of the dehesa of Guadiloba belonging to a woman named María García. This dehesa was situated east of Cáceres, close to the road going towards Trujillo. The following extract leaves no doubt as to the existence of a wood pasture landscape at that time:

> [...] on the road going towards the atalaya towers, [the limit] gives on to a forked holm oak, and farther on the right, big slate rocks situated over a spring, and farther on the right, a round shaped holm oak near a brook valley at the bottom of which there are two forked trees by a stump. And farther upwards in this valley, a big forked holm oak, and on the right, another tall holm oak which is between slate rocks near the brook valley, and farther up the brook valley, a holm oak which is at the summit where there is a boundary mark by a tall oak which is on the crest covered with brooms and oak coppices [...]. [59] (Translated from Spanish)

FIGURE 2. Dehesa landscape in Cáceres, South-west Spain

The document highlights the presence of a landscape which is mainly composed of holm oaks. Most of the trees were ancient, large-sized and pollarded, hence their forked shape. Between those trees, there were spaces without shrub vegetation. This landscape clearly corresponds to a wood pasture. In the treeless spaces, the magistrates of Cáceres used topographic or rocky elements to limit the dehesa. The presence of stumps on the boundary proves that there were some former trees that had been cut down. The holm oak's capacity to bud again from its stumps after having been chopped down explains the existence of isolated clumps of oak coppice on the boundary line.

In the south of Caceres, a second boundary survey, dated 15 March 1406, was made for clarifying the limits between the territories of Cáceres and Mérida:

> [...] another boundary mark has been renewed by putting stones on a rock which is up a hillock, and there is another boundary mark by two cork oaks near an old majada [...] on the right of the majada, farther up towards the summit where there are rocks and a grove of cork oaks, and farther on the right, there is a large boundary mark by the cañada of Valdecintados, and from there, towards the veredas which is coming from the Rincón [...] and farther, on the left, there is another boundary mark by a holm oak with a cross against it, and farther a cork oak near slate rocks and with two crosses against this cork oak [...]. [60]
> (Translated from Spanish)

Here, the boundary marks are mostly associated with cork oaks and secondarily with holm oaks. It is also a wood pasture, composed of sparse trees and areas without shrub stratum. As in the previous case the magistrates defined the limit of the dehesa by using topographic elements (hillocks, slate rocks) in the treeless areas. In the places with neither trees nor rocks they set the borders with stone heaps. Breeding activity in this dehesa is attested by several words associated with the transhumance routes (*cañadas*, *veredas*) or suggesting the presence of a small fenced park used for keeping animals (*majada*).

In the third document, dated 21 November 1457, another boundary survey was made on the territory of Cáceres between the dehesa of Puerto de Carmonita, belonging to Cáceres council, and the dehesas of Mayoralgo and Mayoralguillo, ownership of Diego de Mayoralgo. Those dehesas were on the western part of the territory of Cáceres, more exactly on the eastern hillside of the Sierra of San Pedro:

> [...] and off from the cistus heath, on the right another boundary mark at the foot of a pollarded holm oak and with a cross against this holm oak, and farther, on the right another boundary mark up a small hillock near a holm oak coppice, and from there, farther on the right another boundary mark at the foot of a cork oak close to a few little houses and a cross against this cork oak, and nearby in a brook valley another boundary mark, and farther on the right another boundary mark on the top of a mound near some holm oaks, and on the right another boundary mark made with three pebbles at the foot of a cork oak and a cross against this

cork oak [...]. [61] (Translated from Spanish)

The document also gives us the clearest evidence of a dehesa landscape. It is composed of holm oaks and cork oaks. Sometimes, the magistrates mentioned the presence of gum cistus. This plant colonises the burnt areas. On the border, the particular shape of the trees was sometimes specified (pollarded holm oak). Thus, the trees used as border markers were easily spotted. The magistrates also put crosses on the trees. Transhumance routes going through those wood pastures indicated the importance of breeding activity on them.

These three examples, taken from the territory of Cáceres, demonstrate that the dehesa landscapes were already widespread in the fourteenth century and were always associated with pastoral activity. Indirectly, the documents inform us about the two ways used for thinning the tree coverage and eliminating the shrub stratum: by cutting some trees down (presence of stumps) or by burning the shrubs (presence of gum cistus). The introduction of herds into the forests, however, had certainly accelerated the clearance of the shrub stratum. In this way, the breeders managed to obtain large grazing spaces for their droves. They protected the tall trees because of their usefulness for breeding activity. The trees preserved the moisture of the soil, and so ensured the preservation of grazing spaces. They were also fodder trees. Foliage, young branches and acorns were used to feed the herds. Finally, they supplied shade for the animals.

We can conclude that dehesa landscapes did not appear in the eighteenth century. Wood pastures were still widespread in the Late Middle Ages, at least in Extremadura. In the territory of Plasencia (northern Extremadura), Clemente Ramos has arrived at the same conclusions. [62] From the Middle Ages onwards wood pasture became an extensive landscape pattern in South-west Spain. The spreading of the wood pastures indicates that this landscape form was well adapted for developing livestock breeding activities. Moreover, the soils of Extremadura were poorly adapted for cultivation. The stock breeding industry, and especially transhumance, was the best way to exploit an area which in the Late Middle Ages was still mostly forested and thinly populated. But is that enough to understand the durability of the dehesa landscape over time? Are there any other historical factors to explain the amazing longevity of this landscape to this day?

IV. DEHESAS AS AN ENIGMATIC DURABLE LANDSCAPE

The durability of the dehesa landscape is a real historical enigma. How can one explain the fact that the wood pastures, which were largely diffused thanks to the particular conditions of the Reconquest, are still today an essential component of the landscape in South-west Spain? The situation is all the more surprising given that in the other Mediterranean countries wood pastures have almost disappeared, as in the south of France or in the Italian Peninsula.

The durability of dehesa landscapes can be explained by a combination of several factors. The first factor is the permanence of the border logic over time.[63] After the Moorish conquest in 711, the current Extremadura was included in a border march which extended from West to East between the territories of the Moors and the Christians.[64] At the beginning of the thirteenth century, the Christian Reconquest of Extremadura did not put an end to the border logic. There was only a change in its direction: it became a north to south border between the Portuguese and Spanish. The permanency of the border has influenced the perception of Extremadura a great deal. Indeed, today it is still considered as a marginal territory and as a less populated area of Spain.[65] Insecurity along the frontier with Portugal for a long time ensured only a very small population. In the fifteenth century, the Cáceres council regretted the depopulation of the frequently attacked villages of the border. The village of Aliseda, situated close to the boundary with Portugal, provides a telling example. Armed gangs regularly destroyed, plundered and set the village on fire. In a document of 1426, about two centuries after the reconquest of Cáceres, the council granted tax exemption to the present and future inhabitants of Aliseda in order to attract new settlers.[66] The council also allowed the inhabitants of Aliseda to take shelter in the fortified town of Cáceres in case of Portuguese attacks. Twenty years later, the problem was not yet solved, as another document of 1446 shows.[67] The village of Aliseda was always almost deserted and the rare inhabitants were often threatened. Insecurity affected all the territories near the Portuguese border. In the bylaws of Badajoz (1500), men were allowed to wear a lance and a sword for self-defence.[68] Danger was, for many centuries, a real threat in the border region. It not only limited the populating process, but favoured the development of a rural economy based on stock breeding, which was less vulnerable than cultivation in case of attack or conflict.

However, the border logic alone is not enough to explain the durability of the dehesa landscape. A second factor also needs to be considered: the surprising longevity of the Mesta. Created in 1273, the powerful breeders' organisation survived for almost six centuries. It widely contributed to the conservation of the dehesa landscapes, for three major reasons. First, because the Mesta breeders had a right to go through all the dehesas: through those belonging to military orders of course, but even through the dehesas of the communities. The right of way was a privilege of the Mesta breeders granted by the king. The local ordinances had to recognise it as shown in the fourth article of the bylaws of Badajoz (1500): '[…] *the droves can cross our dehesas, but without stopping, eating or sleeping in them* […]' (translated from Spanish).[69] The right of way strongly limited the change of dehesas into cultivation lands or towards others kinds of use. The second reason is that the Mesta breeders, through renewable rental agreements, sometimes hired the dehesas of communities or private owners. This was the case in Trujillo.[70] So the evolution of dehesas towards another form of exploitation was quite limited. The third reason is due to the importance of the Mesta in the Spanish economy. Between the sixteenth and the eighteenth centuries, the wool of merinos was one of the main sources of Spain's wealth. Wool was exported towards Flanders and England. For that reason, all endeavours on behalf of the political powers to call into question the powerful organisation of breeders was rather a risky venture. The conservation of the dehesas was vital to feed the herds of the Mesta, which did not stop increasing over the centuries: from 2,700,000 sheep in the late fifteenth century,[71] the number increased to 3,500,000 in the early sixteenth century [72] and to 5,000,000 in the late eighteenth century.[73] Nevertheless, progressively the Mesta was a victim of its own success. In the particular context of a growing population and the need for cultivable lands in Spain, the pressure caused by the Mesta's herds had become intolerable in the eighteenth century. This explained the criticisms formulated in enlightened circles, in particular by the Minister Jovellanos who recommended, around 1750, the abolition of the Mesta.[74] But the powerful breeders' association that was inherited from the Reconquest was abolished only in 1836. The decision on behalf of the political powers took place in a particular context: after the war for independence against France and the advent of the bourgeoisie after the Cortes of Cadiz in 1812, the Spanish State wanted to abolish the structures of the Ancien Régime of which the Mesta was one of the strongest symbols. The end of the large privileges of the Mesta made the wool economy less and less profitable. In 1892, only 1,355,000 transhumant sheep remained in Spain,[75] less than a quarter of the herds of the late eighteenth century. But for many centuries transhumance had contributed to the preservation of wood pastures for they were a kind of landscape well adapted to this activity.

The third factor concerns the protective law adopted by the communities of Extremadura to preserve their own dehesas. All of the local ordinances contained number of measures intended to protect the wood pastures. It was absolutely forbidden to cut down trees in the dehesas. In the bylaw of Mengabril (1548, in the territory of Medellín), several fines were imposed as punishment according with the size of the trees or branches illegally chopping down in the dehesa:

> [...] any person caught chopping down, conveying or loading a holm oak tree or coppice in our dehesa will be fined five hundred maravedis in benefit of the council; and for a branch as large as a man's body, the person will pay a fine of three hundred maravedis; and for a branch as large as a man's thigh, two hundred maravedis; and for a branch as large as a man's calf one hundred maravedis; and for a branch as large as a man's wrist twenty-five maravedis; and for smaller branches ten maravedis [...].[76] (Translated from Spanish)

To make it more understandable to everyone, the different size of trees and branches were compared to various parts of the body. The detailed fines demonstrate the great determination of the community to protect the tree coverage (called *vuelo*) of the dehesa. The pasture of the dehesa (called *suelo*) was also protected. It

Vincent Clément

FIGURE 3. Dehesa with black pig in the territory of Cáceres.

was forbidden to cut grass and take it away from the dehesa. Any person who did so had to pay a fine, and the council confiscated his tool as punishment.[77] The exploitation of by-products, like gathering acorns, was subject to a previous permit given by the council of Mengabril.[78] As in Mengabril, other communities of Extremadura took the same measures; there were only slight differences in the size of the fines. So the use of the wood pasture was strictly regulated by the local ordinances of the sixteenth century, which remained in effect until the nineteenth century. Then, the *desamortización* created a new situation. The desamortización was a process of compulsory sale of goods belonging to the Church and to the military orders (Law Mendizábal of 1836), as well as those of the communities (Law Madoz of 1855). Some dehesas disappeared after having been bought, because the new owner wanted either to pay them off by selling the trees, or to convert them into cultivation lands. But in most cases, the constitution of large private properties of dehesas in Extremadura mainly turned towards two new extensive breeding activities: on the one hand, the breeding of brave bulls for corridas and, on the other, the breeding of black pigs (a local breed) to produce a fine ham called *Jamón de Pata Negra* (Figure 3). These new breeding activities resulted in quite a profitable business, and have enabled the conservation of the dehesas in South-west Spain to this day.[79]

CONCLUSION

The origin of the Spanish wood pastures extends back as far as Antiquity. They were present in South-west Spain before the Roman occupation. The wood pastures of Roman times can be considered as a particular form of the *coltura promiscua* system, which was defended by Latin agronomists. Nevertheless, as demonstrated in this article, the current dehesa landscapes are more directly linked with the particular modalities of the Christian Reconquest. After the advance of the Christians, the monk-soldiers of the military orders and the new settlers appropriated the territory. Far from destroying the forest, they shaped an original wooded landscape by clearing the shrub stratum while preserving the tall trees and developing transhumance over long distances. In that way, they managed to exploit the large forest blankets of the border march of Extremadura. The development of stock breeding activity enabled them to turn to good account this thinly populated area, which was unsuited for agriculture. The spreading of the dehesa landscape in South-west Spain began in the thirteenth century, and the dynamic accelerated during the fourteenth and the fifteenth centuries. At the end of the Middle Ages wood pastures extended over the greater part of Extremadura.

The durability of the dehesa is a result of a combination of several factors. The permanency of the insecurity along the border with Portugal and, most of all, the importance of transhumance in the Spanish economy from the Late Middle Ages to the nineteenth century played a decisive part in the conservation of the dehesas in the territories of the military orders. Moreover, the protective measures adopted by the rural communities against the numerous sheep of the Mesta breeders helped to preserve their own dehesas. In the nineteenth century, the abolition of the Mesta in 1836 and the selling of the dehesas to private owners did not lead to the disappearance of the dehesas in Extremadura. New breeding activities (brave bulls and black pigs) have enabled the preservation of the dehesas to the present day.

Two lessons of a more general scope can be draw from the example of the Spanish dehesas. Firstly the Middle Ages are usually known in Europe as a period of intense deforestation favoured by monks and colonists. The shape of the dehesas of the Medieval period clearly challenges this thesis. Secondly the preservation of the dehesa landscapes also proves that the study of the relationship between humankind and the forest cannot be reduced to a systematically negative discourse. If the rural communities in Mediterranean Europe sometimes eliminated the forest to extend croplands because of demographic growth, in other cases, as in Extremadura, they were able to preserve the forest by shaping valuable landscapes. Today, forest engineers, agronomists, geographers, rural historians and politicians recognise the importance of the conservation of the

Vincent Clément

dehesa, which is a cultural landscape that has been shaped over the centuries by rural communities.

NOTES

My grateful thanks to Frédéric Regard, Professor of English Literature at the École Normale Supérieure, for much valuable advice, and to Louise Slater and Georgina Endfield for their assistance in editing this article for publication.

1. Harriet D. Allen, *Mediterranean Ecogeography* (London: Pearson Education Press, 2001), p. 191; André Humbert, *Le monte dans les chaînes subbétiques centrales* (Paris: Publications de la Sorbonne, 1980), pp. 5–8.

2. David E. Vassberg, *Land and Society in Golden Castile* (Cambridge: Cambridge University Press, 1984), p. 151; Roger Dion, *Essai sur la formation du paysage rural français* (Tours: Arrault, 1934), pp. 80–84.

3. The process, known as centuriation, was intended for land distribution to Roman settlers. It consisted of dividing the land along two main streets, which crossed each other at right angles: the '*Cardo Maximus*' (oriented North to South) and the '*Decumanus Maximus*' (oriented East to West). A network of secondary cardines and decumanus created a grid of square plots. Every square plot measured twenty *actus* (around 2,52 hectares). The result of such a division was named '*centuriacio*' (centuriation).

4. Gérard Chouquer, François Favory, *Les paysages de l'Antiquité. Terres et cadastres de l'Occident romain* (Paris: Errance, 1991), p. 227; Philippe Leveau, Pierre Sillières, Jean-Pierre Vallat, *Campagnes de la Méditerranée romaine* (Paris: Hachette, 1993), p. 69; Gérard Chouquer, François Favory, *L'arpentage romain* (Paris: Errance, 2001), p. 457.

5. Columelle, *De l'Agriculture* (Paris: Errance, 2002), book VII, 171.

6. Gérard Chouquer, *L'étude des paysages. Essai sur leurs formes et leur histoire* (Paris: Errance, 2000), p. 76.

7. Pedro Saez Fernández, 'Los agrónomos latinos y la ganadería', in J. Gómez Pantoja (ed.), *Los rebaños de Gerión. Pastores y transhumancia en Iberia Antigua y Medieval* (Madrid: Casa de Velázquez, 2001), pp. 162–3.

8. Christopher Wickham, 'European Forests in the Early Middle Ages: landscape and land clearance', in *L'ambiente vegetale nell'alto medioevo* (Spoleto: Centro Italiano sull'alto medioevo, 1990), 2, p. 486.

9. Wickham, 'European Forests in the Early Middle Ages', p. 488.

10. Pedro Saez Fernández, 'La ganadería extremeña en la antigüedad', in S. Rodríguez Becerra (ed.), *Trashumancia y cultura pastoril en Extremadura* (Mérida: Asamblea de Extremadura, 1993), pp. 44–5.

11. Vincent Clément, 'Une mer des hommes: les limites du monde méditerranéen', in D. Borne, and J. Scheibling (eds.), *La Méditerranée* (Paris: Hachette, 2002), p. 43.

12. Robert Delort, François Walter, *Histoire de l'environnement européen* (Paris: PUF, 2001), pp. 228–9.

13. Columelle, *De l'Agriculture*, book V, 125. See also Maria Gemma Grillotti Di Giacomo, *Atlante Tematico dell'Agricoltura Italiana* (Roma: Societa Geografica Italiana, 2000). There are a number of historical documents showing the consistency, over the centuries, of the *coltura promiscua* system in Italy. For Umbria, see pp. 310–11.

14. Caton, *De l'Agriculture* (Paris: Les Belles Lettres, 1975), p. 10.

15. Columelle, *De l'Agriculture*, book VII, p. 180.

16. Henri Desplanques, *Campagnes ombriennes. Contribution à l'étude des paysages ruraux en Italie centrale* (Paris: Armand Colin, 1969), pp. 236 and 245.

17. Graeme Barker, 'The archeology of the Italian shepherd', *Cambridge Philosophical Society* 35 (1989): 1–19.

18. Joaquín Gómez Pantoja, 'Pastores y trashumantes de Hispania', in Fr. Burillo Mozota (ed.), *Poblamientos celtibéricos, III simposio sobre los Celtíberos* (Zaragoza: Institución Fernando el Católico, 1996), pp. 495–505.

19. Vincent Clément, 'Le territoire du sud-ouest de la péninsule Ibérique à l'époque romaine: du concept au modèle d'organisation de l'espace', in J.G. Gorges and F.G. Rodríguez Martín (eds.), *Économie et territoire en Lusitanie romaine* (Madrid: Casa de Velázquez, 1999), p. 119.

20. Anthony C. Stevenson, Robert J. Harrison, 'Ancient forests in Spain: a model for land-use and dry forest management in Southwest Spain from 4000 BC to 1900 AD', *Prehistoric Society* 58 (1992): 227–47. See also Anthony C. Stevenson, P.D. Moore, 'Studies in the vegetational history of SW Spain. Palynological Investigations at El Acebrón Huelva', *Journal of Biogeography* 15 (1988): 339–61, doi: 10.2307/2845417; and Richard Joffre, Serge Rambal, Jean-Pierre Ratte, 'The dehesa system of southern Spain and Portugal as a natural ecosystem mimic', *Agroforestry Systems* 45 (1999): 57–79, doi: 10.1023/A:1006259402496.

21. Martín Almagro Gorbea, 'El territorio de Medellín en época protohistórica', in Gorges and Rodríguez Martín (eds.), *Économie et territoire en Lusitanie romaine*, pp. 27–8.

22. Iain Davidson, 'Transhumance, Spain and Ethnoarcheology', *Antiquity* 54 (1980): 144–7; Robert W. Chapman, 'Transhumance and Megalithic Tombs in Iberia', *Antiquity* 53 (1979): 150–52.

23. Jonathan C. Edmonson, 'Creating a provincial landscape: roman imperialism and rural change in Lusitania', in. J.G. Gorges and M. Salinas de Frías (eds.), *Les campagnes de Lusitanie romaine*, (Madrid/Salamanca: Casa de Velázquez/Universidad de Salamanca, 1994), p. 22.

24. José María Monsalvo Antón, 'Frontera pionera, monarquía en expansión y formación de los concejos de villa y tierra. Relaciones de poder en el realengo concejil entre Duero y Tajo (1072-1222)', *Arqueología y territorio medieval* 102 (2003): 56.

25. Julián Clemente Ramos, 'La organización del terrazgo agropecuario en Extremadura (Siglos XV-XVI)', *En la España Medieval* 28 (2005): 53–4.

26. Vincent Clément, 'La forêt et les hommes en Castille au XIIIᵉ siècle. L'exemple du territoire de Sepúlveda', *Mélanges de la Casa de Velázquez*, XXX-1 (1994): 271–2.

27. Saez Fernández, 'La ganadería extremeña en la antigüedad', p. 45.

28. Emilio Saez, *Los fueros de Sepúlveda*, (Segovia: Diputación provincial, 1953). See Fuero romanceado, title 169, pp. 119–120. The Fuero of Sepulveda was a code of law specific to the border regions. It served as a legal model in all the Castile Kingdom.

29. François Menant, *Campagnes lombardes au Moyen Âge* (Rome: École Française de Rome, 1993), pp. 206–7.

30. Aline Durand, *Les paysages médiévaux du Languedoc (Xᵉ-XIIᵉ siècles)* (Toulouse: Presses Universitaires du Mirail, 1998), pp. 397–8.

31. José Luis Martín Martín, 'Sur les origines et les modalités de la grande propriété du Bas Moyen Âge en Estrémadure et dans la Transierra de Léon', in D. Menjot (ed.), *Les Espagnes médiévales. Aspects économiques et sociaux, Mélanges offert à Jean Gautier-Dalché* (Nice: Annales de la Faculté des Lettres de Nice, 1983), pp. 81–91; María Dolores Oliva, 'Orígenes y expansión de la dehesa en el término de Cáceres', *Studia Historica. Historia Medieval* 2 (1986): 77–100.

32. Juan Luis de la Montaña Conchiña, 'Poblamiento y ocupación del espacio: el caso extremeño (siglos XII-XIV)', *Revista de Estudios Extremeños* 2 (2004): 574.

33. Montaña Conchiña, 'Poblamiento y ocupación del espacio', p. 594.

34. Vincent Clément, 'Frontière, reconquête et mutation des paysages végétaux entre Duero et Système Central du XIᶜ au milieu du XVᶜ siècle', *Mélanges de la Casa de Velázquez*, XXIX-1 (1993): 117.

35. José Luis Martín Martín, 'Mérida medieval, señorío santiaguista', *Revista de Estudios Extremeños* 2 (1996): 487.

36. Martín Martín, 'Sur les origines et les modalités de la grande propriété', p. 85.

37. Julián Clemente Ramos, 'La evolución del medio natural en Extremadura (1142-1525)', in J. Clemente Ramos (ed.), *El medio natural en la España medieval, Actas del I Congreso sobre ecohistoria en historia medieval* (Cáceres: Universidad de Extremadura, 2001), p. 16.

38. Marie-Claude Gerbet, *Un élevage original au Moyen Âge. La péninsule Ibérique* (Biarritz: Atlantica, 2000), p. 101.

39. Charles J. Bishko, 'The Castilian as plainsman: The medieval ranching frontier in La Mancha and Extremadura', in A.R. Lewis and T.F. McGann (eds.), *The New World Looks at its History* (Austin: University of Texas Press, 1963), pp. 56–7.

40. Pedro Madrigal, *Libro de las leyes, privilegios y provisiones reales del Honrado Consejo General de la Mesta* (Madrid: En casa de Pedro Madrigal, 1586).

41. For example, in 1284, King Sancho VI granted the Order of Alcántara the liberty of taking their herds anywhere in the kingdom. Bonifacio Palacios Martín, *Colección diplomática medieval de la Orden de Alcántara (1157-1494)* (Madrid: Editorial Complutense, Fundación San Benito de Alcántara, 2000), document 361.

42. Daniel Rodriguez Blanco, *La Orden de Santiago en Extremadura (siglos XIV y XV)* (Badajoz: Diputación Provincial, 1985), p. 173.

43. Geronimo López-Salazar Perez, *Mesta, pastos y conflictos durante el siglo XVI* (Madrid: Consejo Superior de Investigaciones científicas, Centro de Estudios Históricos, 1987), pp. 9–10.

44. Gerbet, *Un élevage original au Moyen Âge*, p. 268.

45. Clemente Ramos, 'La evolución del medio natural en Extremadura', pp. 16–20.

46. Martín Martín, 'Sur les origines et les modalités de la grande propriété', p. 87.

47. Thomas F. Glick, *Islamic and Christian Spain in the Early Middle Ages* (Leiden: Brill Academic Publishers, 2005), pp. 106–7.

48. See the transcription of the full text of the Ordinances in José Luis Martín Martín, 'Las ordenanzas viejas de Badajoz (1500)', *Revista de Estudios Extremeños* 1 (2001): 233–60.

49. Vassberg, *Land and Society in Golden Castile*, pp. 28–9. The adjective *boyal* refers to the *buey* (ox).

50. 'Otrosi, que los labradores puedan tener bueyes para arrendar teniendo pues demasiados de su valor y traellos en las dichas dehesas sin pena, y no otra persona [...]', Martín, 'Las Ordenanzas viejas de Badajoz', article 3, p. 247.

51. 'Otrosi, quel boyero sea obligado a estar e residir en la boyada [...]', Martín, 'Las Ordenanzas viejas de Badajoz', article 13, p. 249.

52. 'Otrosi, ordenamos e mandamos que los dichos boyeros al tiempo que se oviere de recabar sea a contento de los labradores, tomando la justiçia e regidores ynformaçión de personas sin sospecha quál es el que más conviene ser boyero y conforme aquello y al contentamiento de los labradores se tome tal boyero por que mejor haga lo que conviene [...]', Martín, 'Las Ordenanzas viejas de Badajoz', article 15, p. 250.

53. As seen in article 25, the peasants used to build small buildings such as cheese dairies (*quesera*) or pigsties (*zahúrda*) at the side of the cañada for the few animals they had. But the cañada, which was usually 75 metres wide, was reserved for the Mesta's breeders. Despite the ban, the farmers frequently colonised the sides of the cañada. That, of course, was another motive of conflicts between the Mesta's breeders and the community. 'Hordenamos y mandamos que ninguna persona vecino destaçibdad y su término no tenga quesera ni çahurdas ni corrales ni

redes de ovejas ni de carneros en cañada, y si la tuviere que se la derriben y deshagan a su costa |… |', Martín, 'Las Ordenanzas viejas de Badajoz', article 25, p. 253.

54. Supremo Consejo de Castilla, *Ordenanzas de la muy noble y muy leal ciudad de Badajoz* (Madrid: Sanz, 1767).

55. ˙Puso despues la Orden de Calatrava por gobernador en esta villa a Don Fernándo Gómez de Guzmán, comendador mayor en ella a quien por sus tiranías y desafueros le dieron sus vecinos sangrienta muerte en el año 1476 |…| bolvieron después a avecimarse algunos cavalleros hijosdalgo, pero la dicha villa les negó el aprovechamiento de pastar con sus ganados en la dehesa boyal del concejo |… |', *Diccionario de Tómas López (1792)*, Biblioteca Nacional de Madrid, sección Manuscritos, Mss. 7294, folios 441–2.

56. José Manuel Llorente Pinto, 'El problema de la sostenibilidad de las dehesas a la luz de la evolución histórica de los terrenos adehesados', *Actas de la II Reunión sobre Historia Forestal, Cuaderno de la Sociedad Española de Ciencias Forestales* 16 (2003): 136.

57. A.T. Grove and Oliver Rackham, *The Nature of Mediterranean Europe: An Ecological History*, (London: Yale University Press, 2001), pp. 201–2; Julius Klein, *The Mesta: A Study in Spanish Economic History (1273–1836)* (Harvard: Harvard University Press, 1920).

58. Vincent Clément, *De la marche-frontière au pays-des-bois. Forêts, sociétés paysannes et territoires en Vieille-Castille (XIᵉ-XXᵉ siècle)* (Madrid: Casa de Velázquez, 2002), pp. 144–50. See also Vincent Clément, 'Les grandes pinèdes au sud du Duero: l'archéologie des paysages au secours d'une origine controversée', in E. Fouache (ed.), *The Mediterranean World Environment and History*, Working Group on Geo-archeology Symposium Proceeding (Paris: Elsevier, 2003), p. 211.

59. Antonio Floriano Cumbreño, *Documentación histórica del Archivo Municipal de Cáceres (1229–1471)* (Cáceres: Institución cultural El Brocense, 1987), document 9.

60. Cumbreño, *Documentación histórica del Archivo Municipal de Cáceres*, document 71.

61. Cumbreño, *Documentación histórica del Archivo Municipal de Cáceres*, document 103.

62. Julián Clemente Ramos, 'Explotación del bosque y paisaje natural en la Tierra de Plasencia (1350–1550)', in R. Uriarte (ed.), *IX Congreso de Historia Agraria*, (Bilbao: Universidad del País Vaco, Departamento de Historia e Instituciones Económicas, 1999), p. 449; Julián Clemente Ramos, 'La organización del terrazgo agropecuario en Extremadura (siglos XV–XVI)', *En la España medieval*, 28 (2005): 66–70.

63. Marcelino Cardalliaguet Quirant, *Sociedad y territorio en la historia de Extremadura* (Cáceres: Diputación Provincial, Intitución cultural El Brocense, 1999), p. 26.

64. These borders were broadly named *thugûr* (singular, *Tâgr*). Extremadura was named *Al-tâgr-al-Djawfî*, which was part of the Lower Marsh. About the relationship between forest and border in the Middle Ages, see: Vincent Clément, 'La frontera y el bosque en el medievo. Nuevos planteamientos para una problemática antigua', in P. Segura Artero (ed.), *La frontera oriental nazarí como sujeto histórico (siglos XIII–XVI)* (Almería: Instituto de Estudios Almerienses, 1997), pp. 329–37.

65. Bishko, 'The Castilian as Plainsman', p. 48.

66. Cumbreño, *Documentación histórica del Archivo Municipal de Cáceres*, document 81.

67. Cumbreño, *Documentación histórica del Archivo Municipal de Cáceres*, document 93.

68. Martín, 'Las ordenanzas viejas de Badajoz', article 6, p. 248.

69. Martín, 'Las ordenanzas viejas de Badajoz', article 4, p. 247.

70. Vassberg, *Land and Society in Golden Castile*, p. 81.

71. Jules Goury du Roslan, *Essai sur l'histoire économique de l'Espagne* (Paris: Guillaumin, 1888), p. 232.

72. Jean-Paul Le Flem, 'Las cuentas de la Mesta (1510-1709)', *Moneda y Crédito* 121 (1972): 68.

73. Angel García Sanz, 'El siglo XVIII: entre la prosperidad de la trashumancia y la crítica antime-seteña de la Illustración (1700–1808)', in G. Anes and A.G. Sanz (eds.), *Mesta, trashumancia y vida pastoril* (Valladolid: Sociedad V Centenario del Tratado de Tordesillas, 1994), p. 139.

74. García Sanz, 'El siglo XVIII: entre la prosperidad de la transhumancia', pp. 154–5.

75. André Fribourg, 'La transhumance en Espagne', *Annales de Géographie* (1910): 235.

76. 'Las ordenanzas de Mengabril (1548)', transcription of Julián Clemente Ramos, *Revista de Estudios Extremeños* 2 (2004): 630.

77. Clemente, 'Las ordenanzas de Mengabril', p. 635.

78. Clemente, 'Las ordenanzas de Mengabril', p. 645.

79. Carlos Gregorio Hernández Díaz-Ambrona, 'La dehesa extremeña', *Revista de Agricultura* 750 (1995): 37–41.

A Native American System of Wetland Agriculture in Different Ecosystems in the Ecuadorian Andes (15th–18th Centuries)[1]

Chantal Caillavet

In affectionate memory of Elinor Melville

WETLAND AGRICULTURE AND RIDGED FIELDS IN LATIN AMERICA

Ridged fields and wetland agriculture are a key element in the culture and ecology of pre-hispanic societies, and in some areas they survived beyond the immediate aftermath of the Spanish conquest. Their importance was first established in a series of groundbreaking publications which appeared from the mid-twentieth century onwards. West and Armillas' study (1950)[2] of the famous 'chinampa' system of wetland agriculture in Mexico paved the way for a series of later studies in archaeology, agronomy and ethno-history. These provided valuable information on the use and functions of the 'chinampas' in the years after Contact. Research that stressed the economic and ecological success of these native systems drew on surprisingly abundant pre-Hispanic documentation, as well as, more predictably, the Spanish Colonial records.[3] Geographers, archaeologists and historians all clarified different aspects of pre-Columbian raised fields and wetland agriculture.[4] At the same time, research on specific wetland techniques clearly lies within the wider frame of reference of the study of pre-Hispanic landscapes.[5]

In terms of South America, pioneering research was carried out by Parsons and Bowen on the ridged fields of the Colombian San Jorge river flats (1966); near Lake Titicaca by Smith, Denevan and Hamilton (1968); and by Broadbent on the Andean highlands near Bogotá (1968).[6] Subsequent surveys of these fossil agrarian forms have established that they can be found in a variety of archaeological areas and climatic and ecological environments, ranging from cold highlands to tropical lowlands.[7] Nevertheless, in marked contrast to Mexico, there is much less historical data on this type of native agriculture in South America. Was this silence a result of these techniques falling into centuries of disuse on the arrival of the European colonisers? Or, despite their ongoing use, did they simply fail to attract enough attention to appear in the literature?

In the absence of any descriptions by the Spanish chroniclers, what other types of historical sources can provide significant information on native traditions

Environment and History **14** (2008): 331–53.

Chantal Caillavet

of land use? The voices of the native peoples were recorded now and again, but indirectly, and only through the filter of Eurocentric forms of recorded history. Specifically, they can be traced in the legal texts that Indians dictated to acculturated Indian translators ('indios ladinos') to satisfy Colonial requirements: wills, declarations during land litigation, and judicial ceremonies associated with land possession. In these circumstances, the witness had to describe the land that he or she was bequeathing or claiming, and would frequently give very detailed information about its location, its key geographical features, climate and crops that were grown there. Place-names, too, were often mentioned. By compiling and comparing archival data of this kind from the northern highlands of Ecuador, the present-day provinces of Pichincha, Imbabura and Carchi (located between 1°N and 1°S), I have identified a range of evidence on the character and functions of raised fields and wetland agriculture (see Figure 1). I was also able to establish the meaning of some local place-names that may be connected to native forms of land use.[8]

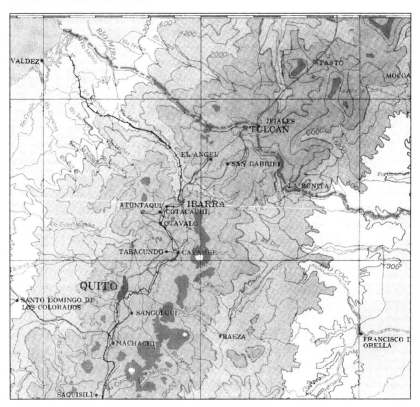

FIGURE 1. Map showing location of study sites

ARCHAEOLOGICAL EVIDENCE OF RIDGED FIELDS IN THE NORTH ECUADORIAN ANDES

My analysis of the available historical data will take into account major innovative work that has been carried out in the fields of archaeology, agronomy and climatology. Archaeologists and geographers have discovered fossil ridged fields ('camellones') in many parts of the north Ecuadorian Andes.[9] Gondard and López carried out a highly interesting and comprehensive survey[10] in which they identified about fifteen sites in cold climates covering roughly 2,000 hectares, at altitudes rising from 2,400 to 3,100 metres. Of these sites, San Pablo – at that time the best preserved – was the subject of archaeological excavations and experimental reconstruction. The scientific results provided us with the most complete information hitherto available on the technique's morphology, functioning and cultural characteristics. The combined ridge and ditch (that is, including both the land above water and the submerged area) has a total width of between 3 and 7 metres. The proximity of still water raises nocturnal temperatures, and mitigates the effects of frost. The mud accumulating in the ditch, that is collected to reinforce the ridge, has a high phosphorus content that will have made up for the lack of animal fertilisers in the northern Andes in pre-Hispanic times.

Experiments carried out at the end of the twentieth century by agronomists and climatologists highlight the positive effect of stagnating water in combating frost. Vacher, Lhomme and Erickson have measured an increase of 1 to 2 degrees in temperature in the area of reconstructed ridged fields near Lake Titicaca. This has had a decisive effect in reducing annual frosts, and, therefore, in saving harvests during key periods of crop growth. Maximum efficiency is obtained through a distribution of 4 metres of platform to between 1.5 and 2 metres of water.[11] The wind-breaks identified on the plain of Cayambe, where screens of trees were planted along the ridges, may also have helped to mitigate the effects of frosts.[12] This hypothesis is consistent with the archaeological evidence that makes use of phytolith remains to reconstruct the area's agrarian landscape.[13]

The best-studied cases – those of the Cayambe and San Pablo plains, north of Quito, in the inter-Andean basin – perfectly illustrate the conclusions reached by a number of geographers, as summarised by Bouchard and Usselmann.[14] In each case, the palaeotechnique of wetlands has a distinct purpose. It may help to resolve problems – permanent or seasonal – such as excess of water in marshlands, or the irregularity of hydraulic supply when a given area is affected by both drought and flooding at different times.

A series of early nineteenth century historical testimonies describe the geographical conditions in these areas. In accordance with modern scientific conclusions, they clearly show how problematic climate and soil type were. In the case of Cayambe in 1808, for example, it was said: 'when there are icy

Chantal Caillavet

spells (as there have been this year), excessive rain or drought, the crops are destroyed and there is great scarcity'. In the same year, Otavalo, north of Quito (see Figure 1), was described in the following terms: 'the harvests in Otavalo are more abundant in less rainy years, because the ground is naturally humid and irrigated by the many streams that descend from the heights, so that great amounts of rainwater are damaging'. Finally, there was a description of San Pablo: 'on account of the great amount of water from the hillsides ... the village near the Lake has several marshy areas, as it is wet even at a considerable distance from the Lake'.[15]

Archaeologists have recently established a chronological framework for the agricultural practice of raised fields. Recent research in vulcanology has led to considerable advances in the dating of population movements and pre-Hispanic settlements in Ecuador.[16] In the Ecuadorian Andes, the most significant event occurred in approximately A.D.1280 when the Quilotoa volcano erupted and scattered ash over a considerable part of the sierra that we now associate with the use of ridged fields. Research based on textural pedofeature characteristics shows that intensive efforts were made to clear away volcanic ash and re-establish the ridged fields.[17] With regard to its subsequent influence, both Villalba working on Cayambe, and Knapp and Mothes studying San Pablo, stress the chronological concurrence of the technique of ridged fields and the use of ceremonial mounds ('tolas') and raised platforms associated with the major highland chiefdoms up to the arrival of the Spaniards.[18]

THE ETHNOHISTORICAL EVIDENCE

What then is the additional information that may be retrieved from the historical documentation? Taking my earlier work on early Colonial sources as my starting point, I will suggest here that a quite specific vocabulary (such as 'camellon' and 'pijal'), as well as more indeterminate terms such as 'marshland' ('cienagas') and 'lakes' ('lagunas'), may refer quite explicitly to a deliberate modification of the landscape and to some form of wetland agriculture. A close linguistic reading of early texts allows us to enlarge our inventory on the basis of case-by-case analysis. In this way I identified various sectors of 'camellones' in the sixteenth century. The area of San Pablo was one such of course, but there was also Cochecarangue (the term for an extensive swathe of territory that ranges from the lower reaches of the valley of Tahuando, Las Monjas, Angochagua and Zuleta and could reach Caranqui). A donation of 1606 by the encomendero of Caranqui of the grazing land ('estancia') of Cochecarangue established its frontier with another one in Caranqui, bordering on 'the village itself and marshland' ('el mismo pueblo y cienagas').[19] At the beginning of the sixteenth century, part of the plain of Cayambe was known as 'the marshland' ('la cienaga'). And in the area near Otavalo there was 'a lake and marshland called Quinchuqui and others called Guabizi', according to Indian title deeds of 1596.[20]

Using the same methodology, I established that the toponym 'Pixal/Pijal' referred to wetland agriculture, and I can now add that another linguistic particle 'Pifo/Biafo' refers to some comparable agrarian form. It is not, however, possible to say whether this was ascribed to another morphological feature, such as distinct types of 'camellones' distributed in a 'chessboard' form, parallel or curvilinear. Up to the 1980s, the varied typology of these remains was visible to the naked eye. An aerial photo dating from 1963 shows very clearly the remains of different types of 'camellones' on the edge of the San Pablo lake (see Figure 2).[21] Three examples are directly related to sites on the edge of the San Pablo Lake where 'camellones' have been specifically identified. In Oyagata, in 1580, a witness attests to having 'two fields of maize, the rest are camellones of potatoes' in a 'place called Abiafu'. A second example comes from San Miguel, 1680, where the witness has 'two sources next to the lake of San Pablo ... and land and marshland ... and a spring called Cuspifu'. Finally, in Cusin, in 1690, land in the 'Rinconada de Cusin' is called 'Intapifo'.

Intriguingly, this toponym also appears in hot lowlands, which raises another question: what is a 'camellon' in warm areas? Why do the same terms ('camellon', 'pifu') refer to arguably distinct techniques of wetland agriculture? In 1560, for instance, cotton is mentioned as being grown in 'Lachipichig Bifu' in the hot valley of Pisque, near the pre-Hispanic settlement of Alchipichí.[22]

While archaeological research has advanced our understanding of wetland agriculture in the northern highlands, similar work has yet to be carried out on neighbouring warmer valley slopes and low-lying areas. The main point of comparison we have is with the Ecuadorian coast – namely 'camellones' in the Guayas basin in the south, and in the Tumaco-la Tolita region in the north – as well as with the hot lands of the San Jorge river in Colombia.[23]

THE ADVANTAGES OF RIDGED FIELD TECHNIQUES IN WARM AND COLD AREAS

With only historical data to count on, it is difficult to reconstruct agrarian techniques in the warmer areas of the northern Ecuadorian Andes. Returning to material I have previously studied,[24] I would argue here that the terms 'camellon' and 'raya' (row or line) and the toponym 'pifu' correspond to the same closely inter-connected logic. 'Camellon' and 'raya' were Spanish words used by the Indians in the sixteenth and seventeenth centuries to designate coca crops, and orchards with fruit such as avocados, guavas and bananas in the tropical valleys of the rivers Coangue-Chota, Ambuquí and Palacara. The toponym 'pifu', on the other hand, was associated with cotton growing on the lower stretches of

Chantal Caillavet

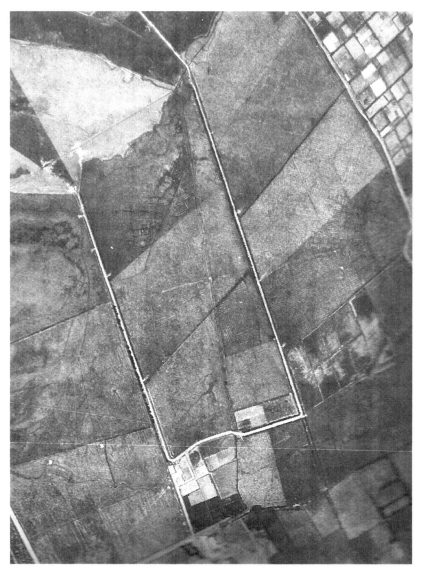

FIGURE 2. Aerial photograph showing remains of 'camellones' near San Pablo Lake.

the River Pisque, so it may be that the similarity was more in form than function. Irrigation is essential for these tropical crops, and they were to be found in economically valuable locations in warm valleys. They were grown on the riverbanks and often exiguous land surfaces known as 'banks' ('playas') that accompany the rivers on their descent through lower ecological levels.

In these tropical fields, the goal is to retain water, and these terms seem to describe both the raised ridge where trees are planted, flanked by water on each side, and the elongated form of the fields. They may refer to wetland orchards ('huertas') or some system of irrigation inherent to plots on riverbanks. Detailed descriptions of coca growing make it clear that there is an association with taller trees that alternate with coca plants, or surround them, probably so as to protect them from the burning Equatorial sun. An example from Pimampiro in 1629 mentions avocado trees growing both within and around the coca plants: 'I have a plot of coca ... with five rows ('rayas') of coca and five avocado trees, and with more than twenty-three avocado trees around the said area'. As to the cotton that is also grown in hot valleys, this was either under the direct control of the Otavalo ethnic group (in the valleys of Intag, Pisque, Guayllabamba, Palacara, Coangue-Chota, Ambuquí), or of the more western ethnic groups around Lita and Cahuasquí, that specialised in its production.[25] It is possible that the terminology refers to a kind that is comparable to irrigation in form, but whose specificity does not emerge from the documentation. Both in cool climates at high altitude, and in hot lowlands, the availability of abundant water (via rain, poor drainage, access to rivers or through artificial measures to retain it) seems to be the common denominator.

The historical data also allows us to identify the distribution of crops by area, taking into account both the region's historical evolution and Spanish Colonial cultural impositions. On various occasions in the sixteenth century, the autochthonous population and its ethnic leaders answered questions for Spanish surveys. Their opinions on agriculture represent a valuable testimony on the Indian approach to risk-taking in agriculture, and their preference for certain crops, whether for their food value or symbolic importance.

For the period 1578–82 archival documentation has survived which includes declarations by Indians from the area of Otavalo. For example, an inhabitant of Sarance (the present-day town of Otavalo at 2,600 metres) said they 'harvest potatoes, squash and beans' and many witnesses declared that the Otavalo area was not ideal for growing maize. Two ethnic leaders ('caciques') from San Pablo, in particular that of Pixalquí (where the 'camellones' mentioned above were located), said that Otavalo had 'a cold climate. Maize doesn't grow'. Other Indians from the same area specified that 'maize does not grow as well in the village of San Pablo as in the other villages'. The principal cacique of the ethnic group blamed the frosts: 'there is no maize ... because although the Indians plant it, most years it freezes and is lost'. The wheat introduced in Colonial times was even more vulnerable to the cold and in 1581 and 1582 there were both droughts and frosts:[26] 'the caciques of Otavalo owe 1500 fanegas of wheat and maize this year and last year ... on account of the great drought which occurred and because their crops froze'.[27]

Frost and drought were the most dreaded climatic conditions. In contrast, no reference was made to problems that might have been caused by excess water, nor by possible drainage problems. When referring to the technique of cultivation of 'camellones' in cold areas, only potatoes and vegetables are mentioned: 'camellones' with potatoes on the edge of San Pablo Lake, in Pijal y Oyagata, 'camellones' of vegetables ('verduras', 'guacamullos') in San Pablo, Caranqui and Guápulo.[28] I now consider that the only documentary reference to the cultivation of maize in 'camellones' in the Quito area, one that comes from a non-indigenous source, does not in fact refer to the technique of ridged fields. The term 'camellón', as used in 1573 by a Spanish official, allowed him to contrast cultivation on flat plains for the wheat and barley coming from Europe, with the undulating ground used for maize. This would have made his description clearer to European readers.[29]

An investigation by Knapp[30] on the present-day distribution of maize and potato crops at different altitudes around the Imbabura volcano helps to address a problem that is of great interest for the sixteenth century and the pre-Hispanic period. He inquires about the preference for potato crops at altitudes above 3,000–3,200 metres, and maize production at lower altitudes of around 2,600–2,800 metres. He examines the classic division into ecological levels, noting that an economic rationale does not provide us with a complete answer insofar as both crops can compete at similar levels, while a combination of calorific value and lower labour requirements favours maize. He provides data on hacienda production in the same areas as those known for traces of 'camellones', San Pablo, La Vega, Zuleta and the neighbouring upper slopes of the Imbabura (Topo, Angla).

I believe we can adopt a similar approach for the sixteenth century. We do not have direct references to the technique of 'camellones' being applied to maize production in cold areas, although Veintimilla's archaeological work suggests intermittent maize production in the 'camellones' on the Cayambe plain.[31] We do, however, have this evidence for potato production. High altitude maize was not highly prized, possibly out of cultural preference or for reasons of agronomic efficiency. This is open to conjecture. The Indians much preferred agricultural products from the hotter lowlands. For the Otavalo ethnic group, as well as other local groups under their control, the areas of maximum economic value were the low-level tropical valley enclaves in the sierra, and the hot lowlands in the western foothills of the Andes. The complementary nature of production in cool highlands/warm lowlands parallels findings from studies of Ecuadorian chiefdoms.[32]

The Indians of the northern highlands express this idea quite explicitly. The distinctions they draw between cold and hot areas is synonymous with that between the poor and the rich. Cotton, capsicum, coca and tropical fruit can only be produced in scarce and privileged areas, generally under the control of the autochthonous elite. For example, the highly prized hot land at Alchipichí (at 1,700 metres), where cotton and coca but also maize was cultivated, was held by the descendants of the principal cacique of the Otavalo ethnic group. This was also the case in the hot valley of Ambuquí, where maize, capsicum and coca were planted. This tropical maize was also grown by the Lita ethnic group in the dense tropical forest ('montaña') of the western slopes of the Andes, as well as by other lowland groups that were linked to the Pasto and Otavalo ethnic groups.[33] Perhaps the varieties of tropical maize were appreciated for their taste or other gastronomical qualities. It is quite clear that one feature was especially appreciated by the Indians: 'the maize is harvested every two months'. In the hot valleys of the south of Colombia that were controlled by high altitude Pasto Indians, the double annual harvest was also highly valued.[34]

It is difficult to estimate productive capacity. As often happens when we study ancien régime conditions, we can indicate the relation between seeds and harvest, but not agricultural extension. Nevertheless, a comparison between four habitats with distinct ecological characteristics is highly revealing, with maize cultivation in the hot lands of Otavalo proving the most productive. Thus we find that in the cold Chambo area, at 3,200 metres, the surface measure of 1 'fanega de sembradura' had a harvest capacity of 8 to 10 fanegas; near Quito, which is also cold, lying at 2,800 metres, 1 fanega produced 20–40; in the warmer valley of Chillos at 2,500 metres, 20–30 and 60–70 fanegas were produced; and in the hot valleys at 1,700–2,000 metres in the Otavalo region, this figure rose to 50–60 and 100.[35] We can only be sure that irrigation was involved in this final example. A logical inference, I think, is that preference was given to maize from the more productive hot lands, while the 'camellones' of cool areas were limited to the cultivation of potatoes and other autochthonous crops (other tubers, lupine, quinoa, beans).

These two autochthonous groups of products were viewed quite differently by the Colonial authorities. Consumption of tubers was left to the exclusive sphere of the autochthonous population, while increasingly, maize was preferred both for new Colonial economic requirements and Indian consumption. Maize fed the local workforce who laboured for the Spaniards, and also fed European cattle which were increasingly numerous from as early as the sixteenth century onwards.[36] The colonisers shifted the pattern of distribution of agricultural land: they assigned a greater area to cereals and grazing, sacrificing autochthonous priorities with regard to tubers, quinoa. Marshy areas were gradually dried out to make way for grazing land, while the amount of land sown with wheat, barley and maize cultivation was expanded.[37]

Another advantage of wetland agriculture is that it seems to be unnecessary to leave land fallow, a practice that was already widespread from the eighteenth century onwards when some land was regarded as infertile. The palaeotechnique of 'camellones' maintained the fertile layer in situ and helped to deter soil erosion.

THE PRODUCTION OF SUBMERGED LAND

The technique of 'camellones' also afforded benefits from the area under water. Archival data show that various aquatic varieties of vegetables were cultivated, although they were not described in detail as they did not form part of Spanish diet. They were green vegetables ('verduras') for strictly Indian consumption. For example, near Caranqui in 1626 we find 'the place called Pigalqui where we have our vegetables as it is marshy land covered in water'. An ethnographic comparison I conducted in 1979–81 revealed the survival of herbs ('yerbas'), in Indian peasant food consumption; these were aquatic herbs ('yuyos' and 'pima' in Quechua), turnip leaves and water cress ('bledo' and 'berro' in Spanish). The Royal administrative surveys, or *Relaciones Geográficas,* of 1573 and 1582 drew attention to the diversity of cultivated varieties and their major role in consumption. A naturalist traveller in the nineteenth century could still observe the cultivation of a variety of aquatic plants in marshland. Reeds ('totora') are also cultivated in water and come up on the colonial market place in the form of mats ('esteras') which are required in the tax assignments of 1551 and 1562.[38]

The flooded part of the 'camellón' will also have been used to raise molluscs and edible fish. The historical information of the sixteenth century is not very detailed, but comparison with the sunken gardens ('huertas') of the Peruvian coast and the Mexican 'chinampas' lead us to this conclusion.[39]

River and lake fish, as well as fish from natural wells within lakes ('preña-dillas'), were abundant and provided surplus production, since the colonial taxation refers to fresh but also dried fish. The description of Otavalo in 1582 offers a nice description of how to fish the abundant and highly appreciated well fish at the very point where the lake and mountain streams intersected. In marshy areas, fish, crustaceans, molluscs and water insects attracted aquatic birds. This is well documented from the sixteenth to the nineteenth centuries.[40] These data refer to the cold areas of Otavalo territory. Archaeological work on the 'camellones' of the San Jorge river in Colombia have identified nine species of fish and many types of birds in the remains of flooded ditches. On the plains, or 'llanos' of Mojos, agricultural and hydraulic remodelling, consisting of large and numerous 'camellones' over an extended area, also reveals an abundance of fishes and birds.[41]

For a direct point of comparison, we may mention the intelligent autoch-thonous management of 'renewable natural resources' on the coast of Peru in the sixteenth century. There, ethno-historical evidence reveals important food reserves with a semi-aquatic milieu of coastal hills ('lomas'), fish farms, the cultivation of reeds and other plants, and the exploitation of aquatic fauna.[42]

The raised field/flooded ditch complex appears to be inextricably associated, a distinctive exploitation of natural resources that was not valued by the Spanish colonisers. It is true that within an emerging market economy, this system, however sustainable, did not correspond to production priorities. I would especially emphasise one cultural factor that was directly involved in the abandonment of these techniques. The colonial policy of 'reducciones' or reductions, that is, the forced relocation of the autochthonous inhabitants into nucleated settlements involved the collapse of the traditional habitat, and the separation of houses from cultivated fields. This exercise of control over the Indian population was accompanied by the firm belief that the way of life in traditional settlements was incompatible with the concept of civilisation, and indeed could barely be considered human at all. In particular, the occupation of marshland was unacceptable to the Spanish mind-set, shaped as it was by traditional contempt for the irrigated agriculture that was practiced by the moors and moriscos. This contempt may also have been related to the fact that Europeans at that time held strong views about the unhealthy properties of marshlands. The only agricultural practices considered worthy of a Christian population were non-irrigated tilling ('labranza de secano') and cattle grazing.[43] A Colonial official summed up the moral value he attributed to this role of 'reduction' for the very area of Otavalo itself: 'I worked in settling them in villages and withdrew them from the mountains and the marshes and ravines where they were living, and put them in proper order'.[44] Virtually from the outset, Colonial initiatives involved almost systematic drainage of still water, although archival documentation usually refers to the larger lakes. The goal was to retrieve cultivable land, and grazing pasture. Thus a lake was drained in the valley of Chota-Coangue, near Pimampiro. And the lake of Iñaquito, which was where the Inca Huayna Capac had maintained his hunting reserves, was drained and given over to the common lands of the city of Quito.[45]

In Spanish eyes, native exploitation of the environment produces an incongruous configuration of the landscape. The agricultural terrain, modified by the autochthonous inhabitants, is one of continuities and inter-relations between lakes, the lake rims with their reeds ('totora'), and the flooded fields whose ditches link up in turn with rivers and lakes. There is no clear borderline between dry and wet, fishing and agriculture. Yet this was synonymous with barbarism to the colonisers who could not accept that Christian human beings were not firmly rooted on dry land.[46] The technique of 'camellones' consists precisely in the 'intricate presence, very close together, of environments with very different characteristics ... handled in ways that are not contradictory but compatible' – this was a logic far removed from Spanish tradition.[47] And this,

I think, explains the almost total indifference of the colonisers to this way of exploiting the environment as well as their likely incomprehension of its economic rationale. Hence the scarcity of discussion in the historical documentation, and the ambiguity of terms like marshland ('ciénagas'), lakes ('lagunas') and wetlands ('tierras pantanosas').

SOCIO-ECONOMIC CONDITIONS AND THE SURVIVAL OF NATIVE TECHNIQUES

Certain historical testimonies prove the survival of the technique of 'camellones' in cold areas throughout the sixteenth century and much of the seventeenth century. The final data comes from the legacy of Indian land in 1649 in the area of Cumbayá to the north of Quito, another dated 1653 from the banks of Lake Pablo, and a third in 1668 in the area of Gualapuro, near Otavalo.[48] This enduring survival of wetland cultivation in cold areas may be compared with the historical evolution and late disappearance of 'camellones' on the sabana of Bogotá.[49]

At lower ecological levels, the survival of autochthonous techniques and cultivation is even more noteworthy. We find data on 'camellones' and 'rayas de coca' and fruit trees in the valley of Ambuquí and the Coangue-Chota all through the seventeenth century and up to beginning of the eighteenth century.[50] In this case, the close link with coca plantations – invariably in the hands of the autochthonous elite – explains their gradual disappearance, as trade in coca was gradually marginalised in the northern Andes by the colonial elite which ascribed a minor economic role to it, in contrast to the central Andes where it has been prevalent right up to the present day. Christian morality, and also the existence of other economic alternatives, combined in the excoriation of coca or the 'devil's weed'. There was considerable competition for resources in the hot valleys, which were quickly monopolised by sugar cane plantations, which had, incidentally, a greater demand for water than was the case for coca and cotton.[51]

It has been possible to establish that on the Ecuadorian coast the tropical 'camellones' of the Daule river plain were in use from the fifteenth to the eighteenth century.[52]

What do the historical data tell us about the topographical distribution of wetland agriculture? And what kind of socio-economic organisation does this type of environmental management imply? Archaeological surveys show a great variety in the morphology of wetland fields, which is adapted to the varying geophysical possibilities. Documentary references from the sixteenth and seventeenth centuries show us distinct toponyms referring to each plot, on the evidence of the owners of 'camellones', themselves. These plots are not contiguous, and the plots group the 'camellones' in very varied numbers: small ones of 2, 6, 8, and more extensive ones of 30, 40 and up to 70. The interpretation of the historical sources suggests variety in topographical organisation: we can imagine some parallel 'camellones' and others breaking the geometry, as suggested in a reference from 1668: 'another plot named Pigudpuela which has about 40 camellones and two more...'. The dimensions could vary enormously, especially in length: for example, in 1606, in Pimampiro, '8 very small rows of coca fields'... 'a coca plot of 4 rows and more land, measuring about a quadra'. A reference to 28 'camellones' that occupy 'two and a half quadras in width and length' gives us an overall surface that may be calculated as 6 hectares, but does not allow us to see the form or the size of each one. The distribution on the same site of small, differentiated groups of 'camellones' with their own toponyms, could correspond to the organisation of perpendicular plots, with a chessboard distribution. In San Pablo, in 1614, we find 'ten camellones named Piroguchi and 5 more camellones named Mimbuara and another one called Calupigal with 6 camellones and another 5 camellones called Ytumiza plus 7 camellones named Lagabiro'.[53]

Insofar as these references can often be found in the context of how a land inheritance was divided up, they seem to indicate the extension of the property as well as the owners who would benefit. Nevertheless these data, which come from the Colonial period, may have already been affected by European-style rules of succession and property laws.

It would be more apposite to ask what kind of human pre-Hispanic settlements were associated with these agrarian techniques, in order to understand what level of socio-economic organisation corresponds to their maintenance and exploitation. The historical texts reveal the existence of dwellings interspersed among cultivated fields and 'camellones', an autochthonous form of settlement surviving the introduction of the forced settlements. For this very area of the northern Ecuadorian Andes, a document of 1649 describes this interlacing of Indian houses and 'camellones' in the Tumbaco area, north of Quito ('and 30 camellones amid thatched huts').[54]

A comparison with hot areas in Colombia, according to the early testimony of one of the discoverers of the province of Antioquia, supports the archaeological evidence for short 'camellones' with dispersed dwellings. Towards 1540, the conquistador Robledo described the dispersed aboriginal habitat, which he associated with dwellings and a hydraulic system: 'and here there was a hut, and another one 2 leagues further on, and in each one (they) had a plot of maize and manioc, and he found big irrigation channels that they had dug'.[55] We may add that this testimony conclusively demonstrates that irrigation channels were used in pre-Hispanic times, as Salomon has shown with ethnohistorical data on Quito and the northern Andes.[56] Bray has assembled the archaeological and geographical evidence on this very question for the northern Andes.[57]

In these cases, agricultural exploitation was in the hands of a family group and seems to have been limited to the production for autarchic consumption.

Chantal Caillavet

It is true that there are not many studies of native family organisation at the moment of Contact. The most likely conclusion points to an extended family providing the labour necessary to build and maintain the ridges and the ditches of the 'camellones' throughout the year, as well as sowing and harvesting. My studies of the demography and settlements in the area of Otavalo in the sixteenth century are consistent with an organisation, partly collective, of agricultural work at the level of the 'parcialidad', that is, the basic ethnic unit of pre-Hispanic societies. These had between 200 and 500 people, but were also sub-divided into family groups of around 10 to 15 members.[58] The Pacific Coast, on the frontier between Colombia and Ecuador, provides a similar picture in the period of the Tumaco-La Tolita culture, with production for local supply, but no evidence of surplus production.[59] Work on the same area and present-day continuities in the management of the environment suggest that maintenance tasks involve large families, but not necessarily a complex socio-political organisation.[60] The most complete study of pre-Hispanic Chimu society, on the Peruvian coast, gives us very interesting conclusions as to functioning, and political hierarchy: the network of inter-connected irrigation channels that form the basis of that state's prosperity are maintained exclusively at a local level, and through family groups, without the need for state control and management. Archaeology reveals that the construction of the irrigation channels was achieved in piecemeal fashion by distinct groups, and that there was no overall plan executed by a higher authority.[61] In Mexico, it has been estimated that 6 to 8 families were necessary to get a 'chinampa' functioning, and a Colonial example shows the spatial and social organisation of the house/'chinampa' complex in 1585 in Huehuecalco, where a modest-sized family orchard was stuck to a building.[62] This exceptional document presents the spatial distribution of the farm holding and allows us to calculate its dimensions, expressed in a Spanish-Nahua system of measure. It can be estimated that the width of three chinampas was rather more than 14 metres (ditches and ridges), which would correspond to an average of 4.2 metres in width by 'chinampa'. The length would be 7.50 metres, orientated east–west on its longitudinal axis.

It is probable, if we look at the documented cases, that during centuries of Colonial rule, this type of small-scale agricultural use was maintained in the Indian domestic economy, in a form that was barely visible to the Colonial market. But it is also appropriate to recall that the data on the economy of the Otavalo ethnic group in the sixteenth century make it clear that agricultural production in hot areas produced a surplus, which led to the exchange of products. Transactions were based on surplus production of cotton, coca and capsicum.[63]

PRESENT-DAY ECOLOGICAL AND ECONOMIC CHALLENGES

In view of the apparent success of this form of management of water and the environment in pre-Hispanic times, various researchers in archaeology and agronomy – beginning with Erickson's pioneering work in the late 1970s – have looked for ways of experimenting with these same ancient techniques. They have sought out comparable agrarian environments, and selected traditional crops with the aim of studying the feasibility of adapting their production to present-day markets. These experiments generally evaluate which crops would benefit from cultivation in 'camellones', how the cycles of production and profitability are structured, take into account labour costs and the process of rehabilitation.[64] One hypothesis, inspired by archaeological work, favours the possibility of maintaining higher population densities: after assessing carrying capacity, they conclude that potato double-cropping or potato/maize double cropping are feasible.[65]

With regard to these patient and systematic experiments, it is helpful to distinguish between highland and tropical ecosystems.

Reconstruction of 'camellones' has been carried out in distinct areas of the hot Ecuadorian coast with archaeological remains. Results show that agricultural production reaches higher levels than usual especially for maize, sweet potatoes, beans and capsicums, and that two harvests can be expected each year.[66] In the similarly hot area of the Mojos in Bolivia, the experimental study of 'camellones' shows that they enable very infertile land to be productive, which is another confirmation of their importance.[67]

With regard to other climatic conditions, experimental cultivation in the high altitude zone of Puno, in Peru and Bolivia, has shown which are the most successful practices in terms of land, water, labour force and seed planting, and has also established the most suitable agricultural calendar.[67] In this case, protecting the crops against frost and guaranteeing the regularity of harvests constitute the most important achievements.

In conclusion, we can say that archaeological and historical research, as well as field experiments by agronomists, has shown that the palaeotechniques of raised fields and wetland agriculture have clear and notable benefits for agricultural production. They aid hydraulic control and the fight against erosion, promote production of fertilisers and raise temperatures, a series of results which taken together can at least guarantee self-sufficiency, if not surplus harvests.

In terms of present-day agrarian development, we can ask ourselves whether these techniques could be redeployed in places where they were once used so successfully.

Quite clearly, it will not always be possible to turn the clock back. The northern Ecuadorian landscape is a palimpsest on which successive generations have left their traces, and land use and landscapes have been progressively

Chantal Caillavet

transformed over the centuries. During the Colonial and Republican periods,[68] land was turned over to grazing pastures and the cultivation of cereals; more recently, floriculture has predominated. On the coast, mangrove swamps and wetland fields are threatened by the growth of shrimp farms that draw in sea water.[69] In other areas, the difficulty lies in the nexus with the market economy. The retrieval of ancient techniques is not a viable option if they cannot be sustained under present-day economic conditions.

Nevertheless, knowledge of native forms of land use, and of the autochthonous landscape, needs to be revived and prized, as proof of human creativity in managing the resources of diversified ecosystems. These techniques could still be of great value today, if applied to new circumstances and re-invented as solutions to quite different problems. This rediscovery would need to be accompanied by an open-minded approach and a spirit of shared enterprise in the fields of commercial development and ecological management.

NOTES

ABBREVIATIONS:

AGI/S: Archivo General de Indias, Sevilla.

AHBC/I: Archivo Histórico del Banco Central, Ibarra.

AHBC/Q: Archivo Histórico del Banco Central, Quito.

AIOA/O: Archivo Instituto Otavaleño de Antropología, Otavalo.

ANH/Q: Archivo Nacional de Historia, Quito.

AM/Q: Archivo Municipal, Quito.

1. A first version of this article was presented at the International Water History Association/ UNESCO, Paris, December 2005.
2. R. West and P. Armillas, '"Las Chinampas de Mexico". Poesía y realidad de los "jardines flotantes"', *Cuadernos Americanos* IX, 2 (1950): 165–182.
3. Teresa Rojas Rabielo, 'Les techniques indigènes de construction des champs artificiels dans la vallée de Mexico', *Techniques et culture* 4 (July 1984): 1–33; 'La agricultura prehispánica de Mesoamérica en el siglo XVI', in M. Miño Grijalva, ed., *Mundo rural, ciudades y población del Estado de México* (Toluca: El Colegio Mexiquense, 1990), 17–40.
4. Andrew Sluyter, 'Intensive Wetland Agriculture in Mesoamerica: Space, Time and Form', *Annals of the Association of American Geographers* 84 (1994): 557–584, doi:10.1111/j.1467-8306.1994.tb01877.x.
5. See for example the contributions of natural scientists, archaeologists and cultural ecologists in David L. Lentz, ed., *Imperfect Balance. Landscape Transformations in the Pre-Columbian Americas* (New York: Columbia University Press, 2000).
6. J. Parsons and W. Bowen, 'Ancient Ridged Field of the San Jorge River Floodplain, Colombia', *The Geographical Review*, no. 56 (1966): 317–378, doi:10.2307/212460; C. Smith, W. Denevan and P. Hamilton, 'Ancient Ridged Fields in the Region of Lake Titicaca', *The Geographical Journal* 134 (1968): 353–367, doi:10.2307/1792964; Sylvia Broadbent, 'A Prehistoric Field System in Chibcha Territory', *Ñawpa Pacha* no. 6, 135–147.

7. William Denevan, 'Hydraulic Agriculture in the American Tropics: Forms, Measures and Recent Research', in K.V Flannery, ed., *Maya Subsistence* (Academy Press, 1982), 181–203.

8. Chantal Caillavet, 'Toponimia histórica, arqueología y formas prehispánicas de agricultura en la región de Otavalo. Ecuador', *Bulletin de l'Institut Français d'Etudes Andines* XII, 3–4 (1983): 1–21; 'Las técnicas agrarias autóctonas y la remodelación colonial del paisaje en los Andes septentrionales (siglo XVI)', in J.L. Peset, ed., *Ciencia, vida y espacio en Iberoamérica*, (Madrid: CSIC, 1989), vol. III, 109–126.

9. On the Cayambe plain: Bruce Batchelor, 'Los camellones de Cayambe en la Sierra de Ecuador', *América Indígena*, vol. 40, no., 4 (1980): 671–689. On the 'vega' of Lake San Pablo: John Stephen Athens, *El proceso evolutivo en las sociedades complejas y la ocupación del período tardío Cara en los Andes septentrionales del Ecuador* (Otavalo: Pendoneros, IOA, 1980); Gregory Knapp, 'Ecology of prehistoric wetland agriculture in some highland basins of Ecuador', 13th International Botanical Congress, Sydney, Australia, 21–28 Aug. 1981; 'El nicho ecológico llanura húmeda, en la economía prehistorica de los Andes de altura. Evidencia etnohistórica, geográfica y arqueológica', *Sarance* 9 (1981): 83–94; 'Soil, Slope and Water in the Equatorial Andes: A Study of Prehistoric Agricultural Adaptation' (Ph.D. diss., University of Madison, 1984).

10. Pierre Gondard and Freddy López, *Inventario arqueológico preliminar de los Andes septentrionales del Ecuador* (Quito: MAG-PRONAREG-ORSTOM, 1983).

11. Knapp, 'Soil, Slope and Water in the Equatorial Andes'; *Riego precolonial y tradicional en la sierra norte del Ecuador* (Cayambe: Ed. Abya-Yala, 1992); Jean Vacher and Jean Paul Lhomme, 'El uso de los camellones y el control de las heladas en el altiplano', paper for the symposium 'Agricultura Prehispánica. Sistemas agrícolas andinos basados en el drenaje o elevación de la superficie cultivada', Quito, July 2003; Clark L.Erickson, 'El valor actual de los camellones de cultivo precolombinos, experiencias del Perú y Bolivia', in F.Valdez, ed., *Agricultura ancestral, camellones y albarradas. Contexto social, usos y retos del pasado y del presente* (Abya-Yala, IFEA, IRD, Banco Central del Ecuador, INPC, CNRS, DRC, Universidad ParisI, Quito, 2006), 315–339; J.P. Lhomme and J.J. Vacher, 'La mitigación de heladas en los camellones del altiplano andino', *Bulletin de l'Institut Français d'Etudes Andines* 32, 2 (2003): 377–399.

12. Fabián Villalba, 'Los camellones de Cayambe', paper for the symposium 'Agricultura Prehispánica', Quito, 2003.

13. César I. Veintimilla, 'Análisis de opal-fitolitos en camellones del sector Puntiachil, Cantón Cayambe, provincia de Pichincha', in Ernesto Salazar, ed., *Memorias del Primer Congreso Ecuatoriano de Antropología*, (Quito: PUCE-MARKA, 1999), Vol. III, 149–181.

14. Jean-François Bouchard and Pierre Usselmann, 'Espacio, medio ambiente y significado social de los camellones andinos', in: F.Valdez, ed., *Agricultura ancestral, camellones y albarradas*, 57–68.

15. AHBC/Q Fondo Jijón y Caamaño. Serie 1ª, Vol. 7, Exp. 22 in Pilar Ponce Leiva, ed., *Relaciones Histórico-geográficas de la Audiencia de Quito (siglos XVI–XIX)* (Madrid: CSIC, 1991), 733, 742, 748; 'cuando hay contratiempos de heladas (como al presente año) muchas lluvias o exceso de sequedad, se aniquilan las sementeras y hay notable escasez'; 'las cosechas en Otavalo son más abundantes en los años menos lluviosos, porque siendo el terreno por su naturaleza húmedo y regado de tantos arroyuelos que descienden de los altos, las muchas aguas llovedizas le perjudican'; 'por las muchas aguas de los montes... participa la población a la parte de la Laguna de algunos pantanos pues a bastante distancia de ella, es cenegoso'.

16. Minard Hall and Patricia Mothes, 'La actividad volcánica del Holoceno en el Ecuador y Colombia Austral : Impedimento al desarrollo de las Civilizaciones pasadas', in P. Mothes, ed., *Actividad volcánica y pueblos precolombinos en el Ecuador* (Quito: Abya-Yala, 1998), 11–39.

17. John Stephen Athens, 'Volcanism and Archaeology in the Northern Highlands of Ecuador', in Mothes, *Actividad volcánica,* 157–189; Project, Universities of York and Stirling (UK), 'Soil management in pre-Hispanic raised field systems, Ecuador' (2002). See also: C. Wilson, I.A. Simpson, and E.J. Currie, 'Soil management in pre-hispanic raised field systems: micromorphological evidence from Hacienda Zuleta, Ecuador', *Geoarchaeology* 17 (2002): 261–283, doi:10.1002/gea.10015.

18. Fabián Villalba, 'Aprovechamiento de campos anegables para la agricultura en la época prehispánica. El caso Cayambe', in Mothes, *Actividad volcánica,* 191–205; Gregory Knapp and Patricia Mothes, 'Quilotoa Ash and Human Settlements in the Equatorial Andes', in Mothes, *Actividad volcánica,* 146–148.

19. AM/Q Censos Libro 15, f. 83r.

20. Caillavet, 'Toponimia histórica', 1–21; 'Las técnicas agrarias', 109–126; AM/Q. Vol. 90, Tierras de Cayambe, 1672–1686; ANH/Q Indígenas 16, Doc. 1687–XI–18 f. 6r, 40–42; 'una laguna y sienaga llamadas Quinchuqui y otros nombres llamada Guabizi'.

21. Photo of the Instituto Geográfico Militar, Quito, ref. n° 6754.

22. ANH/Q Indígenas, Doc.18–11–1728, f. 25v; AHBC/Q Fondo Jijón y Caamaño, 20ª Colección, Doc. 841, f.7r; AIOA/O Paquete especial. Varios años, Caja 1b, Doc. 2 f. 6r; 'dos chacaras de mais la otra son los camellones de papas'; 'dos ojos de manantial junto en la laguna de San Pablo … y tierras y sienaga … y manantial llamado Cuspifu'.

23. Jorge Marcos, ed., *Proyecto arqueológico y etnobotánico 'Peñón del Río'* (Guayaquil: ESPOL, 1980); James Parsons and Roy Shlemon, 'Nuevo informe sobre los campos elevados prehistóricos de la cuenca del Guayas, Ecuador', *Miscelánea Antropológica Ecuatoriana* 2 (1982): 31–36; William Denevan and Kent Mathewson, 'Preliminary results of the Samborondon raised field project, Guayas basin, Ecuador', in J. P. Darch, ed., *Drained Field Agriculture in Central and South America* (Oxford, 1983), 167–181; Jean-François Bouchard, *Archéologie de la région de Tumaco, Colombie* (Paris: Ed. ADPF, 1984); Francisco Valdez, 'La evolución demográfica en los manglares de la costa noroeste del Ecuador en los períodos formativo y de desarrollo regional', *Cultura – Revista del Banco Central del Ecuador,* 24b, vol. VIII (1986): 593–610; Jean-François Bouchard and Pierre Usselmann, *Trois millénaires de civilisation entre Colombie et Equateur. La région de Tumaco La Tolita* (Paris: CNRS Editions, 2003); L.F. Herrera, G. Romero, P. J. Botero and J. C. Berrío, 'Evolución ambiental de la Depresión Momposina (Colombia) desde el Pleistoceno tardío a los paisajes actuales', *Geología Colombiana* 26 (2001): 95–121.

24. Caillavet, 'Toponimia', 'Las técnicas agrarias'.

25. Chantal Caillavet, *Etnias del Norte. Etnohistoria e Historia de Ecuador* (Quito: IFEA – Casa de Velázquez – Abya-Yala, 2000), 115–116, 245–249; Marcos Jiménez de la Espada, ed., *Relaciones Geográficas de Indias* (Madrid: Atlas, BAE, t. 184, 1965), 246; 'tengo una chacara de coca … que tendrá cinco rayas de coca y cinco paltas con mas de veinte y tres arboles de paltas alrededor de la dicha chacara'.

26. Greg Knapp, 'Vertical Agricultural Differentiation in the Andes: A Result of Historical Adaptative Action', Annual Meeting of the Association of American Geographers, Washington, 1984; Pierre Morlon, 'Variations climatiques et agriculture sur l'Altiplano du lac Titicaca (Pérou-Bolivie): une approche préliminaire', in *La Météorologie* (Paris, CNRS–CNFGG, 1991), 10–29; Lhomme and Vacher, 'La mitigación de heladas'.

27. AGI/S Cámara 922A, 3ª pieza, f. 503r/v; 506v, 508r, 644v, 671v, 673r ; AGI/S Cámara 922A, 2ª Pieza, f. 88r, 156r; 'cogen papas çapallos frizoles'; 'tierra fria. No se coge maiz'; 'en el pueblo de San Pablo no se da tan bien el mays como en los demas pueblos'; 'no ay maiz … porque aunque lo siembren los yndios se les yela y pierde los mas de los años'; 'los caciques de Otavalo deven de este año y el pasado mill y quinientas hanegas de trigo y maiz … por la mucha esterilidad que acudio y por averseles helado sus sementeras'.

28. Caillavet, 'Toponimia', 'Las técnicas agrarias'.

29. Jiménez de la Espada, *Relaciones Geograficas*, 212.

30. Knapp, 'Vertical Agricultural Differentiation'.

31. Veintimilla, 'Análisis de opal-fitolitos', 160–165, 175–178.

32. Udo Oberem, 'El acceso a recursos naturales de diferentes ecologías en la sierra ecuatoriana (siglo XVI)', *Actes du XLIIe Congrès International des Américanistes*, vol. IV, (Paris, 1976), 51–64; Frank Salomon, *Native Lords of Quito in the Age of the Incas. The Political Economy of North Andean Chiefdoms* (Cambridge: Cambridge University Press, 1986).

33. Caillavet, *Etnias del Norte*; Chantal Caillavet, 'Masculin–Féminin: les modalités du pouvoir politique des seigneurs et souveraines ethniques (Andes, XV– XVI e siècles)'. in B. Lavallé. ed., *Les autorités indigènes entre deux mondes* (Centre de Recherche sur l'Amérique Espagnole Coloniale, Université de la Sorbonne Nouvelle, Paris III, 2004), 37–102; AHBC/I Juicios, Paquete 6; 1674–1696. Repartición del agua de Ambuquí, f. 2v; Jiménez de la Espada, *Relaciones Geograficas*, 240–244.

34. Chantal Caillavet, 'Entre sierra y selva : las relaciones fronterizas y sus representaciones para las etnias de los Andes septentrionales', *Anuario de Estudios Hispanoamericanos*, t. XLVI (1989): 71–91; AGI/S Quito 25, Quito, 1600, Relación de los caciques de Tulcán, f.4r; AGI/S Quito 60, Doc. 1, 1558, f.1–47.

35. Chambo 1557: AGI/S Justicia 671, f. 225r; Quito 1573: Jiménez de la Espada, *Relaciones Geográficas*, 212; Chillos 1559: AGI/S Justicia 683, f.817r, 838r, 856r, 869r; Otavalo 1578: Cámara 922A, Pieza 3ª, f. 760v.

36. Jiménez de la Espada, *Relaciones Geográficas*, see especially 'Relación de Quito' *(1573)*, 211–213, nº 63, 69, 80, 81 and 'Relación de Otavalo', 235, nº 4, 239, nº 27. For a suggestive comparison with Mexico, and the striking success of European-introduced cattle, see Elinor Melville, *A Plague of Sheep. Environmental Consequences of the Conquest of Mexico* (Cambridge: Cambridge University Press, 1994). Melville highlights the importance of land-carrying capacity in historical contexts.

37. Pierre Usselmann, 'Un acercamiento a las modificaciones del medio físico latinoamericano durante la colonización: consideraciones generales y algunos ejemplos en las montañas tropicales', *Bulletin de l'Institut Français d'Etudes Andines* XVI, 3–4 (1987): 127–135; Daniel Gade, 'Landscape, System, and Identity in the post-Conquest Andes', *Annals of the Association of American Geographers* 82, 3 (1992): 460–477, doi:10.1111/j.1467-8306.1992.tb01970.x.

38. Chantal Caillavet, 'La nourriture dans les projets de développement: le cas d'un village indien de la région d'Otavalo-Equateur', *Bulletin de l'Institut Français d'Etudes Andines* XI, 1–2 (1982): 1–9; AM/I Papeles sueltos nº 216; Caillavet, *Etnias del Norte*, 125; Jiménez de la Espada, *Relaciones Geográficas*; Enrico Festa, *Nel Darien e Nell'Ecuador. Diario di viaggio di un naturalista* (Quito: Abya-Yala, 1993 [Torino, 1909]), 330–331; Alan White, *Hierbas del Ecuador, Plantas medicinales. Herbs of Ecuador, Medicinal plants* (Quito, 1976); Plutarco Naranjo, 'Plantas alimenticias del Ecuador Precolombino', *Miscelánea Antropológica* 4 (1984): 63–82; 'el sitio llamado Pigalqui donde tenemos nuestras verduras por ser tierras pantanosas y encharcadas de agua'; 'yuyos, pima, bledo, berro …'.

39. R. Kautz and R. Keatinge, 'Determining Site Function: a North Peruvian Coastal Example', *American Antiquity* 42, 1 (1979), 86–97; West and Armillas, '"Las Chinampas de Mexico"'; Teresa Rojas Rabielo, 'Xochimilco: cambio ecológico y agrícola', Paper at the 45th International Congress of Americanists, Bogotá, 1985.

40. Jiménez de la Espada, *Relaciones Geográficas*, 237; for example, in 1573, 'donde hay lagunas o rios o pantanos con agua, hay garzas' (Jiménez de la Espada, *Relaciones Geográficas*, 214); in 1808, 'En la laguna (de San Pablo) se mantiene la galical, los patos, chirlillos, zambullidores y garzas, que todos estos son en abundancia' (Ponce Leiva, *Relaciones Histórico-geográficas*, 750); in 1896, Festa, *Nel Darien*, 298–299, 330–331) describes the 'gallinas de agua', 'somorgujas' and ducks of Lake Yahuarcocha, that are drawn by the 'gran botin de crustaceos,

insectos de agua, irrudineos y Planarie', as well as the 'preñadillas, doradillas y sabaletas' of the river Mira and neighbouring streams and sources.

41. Luisa Fernanda Herrera, 'Paleoecología en la depresión Momposina. 21 000 años de cambios ambientales', in Valdez, *Agricultura ancestral*, 227–239; Sneider Rojas Mora and Fernando Montejo, 'Manejo del espacio y aprovechamiento de recursos en la depresión momposina. Bajo río San Jorge', in Valdez, *Agricultura ancestral*, 81–91; Clark Erickson, 'Sistemas agrícolas prehispánicos en los llanos de Mojos', *América Indígena* Vol. XI, 4 (1980): 731–755.

42. María Rostworowski de Díez Canseco, *Recursos naturales renovables y pesca, siglos XVI y XVII* (Lima: IEP, 1981), 25–31, 49–50.

43. Pierre Ponsot , 'Les morisques, la culture irriguée du blé, et le problème de la décadence de l'agriculture espagnole au XVIIIe siècle', *Mélanges de la Casa de Velázquez* VII (1971): 255–256; Caillavet, *Etnias del Norte*, 140–141.

44. AGI/S Justicia 683, f. 80v, 1566; Caillavet, *Etnias del Norte*, 124; 'Yo travaje en poblallos y sacallos de los montes y cienagas y barrancos en que estavan poblados ponyendolos en toda buena orden'.

45. Jiménez de la Espada, *Relaciones Geográficas*, 248, 210–212.

46. Nathan Wachtel, 'Hommes d'eau: le problème uru (XVI–XVII siècle)', *Annales E.S.C.* 33, 5–6 (1978): 1127–1159.

47. Pierre Morlon, 'Informe de consultoría sobre la rehabilitación de camellones en el altiplano de Puno, Perú', Ms, 1990.

48. ANH/Q Indígenas 18, Doc. 1690–VI–30, f. 2v, Tumbaco, 1649: 'en la loma de Apianda para abajo de Pillagua…y en medio de las casas de paxa treinta camellones'; IOA/O EP/J 1ª(1655–6) f. 17v, San Pablo, 1653: 'cuatro camellones llamado Simpia Pigal', 'ocho camellones llamado Lupifu Pigal', 'dos camellones llamado Pirachu', 'ocho camellones llamado Ytambiquincha', 'cuatro camellones llamados Pirachipigal', 'seis camellones llamado Cutpipigal'; ANH/Q Cacicazgos, Libro 29, f.132v, Gualapuro 1668: 'otro pedaso llamado Pigudcapuela que sera cuarenta camellones y dos más'.

49. Inés Cavelier, 'Perspectivas culturales y cambios en el uso del paisaje. Sabana de Bogotá, siglos XVI–XVII', in Valdez, *Agricultura ancestral*, 127–139.

50. For example, in Yromina, in 1625: 'tierras y chacaras de cocales en el lugar llamado Yromina en el término del balle de Amboqui que tendra seis o siete pedaços de tierras y chacaras de cocales ansi de camellones y los demás rinconados' (AHBC/I Juicios, Paquete 16 (1685–1692); Pimampiro, 1629: 'otra chacara de coca que es como tres rayas y dos arboles de paltas'… 'dos rayas de coca con paltas'… (AHBC/I Juicios, Paquete 21 (1605–1699); Ambuquí, 1651: 'en el valle de Ambuquí siete rayas de coca' (AHBC/I Juicios, Paquete 4 (1654–1659); Ambuquí, 1703: 'tengo en el Balle de Ambuquí… once rayas de tierras donde tengo una huerta con seis árboles frutales y coca que las tuve y herede de mis antepasados (ANH/Q Cacicazgos, Libros Empastados, t. 55, f. 27v).

51. Chantal Caillavet, '"La disparition". La coca du Nord andin: où, quand, pourquoi?', in *Recueil en hommage à Nathan Wachtel*, Paris (forthcoming).

52. David Stemper, 'Campos elevados y producción agrícola en los siglos XV a XVIII – Río Daule – Ecuador', in *The Ecology and Archeology of Prehistoric Agriculture in the Central Andes*, 45th International Congress of Americanists, Bogotá, 1985.

53. AHBC/I Juicios, Paquete 21 (1605–1699); ANH/Q Tierras 21, Doc. 1895–25, f. 819r; AHBC/I Juicios, Paquete 2 (1640–1686); 'otro pedaco llamado Pigudpuela que sera cuarenta camellones y dos más…'; 'ocho rayas muy pequeño (sic) chacaras de cocales'; 'una chacara de coca de quatro rayas y mas tierra, abra una quadra'; 'dos quadras y media de ancho y largo tambien'; 'diez camellones llamados Piroguchi y mas cinco camellones llamados Mimbuara y mas otra llamada (sic) Calupigal que son seis camellones y mas otra cinco camellones llamada Ytumiza mas siete camellones llamada Lagabiro'.

54. ANH/Q Indígenas 18, Doc.1690–VI–30, f. 2v; 'y en medio de las casas de paxa treinta camellones'.

55. Herrera et al., op. cit; Hermés Tovar Pinzón, *Relaciones y Visitas a los Andes, siglo XVI*, (Bogotá: Colcultura, 1993), 289; 'e estaba aqui un bohio e a dos leguas otro, e en cada uno habia sembrado su comida de maiz y yuca, e hallo muy grandes acequias de agua, hechas a mano'.

56. Frank Salomon, *Native Lords of Quito*, 56, 60, 63, 166.

57. Tamara L.Bray, *Los efectos del imperialismo incaico en la frontera norte. Una investigación arqueológica en la sierra norte del Ecuador* (Quito: Abya-Yala 2003), 33–34.

58. Caillavet, *Etnias del Norte*, 151–153.

59. Bouchard and Usselmann, 'Trois millénaires…', 20–23; 'Espacio, medio ambiente y significado social de los camellones andinos'.

60. Valdez, 'La evolución demográfica'; 'Drenajes, camellones y organización social: usos del espacio y poder en La Tola, Esmeraldas', in Valdez, *Agricultura ancestral*, 189–225; Alexandra Yépez, 'Visiones y uso actual del espacio en la Laguna de la Ciudad' in Valdez, *Agricultura ancestral*, 341–355.

61. Patricia Netherly, 'The management of Late Andean irrigation systems on the North Coast of Peru', *American Antiquity* 49, 2 (1984): 228, 247–248; 'Out of Many, One: The Organization of Rule in the North Coast Polities', in M.E. Moseley and A. Cordy-Collins, ed., *The Northern Dinasties. Kinship and Statecraft in Chimor* (Washington: Dumbarton Oaks, 1990), 461–87.

62. James Parsons, 'Political implications of prehispanic chinampa agriculture in the valley of Mexico', 45th International Congress of Americanists, Bogotá, 1985; James Lockhart, *The Nahuas After the Conquest* (Stanford University Press, 1992), 61.

63. Caillavet, *Etnias del Norte*, 65–67, 245–246.

64. Clark Erickson, 'Applications of Prehistoric Andean Technology: Experiments in Raised-Field Agriculture, Huatta, Lake Titicaca, Peru, 1981–1983', in I.S. Farrington, ed., *Prehistoric Intensive Agriculture in the Tropics*, (Oxford: BAR, 1985); 'La agricultura en camellones en la cuenca del Lago Titicaca: Aspectos técnicos y su futuro' in *Andenes y camellones en el Perú Andino*, (Lima: Consejo Nacional de Ciencia y Tecnología, 1986), 331–350; 'El valor actual presente de camellones de cultivo precolombino, experiencias del Perú y Bolivia'; C. Erickson and K.L. Candler, 'Raised Fields and Sustainable Agriculture in the Lake Titicaca Basin of Peru', in J. Browder, ed., *Fragile Lands of Latin America : Strategies for Sustainable Development* (Boulder: Westview Press, 1989), 230–248; Ignacio Garaycochea , 'Agricultural Experiments in Raised Fields in the Lake Titicaca Basin, Peru : Preliminary Considerations', in W. Denevan, K. Mathewson and G. Knapp, eds., *Pre-Hispanic Agricultural Fields in the Andean Region* (Oxford: BAR, 1987); 'Recuperación de campos elevados en el altiplano peruano : veinte años después', paper for the international symposium 'Agricultura Prehispánica', Quito, 2003; Pierre Morlon, 'Du climat à la commercialisation : l'exemple de l'Altiplano péruvien', in Michel Eldin and Pierre Milleville, eds., *Le risque en agriculture* (Paris: ORSTOM, 1989), 187–224; 'Informe de consultoría sobre la rehabilitación de camellones en el altiplano de Puno, Perú', Ms, 1990; 'Variations climatiques et agriculture'; *Comprendre l'agriculture paysanne dans les Andes centrales (Pérou-Bolivie)* (Versailles: INRA, 1992); P. Morlon, B. Orlove and A. Hibon, *Tecnologías agrícolas tradicionales en los Andes Centrales : perspectivas para el desarrollo* (Lima: COFIDE-PNUD-UNESCO, 1982).

65. Greg Knapp and Roy Ryder, 'Aspects of the origin, morphology and function of ridged fields in the Quito altiplano, Ecuador', *44th International Congress of Americanists*, Manchester, 1982; Valdez, 'Drenajes'.

66. Samborondón: Kent Mathewson, 'Proyecto Camellones: interim progress report', Museo Arqueológico del Banco Central del Ecuador, Guayaquil, 1980; W. Denevan and K. Mathewson, 'Preliminary results of the Samborondon raised field project, Guayas basin, Ecuador', in J.P. Darch, ed., *Drained Field Agriculture in Central and South America* (Oxford, 1983),

167–181; Peñón del Río: Marcos, Proyecto arqueológico; M. Muse and F. Quintero, 'Experimentos en agricultura: resultados del primer ciclo anual en Cooperativa Las Delicias. Proyecto arqueológico-agrícola, Peñón del Río, Ecuador', paper at the 45th International Congress of Americanists, Bogotá, 1985.

67. Clark Erickson, 'Sistemas agrícolas'; Óscar Saavedra Arteaga, 'El sistema agrícola prehispánico de Camellones en la Amazonía boliviana', 295–313.

68. Piwandes project 2000–2003.

69. On colonial transformations of the agricultural landscape and the redeployment of prehispanic features, see Caillavet, 'Las técnicas agrarias autóctonas y la remodelación colonial del paisaje... ', in Peset, ed., *Ciencia, vida y espacio en Iberoamérica*.

70. Charles S. Spencer, 'Prehispanic Water Management and Agricultural Intensification in Mexico and Venezuela: Implications for Contemporary Ecological Planning', in David L. Lentz, ed., *Imperfect Balance. Landscape Transformations in the Pre-Columbian Americas* (New York: Columbia University Press, 2000).

COLONIAL

Toiling in Paradise:
Knowledge Acquisition in the Context of Colonial Agriculture in Brazil's Atlantic Forest

Rogério Ribeiro de Oliveira and Verena Winiwarter

THE CONTEXT: FIVE CENTURIES OF EXPLOITATION

It is easy to imagine the sense of wonder and excitement of the Portuguese sailors as they set foot on an unknown tropical land after 44 days of sailing, according to the poet Luis de Camões, 'on seas never navigated before'.[1] They had left Lisbon, Portugal, at the end of winter 1499 and had just arrived on a shore of warm and sheltered waters, where numerous rivers flowed into the sea and the dense coastal forest grew almost down to the beach. The impression of this tall forest led the first Europeans to believe that the soil of the land they had discovered was incredibly fertile. Pero Vaz de Caminha, scribe for the fleet of Pedro Álvares Cabral, wrote of this land:

> Em tal maneira é graciosa que, querendo-a aproveitar, dar-se-á nela tudo; por causa das águas que tem! (...). Eles não lavram nem criam. Nem há aqui boi ou vaca, cabra, ovelha ou galinha, ou qualquer outro animal que esteja acostumado ao viver do homem. E não comem senão deste inhame, de que aqui há muito, e dessas sementes e frutos que a terra e as árvores de si deitam. E com isto andam tais e tão rijos e tão nédios que o não somos nós tanto, com quanto trigo e legumes comemos.[2] [It is so gracious, so abundant, (...). They do not farm or raise animals. There are no oxen or cows, goats, sheep or chicken, or any other animal that is accustomed to living with man. They do not eat anything but yams, which are very abundant, and the seeds and fruits of the earth and that which the trees yield to them. And with that diet they are even more vigorous and well-fed than us, with the wheat and vegetable that we eat.] [This and all following translations are by Oliveira.]

As in other instances of Europeans encountering the tropics, the contrast between the landscape of Portugal and that of Brazil, with its profusion of colours and species, lead the colonisers to assume that they had encountered an extremely productive new land, a true paradise on earth. The concept of a paradise is repeated through much of the Brazilian colonial literature. It can be found as late as 1839.

> Entre todas as regiões do globo, talvez a mais apropriada à agricultura seja o Brasil, pois que na sua vasta extensão acham-se climas, terrenos e exposições

de quantas qualidades é possível imaginar, de forma que dificilmente nós po-
deremos lembrar de uma espécie vegetal, ou de uma sorte de cultura, que não
exista já, ou que não possa, para o futuro, introduzir-se neste abençoado país,
tão fecundo e variado em produções, ameno em aspectos e ares, tão regado de
águas, revestido de matas, e aprazível à vista que os primeiros descobridores
não duvidaram avançar que tinham por fim deparado com o paraíso terrestre. [3]
[Of all of the regions of the world, perhaps the most appropriate for agriculture
can be found in Brazil, for in its vast extensions almost unimaginable climates,
lands and conditions can be found, to such an extent that it is difficult to think
of a plant species or a type of plantation that does not now exist, or could not
be introduced in the future into this much blessed country, so fertile and varied
in its production, agreeable in aspect and airs, so abundant with waters, clad in
forests, so delightful to see that the first discoverers had no doubts at all that they
had at last encountered paradise on earth].

Writers like the coffee grower Taunay, from whose work this quote is taken,
convey their edenic visions solely in the introductions of their works, not in their
recipes and practical descriptions, as will be shown. But as faithful subjects of
the Portuguese crown they had to pay reverence to the dominant image to which
their sovereign, the king, subscribed and which justified colonial exploitation
of biological and mineral resources alike.

While some fine scholarship on the colonial history of Brazil exists, the
development of agricultural knowledge in Brazil in the context of colonial
exploitation has attracted less scholarly interest.[4] Warren Dean's *With Broadax
and Firebrand* (1995), Shawn Miller's *Fruitless Trees* (2000) and Padua's *A
Breath of Destruction* (2002) are main texts in the historiography of Brazil's
forests.[5] The authors emphasise political and economical processes of appropriation and discuss the use of environmental resources and the progress of
landscape transformation. They also introduced readers to the first voices against
an agriculture based on destruction of the forest. However, in their narrative,
colonial agricultural writers did not play much of a role. We wish to add to this
knowledge with a case study on colonial agricultural writings in the Atlantic
coastal forest region of Brazil. Figure 1 provides an overview of the original
distribution of this biome on the Brazilian coast.

Agriculture was the central pursuit of all societies before the industrial
revolution. Human sustenance was based on tapping carefully designed biomass
cycles for energy and food demands.[7] Colonialism was no exception to this
pattern. As in other colonies, plantation agriculture played a major role in the
exploitation of the riches of Brazil. The knowledge base for this exploitation will
be discussed after a brief sketch of the history of colonial extraction in Brazil.

Methods to exploit European colonies in the Americas were developed by
trial-and-error in different phases of the colonial enterprise. The first phase of
colonial exploitation in Brazil, which began around 1533, involved the harvesting
of Brazil-wood (*pau-brasil*), much sought after as a red dye. The main driver

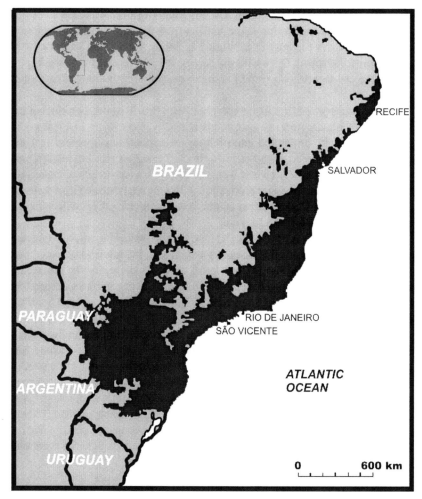

FIGURE 1. Original distribution of Atlantic Coastal Forest biome in Brazil.
(Map adapted from SOS Mata Atlantica[6])

of occupation, colonisation and transformation of the Brazilian territory was sugar production. It had started with the military expedition of Martim Afonso de Souza, sent from Portugal in 1532, which was undertaken to remove French competitors from the coastal areas of Brazil and to promote Portuguese settlement in the region. The fleet brought with it sugarcane plants as well as colonists specialised in the production of sugar.[8]

Sugar production in Brazil had two distinct phases. Up until the seventeenth century, it was the principal colonial activity, with urban areas almost entirely devoted to supporting this industry. The Brazilian sugar economy began to suffer from heavy competition from the Antilles from the mid-seventeenth century onwards, and with the growth of mining in Minas Gerais, the importance of agriculture diminished. Only with the decline of gold production at the beginning of the nineteenth century did the sugar economy regain some of its former intensity.

The first important Brazilian export had been Brazil-wood, a product of the forests, but the more important second commodity, cane sugar, likewise could not have made its unrivalled contribution to colonial development had the Portuguese encountered a land of fewer trees. To make sugar production possible, large quantities of firewood were needed within easy reach of sugar mill furnaces. Any attempt at sugar production without a ready stockpile of forested land had no chance to succeed no matter how favourable other environmental factors such as climate and soil were.[9]

The negligible costs of forests and land led the planters to abuse wood and soil resources. Originally the planters grew cane in the lowland *várzeas'* rich *massapê* soil, famed for fertility and good water retention. By the nineteenth century, however these plains were mostly in use and the planters were deforesting hillsides.[10] The cultivation of sugarcane in Brazil required large land holdings (*latifundios*) and slave labour.

But colonial exploitation schemes had a pronounced impact on the natural systems upon which they were based. The Atlantic Coastal Forest biome was profoundly changed through these schemes.[11] In addition to the space needed to cultivate sugarcane, land was needed for subsistence farm plots of the slaves, with more forest clearing ensuing. These plots played an important role in keeping friction between owners and their slaves at bay. For the slave owners, these small farms represented a significant reduction in maintenance costs, while for the slaves they represented an opportunity to gain a limited degree of economic independence.[12]

Large areas of forest were needed for construction wood and fuel-wood for the boilers. Sugar mills would grind sugarcane day and night for most of the year. The dependence of sugar manufacturing on the Atlantic Forest was enormous, as fuel-wood was essentially the only available energy source for this agro-industry. Extracted from the property itself, or acquired from nearby lands, fuel-wood was deposited near the furnace and used to feed the fires during the long months of sugar production. In addition to fuel-wood to feed the furnaces, other demands for forest products were equally intense, wood was needed for fences, construction, the manufacture and maintenance of ox-carts, and to make crates for shipping sugar. Supplying fuel-wood was always a serious question for the mills, as up to a ton of wood was required for every ton of sugar produced. Generally, eight ox-carts of selected firewood were necessary for twenty cart loads of cut sugarcane; a proportion of 1:2.5. A study based on historical documents about a sugar-mill in Rio de Janeiro that was in operation from 1625 until the end of the nineteenth century gives an estimate of the amount of wood required for sugar production. As this mill processed approximately 6,500 ox-carts of cane per harvest, about 2,600 carts of wood would have been needed for processing. Biomass estimates from neighbouring forests indicate that 10 to 20 hectares of forest were needed per year to supply fuel-wood for a single sugarcane harvest.[13] These numbers all refer to the wood needs of a single mill in a single harvest period. To have an idea of the impact of the sugar industry on the Atlantic Coastal Forest, it must be borne in mind that at the start of the eighteenth century the Capitania of Rio de Janeiro alone had 131 sugar mills in operation.[14]

A new crop was added to the colonial portfolio in the nineteenth century. Around 1830, coffee growers began to alter the landscape in new ways. Coffee beans soon became Brazil's principal agricultural export, economically favouring three south-eastern provinces: Rio de Janeiro, São Paulo and Minas Gerais. In contrast to sugarcane, coffee was generally planted on hill slopes, predominantly in the Paraiba do Sul river valley, the largest drainage basin in south-eastern Brazil.

It is within this spatial and temporal framework of colonial extraction that this paper seeks to analyse the agricultural knowledge of the colonisers and its relation to the Atlantic Coastal Forest. As we will show, indigenous knowledge was ignored whereas the knowledge base of Europe was adapted. While colonial knowledge production was seriously hampered by the political circumstances, several works which survive to date were published, in particular to aid sugarcane and coffee growing. We want to draw attention to contemporary agricultural manuals as a prerequisite for studies of colonial environmental impacts.

INDIGENOUS AGRICULTURAL PRACTICES

The slash-and-burn planting system of Brazil (*roça de toco* or *coivara*) is an Amerindian development. The technique is based on the felling and subsequent burning of the forest, followed by a mixed crop planting regime. Burning is generally performed during the driest months when there is less risk of heavy rain. Ashes are more effectively incorporated into the soil during this dry period, because erosion and leaching would be intense under torrential rains.

An important technique of indigenous agriculture is intercropping up to 15 species together (shrubs, herbs, climbing plants) to reduce plant pathogens and weeds. After a few years – generally three to four – the productivity of the plot will decline due to the exhaustion of soil nutrients by harvesting and erosion. Traditionally, the plots are then abandoned and allowed to rest for at least four years while the farmer cuts another forest area to continue the cultivation

Rogério Ribeiro de Oliveira and Verena Winiwarter

cycle. Leaving the land fallow is an integral part of this technique, allowing the growth of a secondary forest that will aid in reincorporating nutrients lost through harvesting and erosion into the soils. The use of fire is fundamental in shifting agriculture, as it sets free phosphorus and other nutrients accumulated in the living biomass.

Although these agricultural systems have often been falsely considered as primitive, inefficient and environmentally inadequate, under appropriate circumstances they can be highly productive, relatively neutral in terms of their long-term ecological effects, and must be considered as a sophisticated adaptation to nutrient-poor soils.[15] A number of studies have indicated that when this agricultural technique is practiced within certain limits, it is ecologically sustainable for an unlimited period of time.[16] Whether applied by Amerindians or by the mixed European/African populations that succeeded them (the *caboclos, caiçaras, quilombolas*, etc.) for subsistence, the sustainability of this type of agriculture is linked to continued mobility and to relatively low population densities, two framework conditions that were not longer available after European colonists had arrived. Almost none of this knowledge was, therefore, used in plantation agriculture.

At the beginning of the period of colonisation, native peoples were seen as a work force that the colonists could count on for opening trails and clearings, cutting and transporting trees, for constructing canoes, homesteads, mills and forts, as well as for hunting and fishing. Without the labour of the Amerindians, the first colonists would have had very little to eat, but this partnership was anything but peaceful. Indigenous labour was used during the early days of the sugar industry, but was progressively substituted, starting in the sixteenth century, by African slaves who arrived in regular commerce across the Atlantic. The Marquis of Abrantes (see below for details on his work) referred to the participation of the Amerindians in colonial agriculture in the following manner:

> Será de nenhum proveito para a Agricultura a colonização dos nossos indígenas. Estes filhos da Natureza e da indiferença, cujas necessidades são tão limitadas, não têm o estímulo necessário para qualquer tipo de trabalho. Inúteis foram, desde os primeiros tempos da América, as tentativas de colonizar ou de chamar os aborígines ao trabalho.[17] [The colonisation of our indigenous peoples will be of no use to agriculture. These children of nature and of indifference, whose necessities are so limited, do not have the necessary stimulus for any sort of work. Any attempts, since the first moments of America, to colonise or call the aboriginals to work have been useless].

The indigenous knowledge of the Amerindians was not ignored by authors of the agricultural manuals, but while they did see the positive results of indigenous methods, they were not willing to adopt and adapt part of the indigenous system:

> A sua indústria agrícola correspondia à sua natural indolência e aos parcos instrumentos que empregavam. Roçam um pedaço de mato, que queimam

depois, servindo as cinzas de esterco (...) abandonam as plantações à fecundidade natural do terreno e aos métodos naturais e tiravam ricas colheitas desta terra tão mal preparada, porém vigorosa e forte.[18] [Their agricultural activities correspond to their natural indolence and to the few instruments that they use. They clear a small piece of land, which is then burned, with the ashes serving as fertiliser (....) they abandon the plantations to the natural fertility of the lands and to natural methods, and take rich harvests from this land so badly prepared, although vigorous and strong].

The paradoxical image held by the European colonisers was that the new continent was a paradise inhabited by people unfit to use it. As these native populations were oriented towards a system of common lands and self-sufficiency, their agriculture was incompatible with the monocultural model directed towards external markets that the empire sought to establish.

AGRONOMIC KNOWLEDGE IN A COLONIAL CONTEXT

The transition from shifting to permanent agriculture was a consequence of European colonisation. Permanent agriculture in Brazil was introduced in form of mono-cultural sugarcane plantations. As we have detailed above, this development did not lead to reduced forest use; it rather intensified the demand on forest resources. Since sugarcane is a fast-growing grass, it could be left in the ground for multiple cuttings and burned periodically in order to boost production. But because no thought was given to replenishing soil fertility, yields declined over time. When yields decreased, fields were abandoned and given over to pasture in a cycle that took between 12 and 15 years, a type of long rotation that proved unable to sustain soil fertility. New areas would be cleared and cultivation moved on, destroying more forest. Erosion on steeper slopes was also a major problem and contributed to declining yields. Sugar production has been termed a system of profligate use, based on abundant resources of land, forests and labour.[19] The overall ecological effect of this massive change was a conversion of biologically diverse coastal ecosystems into systems favouring a limited number of cultivated and/or exotic plant species, many of which came to be considered agricultural weeds. This same phenomenon occurred also in many other countries under colonial rule, e.g. in Australia.[20]

When land ceased to be used exclusively for subsistence agriculture and regions were transformed into agro-industrial landscapes, new practices had to be adopted to compensate for the rapid exhaustion of soils. Subsistence agriculture had relied on shifting populations and agricultural areas. Permanent agriculture was based on the availability of immense reserves of forested land that were used as sources of wood and bio-fuel, and to substitute the lands damaged by erosion and/or having reduced fertility. The continuation of sugar cultivation and later that of coffee both depended on the availability of new forest areas as the easiest way to overcome the reduction of soil fertility due to depletion of soil nutrients. In this situation, colonial estate owners started to codify their knowledge in order to sustain, if not their soils, so at least their income basis.

Although colonial sugarcane production was exploitative, colonisers were not oblivious to the effects of nutrients and can be assumed to having principally been interested in remedies that could boost their production. Agronomic knowledge was organised and systematised to adapt it to the tropical characteristics of Brazil, but with a long delay. The first Brazilian agricultural manual appeared three centuries after the introduction of sugarcane cultivation. All information available to the farmers up until that time was based on oral transmission among colonisers, on personal experience, or on imported European manuals. The oral tradition of indigenous peoples was ignored in the process of knowledge acquisition and adaptation, as it would not have been useful for sugarcane plantations. There were no positive references in agricultural manuals from the colonial period to any of the techniques employed by the Amerindians, and the dominant vision adopted by Europeans was that the coastal forest soils were rich and that the agricultural knowledge of the indigenous peoples was primitive and rudimentary, according at least to one manual writer, the Marques de Abrantes.[21]

Most European agricultural manuals contain extensive references to earlier works, building a sizeable body of (sometimes contradictory) knowledge which was available at least partially to literate landowners all over Europe.[22] Not all information was actively applied. To give but one example, the technique of planting leguminous crops to add nitrogen to the soil was known to the Ancient Romans but, like crop rotation, was not regularly practised by farmers in the Middle Ages.[23] One is left with speculation as to the reasons for such 'forgetting', but it should be borne in mind how much the organisation of agricultural operations mattered for the applicability of agricultural knowledge. Some methods might simply have ceased to be feasible for economic and labour reasons.

The processes of acquisition and distribution of agronomic knowledge by Brazilian authors were very slow during the colonial period, although the number of published works was significant – especially considering that the country's first printing press was only installed in 1808, approximately 280 years after the start of sugarcane cultivation. It is also true that Brazilian literary production circulated very slowly in the country, and openly communicating and exchanging ideas was no habit among the colonial agricultural authors. References to Brazilian authors were rarely made. Emphasis was being put almost exclusively on foreign writers. A good example of this attitude is the aforementioned manual written by Miguel Du Pin de Almeida, the Marquis of Abrantes, in 1834. The author, an estate owner himself, demonstrated considerable knowledge of the classical writers and provided extensive descriptions of state-of-the-art agriculture in various European countries (like France, Germany, Austria, Switzerland,

Italy, Spain, Sweden and Great Britain) and the United States. His work contains extensive reviews of the principal agronomic writers from each of these countries. But the Marquis of Abrantes referred only superficially to information produced by Brazilian authors, in general limiting himself to referring to a periodical he had founded, the *Annals of the Agricultural Society*.[24] There are exceptions to this pattern, but in general, locally acquired knowledge was not valued as much as written expertise from Europe, at least not in the production of written knowledge. It is also probable that some Brazilian landowners had access to agricultural manuals written in Portugal, 'learned' agriculture in Brazil was dominated by knowledge written for vastly different ecological systems.

THE EIGHTEENTH CENTURY: AGRICULTURAL ENCYCLOPAEDIAS

Nevertheless, there are books discussing subjects closely related to tropical soils. A good example is the work entitled *Memórias de agricultura premiadas em 1787 e 1788*, published by the Royal Academy of Sciences of Portugal. This book is a collection of treatises on various agronomic topics and they were not addressed to tropical agriculture. Although it largely concerned crops that were rarely cultivated in Brazil at that time, such as wheat and grapes, there is evidence that it had been used in Brazil. [25] In it, numerous references concerning soil management by classical authors such as Virgil and Columella are reproduced. Various articles recommend techniques for replenishing soil nutrients through incorporating animal or plant wastes into the soil, and describe alternative fertilisation techniques for localities where animal manure might not be available. One author stressed the use of ashes and of 'exposing the soil to the atmosphere numerous times, which is sufficient to fertilise it'.

Most soils of the Atlantic Coastal Forest are acidic and usually deficient in potassium. The use of ashes would, therefore, have resulted in significant benefits to the crops, had this technique been used, but this cannot be ascertained. In one of the chapters of the manual, the author lists arguments against the use of fallowing as a technique of recuperating soil fertility. According to this treatise, the farmer who uses fallowing will not produce large harvests and could even promote the progressive sterilisation of his lands: 'to leave the soil fallow is an abuse, which is detrimental to the farmer and to the state'. [26] No references were made in nineteenth-century manuals to the use of fallowing in Brazil, which may well be a reflection of the influence of this concept. The stock of virgin soils is another reason which made the use of fallowing unnecessary.

The first Brazilian agricultural manual, *Cultura e Opulência do Brasil* (Culture and Opulence in Brazil), was written in 1711 by the Jesuit André João Antonil, predating the Portuguese collection.[27] But two weeks after its publication the entire edition (except for three or four books) was confiscated and burned by order of the crown, in an attempt to keep the riches of the colony secret. Only in 1798 (that is, 87 years after its first publication) was Frei José Mariano da Conceição Velloso able to produce a partial edition of the book.

Antonil must have been aware of the habits of sugarcane planters, as he offers information on how planters valued different types of land. In the early chapters of his book, Antonil demonstrated an understanding of Brazilian soils and set up a classification system to match crops with different soil types. He considered dark types of clay, *massapê*, 'strong dark soils', as ideal for the cultivation of sugarcane. The *massapê* is a dark soil derived from cretaceous sediments; a heavy clay that can retain a great deal of water. Under heavy rains this material forms a compact mud capable of bogging down the ox carts used to transport sugar or sugarcane. These thick clays made ploughing impractical in cane fields, for it was very difficult for the draft animals to turn this viscous mud.[28] Antonil next refers to sandy/clay soils, called *salões*, 'suitable only for short cultivation, as it soon becomes weak', followed by *areiscas*, 'a mixture of sand with salão that is suitable for manioc and vegetables, but not for sugarcane'. [29] The *salões* is a reddish substrate derived from the decomposition of crystalline rocks; a lighter soil, but one with a lower water retention capacity. In rainy years the *salões* was better for planting sugarcane than *massapê*, but sugarcane growers in general seem to have valued it less than the thicker clay. Figure 2 shows a geomorphologic scheme of this soil classification system, showing that Antonil's way of describing the soils he encountered is consistent with the basic geomorphological situation in the Atlantic coastal region of Brazil.

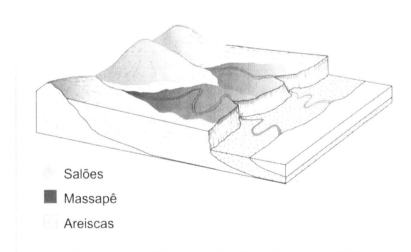

Salões

Massapê

Areiscas

FIGURE 2. The scheme of soil classification system used by the Jesuit André João Antonil as a geomorphological soil classification [Drawing by Marcelo Motta]

According to Antonil, 'new' lands (where forests had been cut only recently) would produce a thick crop of sugarcane for the first and second harvest, but this harvest would not be suitable for making sugar as the cane would be very 'watery'. Antonil recommended the use of animal manure, but only for cultivating tobacco. To fertilise the sugarcane plantations, plant wastes should be burned on the land: 'burning should be done either in the early morning or at night when the wind is calmer and better serves to make the land more fertile'.[30]

When Antonil's book was written, sugarcane cultivation was at its apogee, covering a large part of the Atlantic Forest biome and leading to drastic alterations in the extension and structure of this forest. In contrast to later writers, Antonil viewed this resource as inexhaustible:

> O alimento do fogo é a lenha, e só o Brasil, com a imensidade dos matos que tem, podia fartar, como fartou por tantos anos, e fartará nos tempos vindouros, a tantas fornalhas, quantas são as que se contam nos engenhos da Bahia, Pernambuco e Rio de Janeiro, que comumente moem de dia e de noite, seis, sete, oito e nove meses do ano. [The fires are stocked by fuelwood, and Brazil, with the immensity of the forests that it contains, can count on abundance, abundance in the future, for all its furnaces, as many as there are sugar-mills in Bahia, Pernambuco and Rio de Janeiro, that commonly grind cane day and night for six, seven, eight, nine months a year].[31]

Generally, during early colonial times, the sentiment was that the Brazilian forest was of little intrinsic value, although Brazilian native plants like cashews, papayas, passion fruits and pineapples were quickly adopted and spread throughout the world.[32] But the view of Brazil's forests changed substantially over time. Between 1798 and 1806 Frei José Mariano da Conceição Velloso produced (as writer, coordinator and translator) an important edition of works specifically oriented towards scientific, agricultural and industrial development in Brazil[33]. His eleven volume encyclopaedia *O Fazendeiro do Brasil (...) segundo o melhor que se tem escrito a este assunto* - ['The Brazilian Farmer (...) according to the best that has been written about this subject'] boldly claimed to cover new ground. He covered subjects ranging from the manufacture of sugar, the cultivation of numerous species, including vegetable dye sources, coffee and cocoa, to the preparation of milk and its derivatives.[34] This publication, compiled by Velloso with articles by more than forty foreign authors, was specifically designed to instruct Brazilian farmers. He used most of the space of the first book of his encyclopaedia to convince the Emperor of the damage wrought by the destruction of forests. Frei Velloso also, as a secondary goal, advocated the introduction of new crops in the country. Another author, Baltasar da Silva Lisboa, was one of the first to explicitly denounce the destruction of the Atlantic Coastal Forest in 1786. In fact, most of these agricultural publications contained expressive accusations against the predatory techniques used by the farmers.

Rogério Ribeiro de Oliveira and Verena Winiwarter

In the preface of Tome I, Velloso clearly addresses the non-sustainability of a cultivation system based on the heavy exploitation of the Atlantic Coastal Forest:

> Mas porventura a Natureza será tão liberal na produção destas matas preciosas, que suposta a sua abundância nos reais domínios de Vossa Alteza, possam satisfazer as nossas necessidades presentes e à dos vindouros, sem economia alguma e sem o receio de virem a faltar no futuro? Certamente a devemos recear, pela continuação do presente sistema praticado no Brasil, onde no futuro pode tornar-se difícil este caro e precioso donativo da Natureza. [...] A pobre Natureza vigente, que supre a todas as nossas necessidades é anualmente assassinada nestas máquinas açucareiras, pela indiferença de seus donos. [...] Não há a abundância de matas que se apresenta à primeira vista.[35] [Is it possible that nature was so liberal in the production of these precious forests, which grace with abundance the royal dominions of your Majesty, that it can satisfy our present needs and those to come, without any economy of their use and without fear of scarcity in the future? We must certainly fear, with the continuation of the present system practised in Brazil, a future where this rich and precious gift of nature will become scarce. [...] Poor nature, which supplies all our needs but is regularly assassinated by these sugar-machines, by the indifference of their owners. [...] There is no longer the abundance of these forests as we have seen at first glance.]

Velloso had realised the inadequacy of the sugar-mill system and called attention to the very real possibility of production coming to an end if the forests were indiscriminately felled. He also refers to the idea of paradise comparing it to the richness of nature using terms like 'lost paradise' and 'there's no country in the world with a flora like Brazil'.[36] On the other hand, he paints with strong colours the destruction of the 'paradise', that is, the wrong use of natural resources by the farmers:

> As matas são finitas. Quantos engenhos de açúcar não têm deixado de existir pela falta deste combustível? A Ilha do Governador, no Rio de Janeiro, foi chamada antigamente de sete engenhos; hoje tem apenas um, insignificante. Quantas fazendas se acham reduzidas a taperas, porque seus matos se converteram em sapezais e setais, pelo errado princípio da sua agricultura? [...] Cartago e Tróia não viram certamente maiores montes de cinzas quando foram arrasadas do que se vê nas roças do Brasil e com que se destroem essas importantíssimas e belíssimas matas atualmente.[37] [The forests are finite. How many sugar mills have not disappeared because they ran out of fuel-wood? The community of Ilha do Governador, in Rio de Janeiro, used to be called Seven Mills; but today there is only one insignificant mill left. How many farms have been reduced to abandoned manors because their forests were converted into *sapezais* and *setaes*, by the errors of their agriculture?[38] [...] Carthage and Troy did not turn into larger mounds of ash when pillaged than those you can see in the farms of Brazil, where they destroy these important and beautiful forests even today].

Velloso demonstrated a clear concern about felling forests to fulfil the energetic needs of the sugar-mills:

Não há outra lavoura, outro cultivo no Brasil senão derrubar matos. Que extensão de terra não tem sido descortinada por proprietários de engenhos para a construção das suas fábricas, para a plantação das suas canas, para a combustão das fornalhas de caldeiras e para a fabricação de suas caixas? Quantos lenhos preciosos não foram vítimas de suas mal construídas fornalhas?[39] [There is no farming, no cultivation in Brazil without cutting down the forests. What extensions of land have not been cleared by the owners of sugar plantations in order to run their mills, plant their sugarcane, to supply fuelwood for their furnaces and to build their packing crates? How much wood has been wasted heating their badly constructed furnaces?]

Realising the necessity of increasing productivity without destroying the forests, Velloso proposed a number of alterations to the productive chain: '[...] instead of axes and machetes, the plough; instead of the ashes of the precious and necessary woods, dung and all other types of fertilisers should be used'.[40]

He also linked the preservation of the forests to the restoration of soil fertility, with direct benefits for the cane fields:

Que proveitos não resultariam dessa mudança sábia e prudente? (...) Os campos vastíssimos até aqui reputados como infecundos se voltariam fertilíssimos; as terras que se dizem cansadas e estão reduzidas a sapezais e setais tornariam a dar copiosas searas e ótimos frutos: não se precisaria de tanta extensão de terra para se fundarem fazendas lucrativas.[41] [What benefit would not result from this wise and prudent change? The vast fields that are considered infertile would return to fertility; the lands that are considered worn out and reduced to grasslands and scrub would go back to yielding copious harvests and wonderful fruit: one does not need huge extensions of land to have lucrative farms].

Many of the texts concerning sugarcane production translated by Velloso in the two volumes of his encyclopaedia *O Fazendeiro do Brasil* make reference to the use of fertilisers, linking soil fertility restoration and forest conservation. Bryan Edwards (in his chapter 'Civil and Commercial History of the English Colonies' in the encyclopaedia) discusses fertilising through the addition of ashes, vegetable substances, garbage, lime, cane leaves, etc. to the soil. For his encyclopaedia Velloso selected and translated more than forty studies from various countries, the majority from English or American authors and from the Antilles. The texts show commonalities in terms of their concern about the loss of productivity due to soil exhaustion, their efforts to promote fertiliser use, and their calls for preservation of forest lands. In this collection, Velloso's objective was to present alternatives to forest destruction for Brazilian farmers. As a botanist, he was familiar with the botanical tradition to include the study of the soil into plant descriptions, which is already found in Theophrastus' *Causis Plantarum*, the oldest botanical work of Europe.

THE NINETEENTH CENTURY: AGRICULTURAL MANUALS

During the period from around 1800 to 1860, many agricultural manuals became available in Brazil. [42] This burst in agricultural publications was related to economic growth in the country, and to the expansion of coffee cultivation, but also reflects the worldwide increase in book production during this period. Manuals had to compete in a market. In order to be interesting for readers, they chose different approaches. One can find different motivations, goals, various degrees of in-depth coverage of issues, sets of techniques covered and so on. Each editorial production was sold as 'unique' and 'novel'. The manuals drew attention to different problems. For instance, Taunay's manual focuses particularly on the administration of slaves and related problems. This seems to him a more important issue than the technical information about cropping or grazing. He strongly favours what he considers a more 'rational' way of administrating the slaves. Slavery was an important topic in Brazilian manuals. Jean-Baptiste Alban Imbert and Frei José Mariano da Conceição Velloso both wrote manuals about the medical treatment of slaves under the title of 'agriculture manual'. The manuals of João Rodrigues de Brito and Marques de Abrantes try to convince their readers with mathematical arguments of the necessity of using new techniques and equipments in sugar cane mills. The manual by Baron of Pati do Alferes is very rich in things concerning practical aspects of farming administration.

Carlos Taunay (1791–1867) was the son of Nicolas Taunay, an important member of the French Artistic Mission brought to Rio de Janeiro by D. João VI. Coming from a military background, Taunay wrote a *Manual for the Brazilian Farmer* in 1839. His experience in agricultural affairs was derived from his managing a coffee plantation in the mountains near Rio de Janeiro. Although this book was largely directed towards the administration of slaves, he also proposed innovative techniques capable of increasing productivity of rural properties and dealt with the cultivation of numerous crops, such as coffee, sugarcane and plants used in dyeing, as well as teas and other crops. Taunay expressed his belief that the fertility and abundance of Brazil only needed a rational approach in order to reach their full development. Taunay, echoing Velloso's earlier conviction, considered the plough the principal and most admirable of all agricultural machines, although neither the plough nor manure were regularly used in colonial agriculture. As we have seen, ploughing on *massapê* soils was not feasible, the esteem of the plough hence cannot be considered as stemming from practical experience but is rather a claim founded on the appreciation of the European agricultural model.

Taunay attributed the absence of the plough to 'the facility and simplicity of cutting down virgin forests in Brazil and abandoning the tired land again until the forests grow back'. [43] In terms of the continued use of the land, Taunay affirms that: 'the difference in the quality of the lands influences only the longer or shorter time it can be cultivated. With the exception of very privileged lands, which can give yields for seventy or eighty years in equal abundance, cultivated lands are soon abandoned for newly cleared plots. Only near the larger cities, where the land is greatly sub-divided, is the farmer obliged to use the same plot continuously. As the use of fertiliser is little known or practised, the results are much inferior to the almost spontaneous production from recently cleared forest land'.[44]

Like Antonil, Taunay attempted to classify the soil types found in Brazil, dividing them into three basic classes: heavy clays (*massapê*), sandy/clay (*salões*) and sandy soils, and he remarked that the sugar-mill owners considered the latter soil to be very weak.

Taunay wrote that the sandy soils 'produced almost nothing, but under the influence of the '*meteoros*' and of the atmosphere, vegetation could progressively cover these lands and make them appropriate for cultivation'.

Studying Taunay's Manual I: The role of atmospheric nutrients in tropical agriculture

The colonial literature frequently alludes to fertility derived from the '*meteoros*' and the '*adubos meteóricos*'. Taunay is no exception. The terms 'meteoros' and 'adubos atmosféricos' (atmospheric fertilisers) appear in the pages of his book as a source of renewed fertility for the soil. That the rains make soil fertile is an old conviction. Such lore can be found in many of the Portuguese and European manuals of Early Modern times and are also documented for Classical antiquity, e.g. in the works of Columella (first century C.E.). In a section of the previously mentioned book *Memórias de agricultura premiadas em 1787 e 1788*, the role of nutrients derived from the atmosphere is emphasised: 'rainwater is best for the soils as it gathers all the atmospheric fertilisers (adubos atmosféricos), especially those that originate from lightning'.[45] The Marquis of Abrantes recommended planting trees in pits opened a full month before planting, so that they could 'receive during that interval the benefits derived from the atmosphere'.[46]

The beneficial influence of atmospheric precipitation can be attributed to nutrients deposited in the form of dust, aerosols and above all rainfall that, in tropical systems with very poor soils, can represent a significant source of nutrients. One study undertaken in an area of Atlantic Coastal Forest (situated near the farm plots of a traditional population) compared the uptake of nutrients into that ecosystem by way of two vectors: a) the production of leaf litter; and b) by rainfall and dry deposition.[47] In this study, rain and litter production were monitored over one year, and samples were collected for chemical analyses. The total input amount of these nutrients is shown in Table 1; Figure 3 compares the percentages of nutrient input from leaf litter and from the atmosphere.

Rogério Ribeiro de Oliveira and Verena Winiwarter

Table 1. Total input of nutrients from leaf litter and from atmospheric sources at Ilha Grande, Rio de Janeiro (values in kg.ha⁻¹yr⁻¹). [45]

	N	P	K	Na	Ca	Mg
leaf litter input	105,0	1,5	25,1	17,6	269,9	49,7
atmospheric input	3,6	6,6	100,1	115,1	13,0	97,2

Figure 3 shows that these atmospheric inputs are quite significant for the nutrient balance within the ecosystem, whether for native forests or for any type of agriculture practiced there. In situations of low soil fertility, as in the majority of the soils in the Atlantic Coastal Forest, these inputs mattered for agricultural production. An old claim by agricultural writers proves to be particularly true under tropical circumstances. We have no means to tell if the Brazilian writers emphasised the importance of rainfall due to practical experience or due to adherence to tradition, but in their agro-systems, rainfall did indeed have measurable effects.

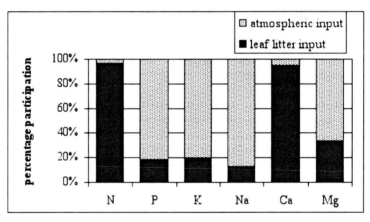

FIGURE 3. Distribution of the percentage of nutrient input from leaf litter and from atmospheric sources in the Atlantic Coastal Forest ecosystem for the elements Nitrogen, Phosphorus, Potassium, Sodium, Calcium and Magnesium. All but Sodium are among the seven main plant nutrients. [45]

Studying Taunay's Manual II: Soil erosion

Taunay was also intimately familiar with the problem of soil erosion in coffee plantations. His plantation was situated in the Maciço da Tijuca, Rio de Janeiro, a very steep landscape, with topographic gradients varying between 30 and 50 per cent.[48] According to him, the loss of humus in unprotected soils was responsible for transforming the soil in *caput mortuum*. Taunay was the first author to recommend planting coffee along hill contours, which was due to an influence, he revealed, of Dutch farming technology at their Asian colonies. Up until that time coffee had been planted in vertical lines on hillsides, a technique that increased erosion but was favoured because it allowed better oversight of the slaves. According to Taunay in some plantations where coffee bushes had been planted randomly, the erosion was avoided. Those farmers who did not plant their coffee plants in regular spacing and alignment did so to avoid having the soil washed off the roots by the rains. But this was an exception. A new front was opened up for the occupation and alteration of the Atlantic Coastal Forest biome. In contrast to sugarcane, coffee was generally planted on hill slopes, principally in the extensive Paraíba do Sul river valley (the largest drainage basin in southeastern Brazil), thus initiating an erosion cycle of enormous proportions. A study on sedimentation in an auxiliary water shed region of the Paraíba River, for example, indicated that during the Pleistocene-Holocene transition (13,000 to 8,000 B.P.) the sedimentation rate there was approximately 300 m³/year; but during just 100 years of coffee cultivation (from 1830 to 1930) the deposition rate in the same location had more than doubled, to almost 750 m³/year.[49] This cycle of deforestation and erosion on an enormous scale created entire landscapes of infertile and eroded lands without any forest cover. The lack of erosion control in colonial coffee farming had led to a regional landscape of eroded lands with severe gully and sheet erosion problems (Figure 4).

FIGURE 4. Gully and sheet erosion in a slope of Rio de Janeiro State where coffee plantations were established in the nineteenth century. [Photo R.R.Oliveira.]

THE MANUAL OF PATI DO ALFERES: THE USE OF INDICATOR
PLANT SPECIES

The extent to which the agricultural manuals were acquired and read by contemporary plantation owners is hard to ascertain. But at least one manual, written by the Baron of Pati do Alferes in 1847 could be considered a best-seller of its time. The first edition was sold out in less than a year and it was followed by four other editions. This work explained how a plantation should be established and administered. It covered techniques useful for cultivating a large number of crops, and provided information concerning animal husbandry. At the end of the book there are some studies from different authors concerning the characteristics and methods of cultivation of a number of crops, such as tea, indigo, cotton and coffee. This was the first agricultural manual derived from experiences accumulated during the boom in coffee cultivation. As in the case of sugarcane, the coffee plantation owners depended on the continuous availability of workers and on virgin lands that could be incorporated into the production process. The Baron was also worried about soils and was the first author to recommend the use of intercropping of coffee with other species like castor oil plant (*Ricinus communis* L.) or *melão de S. Caetano* (*Momordica charantia* L.) as a protection against long dry spells. [50]

In comparison with earlier authors he also demonstrated considerable knowledge of the Atlantic Coastal Forest.[51] In one of the book's chapters he offered a list of the native species and their farm uses (fashioning struts, roof wood, water wheels, planks for doors and windows, etc.). Although he did not have the botanical knowledge of Velloso, he had a great deal of practical experience. This empirical bent is noticeable in the way this author recognised the quality of the soils and the species that grew on them:

> O conhecimento das terras boas ou más é sem dúvida um grande alcance em que está o lavrador a outro meio que as não conhece. As terras apreciam-se avistando as suas florestas ao longe, e principalmente nos meses da primavera. A folhagem de seus arbustos, a configuração de seus galhos, a altura deles, faz distinguir a sua qualidade nessa distância. Entrando em suas matas, ao primeiro golpe de vista conhece-se pela madeira a qualidade da terra, se boa, se média ou má. [Recognising good or poor soils is without doubt a great advantage to the farmer who can, in relation to one who cannot. The lands reveal themselves through their forests when seen from afar, principally in the months of spring. The foliage of the shrubs, the configuration of the branches, their height, all reflect their quality even from a distance. Entering into the woods, at first glance one can recognise in the trees the quality of the land, whether it is good, medium, or poor.][52]

By examining the common plant names published in the Baron of Pati do Alferes' book, it was possible to gain a fair idea of the tree species to which he referred, although one has to take into account that popular names change over time and are different in different regions. Based on published studies of the modern flora in the region of the mid-Paraiba valley, it was possible to identify with some precision the modern species designation of the Baron's list. A large majority of the references are to trees, followed by shrubs and herbs. It was not possible to detect a pattern of successional classification among these plants, as the species mentioned as soil indicators by the Baron of Pati do Alferes can occur in both climax and secondary forests.[53] It is not frequent to find quotations or direct references to particular species of plants in the colonial and post-colonial agronomic literature and Baron of Pati do Alferes shows a great familiarity with the Brazilian flora. He describes also the best woods to be used as furniture, houses and rural machines. Even a botanist like Velloso did not make such association (flora and quality of soils) and referred to the plants in an unspecific way in his treatise of agronomy. Probably the use of this association is an appropriation of native knowledge. But more than a simple list of bioindicators or useful species, Baron of Pati do Alferes refers to them as values to be preserved from destruction.

The use of indicator plants to judge soil quality adapts, for the first time in Brazilian agricultural manuals, a tradition that reaches back to ancient Rome. Columella, as well as Pliny and Cato presented lists of plants useful for this purpose.[54] In their work, as in Pati do Alferes, many species that occur in pristine environments were used to distinguish between fertile and infertile soils.

CONCLUSION

To sum our findings up, the writers presented here are in agreement with regard to the threatened fertility of Brazilian soils, in spite of the recurrent topical invocation of the new world being a green paradise. As all the works were written in the context of colonial politics, such statements of concern about soil fertility, which differed from the official perception, have to be evaluated as departures from it, and hence as inherently political.

A majority of the authors attempted to establish classification systems for the foreign soils they encountered. Most of them were using the system that the Jesuit Antonil had developed or taken from native informants. Later, many authors used systems of soil classification based on this threefold system (*massapê*, *salão* and *areisca*) or similar to it. [55] It is important to note that this system of soil classification does not in any way resemble European classifications. A number of authors tried to understand the natural and artificial processes involved in soil fertility and prescribed procedures to increase it. The use of cattle manure, ashes, marl, the influence of the 'meteoros' and the 'atmospheric fertilisers' appears recurrently in most colonial manuals.

Rogério Ribeiro de Oliveira and Verena Winiwarter

TABLE 2. Indicator species for land quality according to the Baron of Pati do Alferes (list compiled and correlated to modern scientific names by R.R.O.).

old vernacular name	scientific name	family	habitus
good soils			
óleo vermelho**	*Myroxylon peruiferum* L.f.	Leguminosae	arboreal
jacarandatã *	*Machaerium pedicelatus* Vogel	Leguminosae	arboreal
guarabú	*Goniorrhachis* sp	Leguminosae	arboreal
guararema *	*Gallesia integrifolia* (Spreng.) Harms	Phytolaccaceae	arboreal
guarapoca *	*Raputia magnifica* Engl.	Rutaceae	shrub
canela-sassafrás*	*Ocotea pretiosa* (Nees) Mez	Lauraceae	arboreal
cedro	*Cederela fissilis* Vell.	Meliaceae	arboreal
jequitibá *	*Cariniana legalis* (Mart.) Kuntze and *C. estrelensis* (Raddi) Kuntze	Lecythidaceae	arboreal
laranjeira *	*Sloanea monosperma* Vell.	Elaeocarpaceae	arboreal
arco de pipa **	*Erythroxylum pulchrum* A. St.-Hil.	Erythroxylaceae	shrub
pau-paraíba **	*Schizolobium parahyba* (Vell.) S.F. Blake	Leguminosae	arboreal
canela de veado **	*Senepheldera multiflora* Mart.	Euphorbiaceae	arboreal
sucupira	*Boldichia* sp	Leguminosae	arboreal
tenguassiba **	*Zanthoxylum rhoifolium* Lam.	Rutaceae	arboreal
medium soils			
peroba *	*Aspidosperma polyneuron* Müll. Arg.	Apocynaceae	arboreal
cabiúna *	*Machaerium incorruptibile* Allemão	Leguminosae	arboreal
canjerana **	*Cabralea canjerana* (Vell.) Mart.	Meliaceae	arboreal
canela de brejo **	*Nectandra lanceolata* Nees	Lauraceae	arboreal
canela preta **	*Nectandra membranaceae* (Sw.) Griseb.	Lauraceae	arboreal
canela cheirosa *	*Ocotea corymbosa* (Meisn.) Mez	Lauraceae	arboreal
ipê	*Tabebuia* sp	Bignoniaceae	arboreal
taquaracaú **	*Guadua tagoara* (Nees) Kunth	Poaceae	herbaceous
taquarapoca **	*Merostachys riedeliana* Rupr.	Poaceae	herbaceous
poor soils			
tapinhoã *	*Mezilaurus navalium* (Allemão) Taubert ex Mez	Lauraceae	arboreal
bicupari *	*Garcinia gardneriana* (Planch. & Triana) Zappi	Guttiferae	arboreal
milho-cozido	*Licania* sp		arboreal
negro-mina **	*Siparuna apiosyce* (Mart. ex Tul.) A. DC.	Monimiaceae	shrub
caeté	*Calathea* sp	Marantaceae	herbaceous
taqura-de-lixa	*Merostachys* sp	Poaceae	herbaceous

* species normally associated with climax forests
** species that occur in secondary forests

Colonial plantation agriculture, both for sugarcane in the earlier days and later for coffee, had detrimental effects on Brazilian ecosystems. In contrast to a dire reality, viewing the tropical land as paradisiacal was deeply engrained in the coloniser's perception, and was in compliance with the wishes of the crown. Most of the agricultural authors of the colonial period, such as Velloso, the Marquis of Abrantes, Taunay, and the Baron of Pati de Alferes are good examples of the ambivalence between edenic pretension and the requirements of practical recommendation. Local, if not indigenous, knowledge did permeate the coloniser's view of the world. The threefold distinction of soils which Antonil had first presented was obviously developed not through recourse to classical European agricultural literature, but inferred from local experience. Although the introductions of many works reproduce the colonial ideal, the main body of the texts does not. All of them were very concerned with the preservation of forests. The authors clearly perceived that the production systems of sugar or coffee cultivation could not be maintained indefinitely if the forests were destroyed. While claiming erudition and making recourse to European knowledge, the authors – out of their practical experience – understood that there was a clear relationship between forest conservation, soil fertility, and the maintenance of water resources. In this respect, the Brazilian colonial agronomic literature constitutes a first instance of insight into the need of preservation of Brazil's biodiversity against the devastation wrought by colonial exploitation.

ACKNOWLEDGMENTS

We are very grateful to Dr. Carlos Engemann for reviewing the manuscript. This co-operation grew from a research fellowship of R.R.O at the IFF Vienna. We are grateful to both our organisations (IFF Vienna and PUC-Rio) for the support of this exchange. R.R.O. has a research grant from National Council for Scientific and Technological Development (CNPq).

NOTES

1. Luis de Camões, *Os Lusíadas*, Publicações Europa-América, 1997 (original edition 1572).
2. Used edition: Pero Vaz de Caminha, *Carta a El Rei D. Manuel* (São Paulo: Dominus, 1963), 9–10.
3. Carlos Augusto Taunay, *Manual do Agricultor brasileiro*; ed. Rafael de Bivar Marquese (São Paulo: Companhia das Letras, 2001), 33 (original edition 1839).
4. See e.g. Warren Dean, *A ferro e fogo: a história e a devastação da Mata Atlântica brasileira* (São Paulo, Companhia das Letras, 1996). The original edition of this book is Warren Dean, *With Broadax and Firebrand: The Destruction of the Brazilian Atlantic Forest* (Berkeley: University of California Press, 1995).
5. Warren Dean, *A ferro e fogo: a história e a devastação da Mata Atlântica brasileira*; Shawn W. Miller, *Fruitless trees: Portuguese conservation and Brazil's colonial timber* (Stanford:

Stanford University Press, 2000; José Augusto Pádua, *Um Sopro de Destruição: Pensamento Político e Crítica Ambiental no Brasil Escravista (1786–1888)*. (Rio de Janeiro: Jorge Zahar Editor, 2004).

6. SOS Mata Atlântica. *Atlas dos Remanescentes Florestais*. http://mapas.sosma.org.br/ (accessed July 21, 2009).

7. Rolf Peter Sieferle, *The Subterranean Forest: Energy Systems and the Industrial Revolution* (Cambridge: The White Horse Press, 2001).

8. Basilio Magalhães, *O açúcar nos primórdios do Brasil colonial*. (Rio de Janeiro: Imprensa Nacional, 1953), 33.

9. Shawn W. Miller, 'Fuel wood in Colonial Brazil. The Economic and Social Consequences of Fuel Depletion for the Bahian Reconcavo, 1549–1820', *Forest & Conservation History* 38 (1994): 181–92.

10. Peter L. Eisenberg, *The Sugar Industry in Pernambuco: Modernization without Change, 1840–1910* (Berkeley: University of California Press, 1974), 35–6.

11. Dean, *A ferro e fogo: a história e a devastação da Mata Atlântica brasileira*, 183–205.

12. Carlos Engemann, 'Vida cativa: condições materiais de vida nos grandes plantéis do Sudeste Brasileiro no século XIX', in João Fragoso *et al*. (ed.), *Nas Rotas do Império* (Vitória: EDUFES, 2006): 423–45.

13. Carlos Engemann *et al.* 'Consumo de recursos florestais e produção de açúcar no período colonial – O caso do Engenho do Camorim, RJ', in Rogério Ribeiro de Oliveira (ed.), *As marcas do homem na floresta: História ambiental de um trecho urbano de Mata* Atlântica (Rio de Janeiro: Ed. PUC-Rio, 2005): 119–40.

14. Mauricio de Almeida Abreu, 'Um quebra cabeça (quase) resolvido: os engenhos da capitania do Rio de Janeiro, séculos XVI e XVII', *Scripta Nova. Revista electrónica de Geografía y Ciencias Sociales* 10 (2006): 1–23.

15. David McGrath, 'The role of biomass in shifting cultivation', *Human Ecology* 15 (1987): 221–42, p. 231.

16. Cristina Adams, 'As Roças e o manejo da Mata Atlântica pelos caiçaras: uma revisão', *Interciencia* 25 (2000): 143–50; Alpina Begossi, Natália Hanazaki and Nivaldo Peroni, 'Knowledge and use of biodiversity in Brazilian hot spots', *Environment, Development and Sustainability* 2 (2000): 177–93.

17. Miguel Calmon Du Pin de Almeida (Marquês de Abrantes), *Ensaio sobre o fabrico do açúcar* (Salvador: Tipografia do Diário, 1834), 25.

18. João Pedro Gay, 'História da República Jesuítica do Paraguai', *Revista do Instituto Histórico e Geográfico Brasileiro* 26 (1863): 12-3.

19. Michel Williams, *Deforesting the Earth: From Prehistory to Global Crisis* (Chicago: The University of Chicago Press, 2003), 351–2.

20. Peter Griggs, 'Deforestation and Sugar Cane Growing in Eastern Australia, 1860–1995', *Environment and History* 13 (2007): 255–83.

21. Almeida, *Ensaio sobre o fabrico do açúcar*, 1834, 25–6

22. Verena Winiwarter, 'Soil Scientists in Ancient Rome', in *Footprints in the Soil: People and Ideas in Soil History*, ed. Benno P. Warkentin (Oxford: Elsevier, 2006), 3–7.

23. Antonio de Oliveira Marques, *Introdução à história da agricultura em Portugal: a questão cerealífera durante a Idade Média* (Lisboa: Edições Cosmos, 1978).

24. Miguel Calmon Du Pin de Almeida (Marquês de Abrantes), *Ensaio sobre o fabrico do açúcar* (Salvador, Tipografia do Diário, 1834), 1–20.

25. Manoel Joaquim Henriques Paiva, 'Memória Químico Agronômica', in *Memórias de agricultura premiadas em 1787 e 1788*, ed. Academia Real de Ciências (Lisboa: Oficina da Academia Real de Ciências, 1788), 185. This book is available in many Brazilian libraries.

26. Constantino Botelho de Lacerda Lobo, 'Quais são os meios mais convenientes de suprir a falta de estrumes animais nos lugares onde é difícil havê-los', in *Memórias de agricultura premiadas em 1787 e 1788*, ed. Academia Real de Ciências (Lisboa: Oficina da Academia Real de Ciências, 1788), 201–2.

27. André João Antonil, *Cultura e opulência do Brasil.* (Belo Horizonte: Itatiaia/Edusp, 1982).

28. Stuart Schwartz, *Sugar Plantations in the Formation of Brazilian Society: Bahia, 1550–1835* (Cambridge: Cambridge University Press, 1985), 106–7; Stuart Schwartz, *Segredos internos: engenhos e escravos na sociedade colonial* (São Paulo: Companhia das Letras, 1999), 101–02.

29. Antonil, *Cultura e opulência do Brasil*, 101 – 2.

30. Antonil, *Cultura e opulência do Brasil*, 105 – 7.

31. Antonil, *Cultura e opulência do Brasil*, 115.

32. Williams, *Deforesting the earth*, 350–1.

33. Velloso was also an important botanist and the director of the publishing house *Regia Officina Typographica, chalcographica, tipoplastica e Litteraria do Arco do Cego* in Lisbon. His most famous scientific work is the *Flora Fluminensis* (1825), a substantial publication with three volumes of text and fifteen volumes of copper plates. This publication comprises descriptions of seventeen hundred species, many of them previously unknown to science.

34. Frei José Mariano da Conceição Velloso, *O Fazendeiro do Brazil* (Lisboa: Regia Officina Typografica, 1798).

35. Velloso, *O Fazendeiro do Brazil*, tome 1, 16-9.

36. Velloso, *O Fazendeiro do Brazil*, tome 1, 10-2.

37. Velloso, *O Fazendeiro do Brazil*, tome 1, 22.

38. Velloso is refering to these two exotic Graminae (=Poaceae): *Imperata brasiliensis* Trin. (sapê) and *Tristachya leiostachya* Nees. (capim seta), species that are typical for degraded lands.

39. Velloso, *O Fazendeiro do Brazil*, tome 1, 28.

40. Velloso, *O Fazendeiro do Brazil*, tome 1, 27–8.

41. Velloso, *O Fazendeiro do Brazil*, tome 1, 37.

42. They are: André João Antonil, *Cultura e opulência do Brasil.* (Rio de Janeiro: Typ. Imp. e Const. de J. Villeneuve, 1711. (Velloso's 1837 edition of this book); Frei José Mariano da Conceição Velloso, *O Fazendeiro do Brazil – melhorado na economia rural dos gêneros cultivados e de outros, que se pode introduzir, e nas fábricas que lhe são próprias, segundo o melhor que se tem escrito a este assumpto* (Lisboa: Regia Officina Typográfica, 1743); Baltazar da Silva Lisboa, *Discurso histórico, político e econômico, e estado atual da filosofia natural em Portugal, acompanhado de algumas reflexões sobre o Estado do Brasil.* (Lisboa 1786); João Rodrigues de Brito, *Cartas econômico-políticas sobre a agricultura e comércio na Bahia.* (1ª ed.: 1821, Salvador, 1985); Miguel Calmon Du Pin de Almeida (Marquês de Abrantes), *Ensaio sobre o fabrico do açúcar* (Salvador: Tipografia do Diário, 1834); Jean-Baptiste Alban Imbert, *Manual do fazendeiro ou tratado doméstico sobre as enfermidades dos negros* (Rio de Janeiro: Typ. Nacional e Const. De Seignot-Plancher e Cia., 1834); João Joaquim Ferreira de Aguiar, *Pequena Memória sobre a Plantação, Cultura e Colheita do Café* (Rio de Janeiro: Imprensa Americana, 1836); Carlos Augusto Taunay, *Manual do agricultor brasileiro*; organização Rafael de Bivar Marquese (1839) (São Paulo: Companhia das Letras, 2001); Agostinho Rodrigues Cunha, *Arte da Cultura e Preparação do Café* (Rio de Janeiro, Typ. Universal de Laemmert, 1844; Francisco Peixoto de Lacerda Werneck (Barão de Pati do Alferes), *Memória sobre a fundação de uma fazenda na província do Rio de Janeiro* (Rio de Janeiro: Fundação Casa de Rui Barbosa. Original edition: 1847); Carlos Augusto Taunay and Antônio Caetano da Fonseca, *Tratado da cultura do algodoeiro ou arte de tirar vantagens dessa plantação* (Rio de Janeiro, 1862); Frederico Leopoldo Cesar Burlamaqui, *Monographia do Cafeeiro e do Café* (Rio de Janeiro, Typ. N. L. Vianna e Filhos, 1860); Antônio Caetano

da Fonseca, *Manual do agricultor de gêneros alimentícios* (Rio de Janeiro: Ed. Eduardo & Henrique Laemmert, 1863).

43. Taunay, *Manual do agricultor brasileiro*, 44.

44. Taunay, *Manual do agricultor brasileiro*, 44–5

45. Constantino Botelho de Lacerda Lobo, 'Quais são os meios mais convenientes de suprir a falta de estrumes animais nos lugares onde é difícil havê-los', in *Memórias de agricultura premiadas em 1787 e 1788, Academia Real de Ciências* (Lisboa: Oficina da Academia Real de Ciências, 1788).

46. Almeida, *Ensaio sobre o fabrico do açúcar*, 97.

47. Rogério Ribeiro de Oliveira and Ana Luiza Coelho Netto, 'Captura de nutrientes atmosféricos pela vegetação na Ilha Grande, RJ', *Revista Pesquisas* (2001): 31–49.

48. Manoel do Couto Fernandes; André de Souza Avelar and Ana Luiza Coelho Netto, 'Domínios Geo-Hidroecológicos do Maciço da Tijuca, RJ: Subsídios ao Entendimento dos Processos Hidrológicos e Erosivos', *Anuário do Instituto de Geociências – UFRJ* 29 (2006): 120–146.

49. Marcelo Dantas and Ana Luiza Coelho Netto, 'O Impacto do Ciclo Cafeeiro na Evolução da Paisagem Geomorfológica do Médio Vale do Rio Paraíba do Sul', *Cadernos de Geociências*, 15 (1995): 65–72.

50. Barão de Pati do Alferes / Luiz Peixoto de Lacerda Werneck, *Memoria sobre a fundação e costeio de uma fazenda na província do Rio de Janeiro* (Rio de Janeiro: Eduardo & Henrique Laemmert, 1878): 201–2.

51. José Augusto Pádua, '"Cultura esgotadora": agricultura e destruição ambiental nas últimas décadas do Brasill Império', *Estudos Sociedade e Agricultura* (1998): 134–63.

52. Barão de Pati do Alferes, *Memória sobre a fundação de uma fazenda*, 96–8.

53. Gilson Roberto de Souza, *Florística do estrato arbustivo-arbóreo em um trecho de floresta atlântica, no médio Paraíba do Sul, município de Volta Redonda, Rio de Janeiro* (Rio de Janeiro: Universidade Federal Rural do Rio de Janeiro, 2002).

54. Verena Winiwarter, 'Prolegomena to a History of soil knowledge in Europe', in John McNeill and Verena Winiwarter (eds.), *Soil and Societies: Perspectives from Environmental History* (Cambridge: White Horse Press, 2006): 177–215.

55. The authors that used this or a similar system were: Taunay, *Manual do Agricultor brasileiro*, 1839; Barão de Pati do Alferes, *Memória sobre a fundação de uma fazenda*, 1847; Almeida, Ensaio sobre o fabrico do açúcar, 1834; Taunay and Fonseca, *Tratado da cultura do algodoeiro*, 1862.

Contestation over Resources:
The Farmer–Miner Dispute in Colonial Zimbabwe, 1903–1939

Muchaparara Musemwa

INTRODUCTION

Most studies on the so-called 'disputed territories' in white settler societies in southern Africa have focused on how these landscapes became sites of struggle mainly between black and white over ownership and control of resources such as land.[1] For example, historians such as Carruthers and Ranger have individually examined the extent to which national parks (Kalahari Gemsbok and Matopos, respectively) have been 'sites of vigorous contests' over land between black and white over possession, representation and control.[2] Yet, less examined, if at all, are struggles within white settler societies themselves or between white and white over resources from the moment of colonial occupation and beyond. Specifically, the history of the long-standing dispute between miners and farmers over timber, water, grazing rights and land damage caused by mining operations on farms on the Gold Belt in colonial Zimbabwe (Southern Rhodesia), has received astonishingly little attention from either historians or other social scientists. What has been the meaning and significance of resources such as land, timber, grazing and water to different white settler economic classes in those environments where they dislodged indigenous Africans?

This article answers this question by exploring this long-drawn out controversy and examines how it partly gave rise to a formal state-sanctioned conservation regime in Southern Rhodesia. Its central proposition is that competition over access to land, timber, grazing and water, and over the control of that access, characterised the relations between these two sets of capitalist classes. The article also argues that the controversy between farmers and miners in colonial Zimbabwe was not about a conflict between environmental concerns and economic development interests, as occurred in post-World War Two United States, for example.[3] We hardly begin to witness the origins of an environmental movement as a result of this conflict. When farmers raised concerns about the wanton destruction of the surface of their farms, or excessive felling of trees, this was not an expression of what David Lowenthal, commenting on a different case, has called, 'precautionary stewardship' of the environment designed to stop 'entrepreneurial practices harmful to soils, vegetation, wildlife, even climate'.[4]

Environment and History **15** (2009): 79–107.

Both farmers and miners were equally destroyers of indigenous forests in co-lonial Zimbabwe. While miners cut down extensive forests for timber required for mine props and for fuel, farmers, in order to cultivate the most fertile soil, equally 'had to do heavy stumping and clearing of indigenous trees'.[5] Tobacco farmers also required timber for processing flue-cured tobacco.[6] In fact, just before the beginning of World War One, some gambling farmers in Southern Rhodesia became notorious for 'mining the soil and plundering its assets'[7] for over a generation as settlers sought to 'get-rich-quick'.[8] This was, therefore, a struggle over the ownership of the means of production by two competing types of capitalism, a characteristic intra-class as well as intra-racial conflict – a struggle in which indigenous Africans were decidedly excluded as the two parties contested over the spoils of colonisation. Thus, it was about capitalist greed. This dispute represents an ecological narrative of settler colonisation and entrenchment in Southern Rhodesia.

This article builds upon prior scholarship, albeit limited in scope, on the miner–farmer dispute. In addition to a few scattered and passing comments in the works of Hone, Murray and Phimister, the farmer–miner dispute received a relatively more concentrated focus in Lee's doctoral thesis.[9] However, Lee's work on the dispute has remained hidden due to the unpublished state of her thesis. Even more importantly, it is limited both in its depth and coverage of the dispute and its temporal scope, as it goes only up to 1923, the year when settlers attained 'Responsible Government'. Lee did not devote an entire chapter to the dispute, but treated it as merely one of many dimensions of the evolving complex relations between the various institutions forged in the nascent stages of colonial development.[10] Poignantly silent in her analysis, as in the other works, are the long-term environmental and ecological implications of the farmer–miner struggle over access to, distribution, and control over natural resources and space. Finally, Kwashirai is the only historian recently to have explored the farmer–miner dispute at some considerable length.[11] Focusing on the Mazoe District of colonial Zimbabwe, he used the farmer–miner controversy to exemplify the settler society's lack of commitment to attend to the incipient crisis of deforestation and soil erosion as well as the importance of profound changes in both individual and collective mindsets towards the preservation of natural resources. Unlike Kwashirai's extremely relevant and pertinent study which concentrates on just one district, my article examines the farmer–miner dispute writ large, covers a wider geographical scope and frames the wrangle over resources as an intra-class and intra-racial clash.

This study also draws on the important insights of American scholarship which has explored disputes that resonate with issues at the centre of the farmer–miner controversy in colonial Zimbabwe. For example, Kelley's essay, 'Mining on Trial' discusses the conflict between farmers and miners that ensued after the latter group went and settled in the valleys of California and established an agrarian empire. While the 'older economic interests' were chagrined by this 'rapid rise of farm power' and struggled to defend their position, farmers also fought against the impact of hydraulic gold mining in the northern Sierra Nevada Valley.[12] Similarly, Wirth's book *Smelter Smoke in North America: the Politics of Transborder Pollution* examines the US–Canada Trail smelter conflict along the Washington State–British Columbia boundary, 1927–1941.[13] This dispute embroiled United States farmers against the Consolidated Smelter and Mining Company's (CSM) milling complex in Trail, British Columbia. The company emitted fumes of sulphur dioxide which damaged the farmers' cropland in Northport, Washington. Both parties were locked in a long judicial combat which culminated in a legal precedent which laid down the principle 'the polluter pays'. As in Wirth's trans-boundary dispute case study, Aiken argues that farmers with lands downstream from the Bunker Hill Smelter were one vocal group that vehemently grumbled about how lead debris, smoke and fumes emanating from the Bunker Hill damaged their property and endangered their livestock.[14] Morse's book, *The Nature of Gold: An Environmental History of the Klondike Gold Rush* vividly exposes the deleterious consequences (for animals, people and the land) incurred in the wake of the gold rush.[15] Finally, Montrie's book on opposition to coal surface-mining graphically demonstrates how citizens all across the region waged an inexorable decades-long struggle to protect their farms and communities from the depredations of strip-mining.[16] These works provide a firm base upon which to anchor my study.

PERIODISATION AND FOCUS OF STUDY

The study covers the period extending from 1903 when: (a) concessions were granted to the hitherto unrecognised category of miners – otherwise known as 'small workers' or 'tributors'; (b) a consolidated Mines and Minerals Ordinance was introduced; and (c) the Gold Belt areas, once the preserve of large mining companies only, were not only opened up to small workers but were also thrown open to farming under title, thus marking the establishment of agriculture as an indispensable pillar of the colonial economy. The study ends in 1939 when the Report of the Commission to Enquire into the Preservation of Natural Resources of the Colony of Southern Rhodesia [chaired by Sir Robert McIlwaine] was released and tabled.[17] It is a convenient cut-off point because the recommendations of this Commission resulted in the passage of the Natural Resources Act of 1941, thus firmly putting in place natural resources conservation machinery. Within this broad temporal framework, the article very briefly traces the origins of agriculture between 1890 and 1902. Against this background the study shifts to an exploration of the collision course between agricultural and mining

Muchaparara Musemwa

interests and examines the significant signposts, particularly the 1925 and 1933 conferences held between the two sectors to try to resolve the conflict.

The article focuses mainly on the 'Gold Belt' as this was, by and large, the terrain of conflict and engagement between the mining and farming industries. This was the area imagined by the British South Africa Company (BSAC) to be richly endowed with gold deposits. Although geographically ill-defined even by the Company, for the purposes of this study, I adopt Murray's delineation of the Gold Belt. Murray suggested that the Gold Belt roughly stretched from Bindura and more particularly Chegutu (formerly Hartley), through Kadoma (Gatooma), Kwekwe (Que Que), Mvuma (Enkeldoorn), Gweru (Gwelo), down south to Filabusi and Gwanda.[18] The study also extends its focus to areas beyond the Gold Belt where base minerals were mined, i.e. in the Midlands province. Specifically, Shabani, Belingwe, and Mashaba are the areas in which asbestos was to be found, and Selukwe was richly-endowed with both chromium (chrome) and gold deposits. Not only was the Gold Belt rich in minerals, it also had abundant forest resources. A contemporary South African traveller, S.J. Du Toit attested to the splendour of the Gold Belt thus giving us a sense of the evocative setting in which and over which farmer–miner hostilities were to unfold over four decades. Describing the 25-mile range stretching from 'the village of Gwelo to the Selukwe gold-fields' in 1897, for example, Du Toit noted:

> The whole region is densely wooded, especially with the wild loquat, which generally grows as thick as our pine forests; it bears a very palatable fruit; its wood is excellent for fuel and for timber, which is of great advantage to the gold-fields. For here is a gold-bearing region fifty miles long and twenty-five miles broad ... The prospects of these fields are exceptionally good, on account of the extensiveness and the richness of the reefs, and especially on account of the mining facilities; here is a supply of fuel and timber for years (the wood of the wild loquat has been found in the old mines, sufficient proof of its durability, and also on account of its favourable situation.[19]

Thus, claims, counterclaims and struggles to shore up diametrically opposed ownership, production and consumption patterns became the crux of resource politics between white miners and farmers on the Gold Belt during the first few decades of white settler occupation.

ABSENCE OF AFRICANS FROM THE FARMER–MINER STRUGGLES

It needs to be stated from the outset that Africans did not feature prominently in these farmer–miner struggles. For the most part, Africans merely entered into the conflict as 'culprits' – accused by white farmers of disrupting farming operations by 'removing fences' – or of being 'pilferers of crops',[20] or as mere labourers for both industries, or as functionaries in guiding prospectors to disused sites of 'ancient workings' of mining 'for a small reward'.[21] As the historian Rolin proclaimed, '[t]he farmer sees him (the African) as a pilferer and cattle-thief'.[22] In the entire controversy, the only matter on which the two antagonistic industries elected to agree was that Africans should not be given mining licences, to prevent them from competing with miners or from contributing to environmental degradation on farmlands.[23] For example, Abrahamson of the Rhodesia Mining Federation stressed the need for an amendment of the Mining Law, 'to prevent the issue of licences to natives'.[24] Thus, as Mason pointed out: 'In Southern Rhodesia, almost to a man, the farmers and miners who meant to make the country their home believed that the social gap between themselves and the African ought to be maintained; they might wish to be kind masters – most of them did – but social equality of any kind was something too remote to consider'.[25]

But this exclusion is also largely explained by the historical fact that Africans were driven out of most fertile lands and resource-endowed areas such as the Gold Belt and areas along the Bulawayo-Harare-Mutare railway line, and dumped in less fertile and perennially dry reserves such as Shangani, Gwaai and Tsholotsho, as early as 1893, after the defeat of the Ndebele by the BSAC. This set in motion further land alienation, especially after the suppression of the 1896–97 joint Shona-Ndebele *Chimurenga/Umvukela*.[26] More productive land was designated as white land, while Africans were forcibly settled on unproductive lands. The Southern Rhodesia Order in Native Reserves Order in Council of 1898 legalised the dispossession of Africans of their land. By 1914, the African population of 752,000 possessed only 21,390,080 acres of land compared to the 19,032,320 acres that had already been seized and allocated to 23,730 of the white settlers.[27] Further expropriation of land under the Land Apportionment Act of 1930 moved Africans to environments less attractive to European settlers. Africans numbering 1.1 million were allocated only 22 per cent of their land, while 51 per cent was allocated to a minority of 50,000 whites. The remaining percentage comprised forest areas, unassigned areas and Native Purchase Areas.[28] Land was placed in a hierarchy based on fertility and amount of rainfall an area received, and agricultural productivity. Five regions were demarcated. Regions 1 to 3, which receive more rain and are comparatively the most fertile, were allocated to white commercial farmers. Regions 4 and 5, both dry and unsuitable for crop cultivation, were the areas where reserves for African settlement were carved out. Once Africans were out of the way, it was left to classes within the settler community to contest the allocation of the spoils amassed from African land dispossession.

Muchaparara Musemwa

ORIGINS OF SETTLER AGRICULTURE AND THE ROOTS OF ITS CONFLICT WITH MINING: 1890–1902

The farmer–miner dispute had its origins in the manner in which early settler capitalist development in colonial Zimbabwe was established and structured. The mining industry commanded a prime status over agriculture during the first two decades of colonial occupation in Southern Rhodesia. Lured to Southern Rhodesia by the prospect of finding the 'Second Rand' in and beyond the Gold Belt, the BSAC wasted no time in setting up the mining sector by creating the requisite enabling legislative, administrative and political setting for the quick realisation of profits and also to meet expenditure incurred in the process of administering the colony.[29] However, it was not long before the BSAC admitted that no greater Witwatersrand lay under the sub-soil of Rhodesia – it was no gold-bearing country – and that 'its proper economic basis must after all be farming'.[30] This, however, did not immediately change the fortunes of the farming sector, for the BSAC was reluctant to be weaned completely from mining because of its attractive financial returns. Hence it continued to enjoy preferential treatment.

Before 1897, white settler agriculture in Southern Rhodesia was virtually non-existent. It began as a 'scratch affair, its main attraction being the low price of land'.[31] Indeed, 'farms were claimed, located, granted, and surveyed, but few of them were farmed'.[32] Of the less than 250 individual settlers who together owned six million acres, few welcomed the 'hardships of farming'.[33] The majority, of course, were happier to simply eke out a living without necessarily turning to extensive cultivation. While some pursued the elusive fortune to be found in gold, others resigned to cutting down timber and selling it as firewood or using their oxen and wagons for transport riding.[34] It is, perhaps, not surprising, therefore, that once they became established, the privileges enjoyed by miners over wood became one of the sources of conflict, given its importance as a source of livelihood. Despite the outbreak of rinderpest and the *Chimurenga/Umvukela*, which almost led to collapse of the emerging agricultural industry, it was, paradoxically, these very obstacles to the potential success of agriculture which provided the impetus to agricultural transformation between 1897 and 1903.[35]

The BSAC increasingly became responsive to the needs of, and indeed encouraged the development of, commercial agriculture. Simultaneously, it also granted far-reaching concessions to the mining sector. Following the legislative concessions of 1903–4, a hitherto unrecognised entity, the individual 'small worker' or 'producer' as opposed to large mining companies had 'rapidly become an important, and in some ways, typical structural component of the mining industry'.[36] From 1903, 'began the era of the "small workers", and it is thanks to them that the mining industry has been given new impetus', proclaimed Henri Rolin in 1913.[37]

It must be emphasised that both the individual small worker and the individual small farmer emerged at the same time, albeit with inequitable rights to space and resources on the Gold Belt. This point is significant because it was not large mining capital versus large agricultural capital that locked horns in this protracted conflict. It was largely the small worker against the small farmer. This is not to suggest that both forms of large capital were totally marginal to the dispute. Both usually surfaced in support of the respective small capitals as and when it suited their own interests, which, however, were often different from those of the latter. Therefore, differentiation of the two types of capital is of the essence, for neither mining nor agricultural interests comprised a monolithic entity.

Mining companies and small workers had different issues over which they came into conflict with the agricultural sector. These issues also differed in degree, substance and importance. Indeed, both the large mining companies and the Company hailed the development of agriculture because it added value to their expansive land holdings as well as providing a market. But mining companies always feared the likelihood that a government that was sensitive to white farmers' needs would tax the mining sector in order to fund the development of agriculture. As such, large mining capital was antagonistic and rather cautious of yielding too much to commercial agriculture. However, the success and productivity of large mines was not predicated on exclusive rights such as those to free wood and grazing. The issue was far more significant for small workers such that 'much of the bitterly-fought "miner–farmer" contest really only concerned them and the agricultural sector'.[38] Large mining companies and small workers went on the defensive when farmers' demands to end what they conceived as 'disabilities' appeared to be eroding their vested rights.[39] It is to this evolving contest that I now turn.

THE DEVELOPMENT OF THE FARMER–MINER CONFLICT: 1903–1922

The mining industry was, among other forms of support, bolstered by a labyrinth of laws which gave prospectors and miners an assortment of rights to assist them in the discovery and development of mineral deposits. The Mines and Minerals Ordinance (1895), for example, bestowed upon prospectors and miners 'preferential access to water, wood and grazing; ... right of entry on to farmland; and still greater privileges on land falling inside the vaguely-defined gold-belt'.[40] In order to prevent the interruption of prospecting and mining operations, particularly by farmers, a 'universal prohibition against the selection of land on the Gold Belt' was put in place.[41] In other words, no land on the Gold Belt could be bought or sold, as the Chartered Company feared that farming would seriously interfere with mining.[42] Convinced beyond any reasonable doubt that the law 'expressly subordinated agricultural to mining interests', farmers did not just sheepishly resent prospectors' and miners' protected positions but went on the offensive and challenged them together with the law which cushioned them.[43]

Although the farmer–miner conflict became more pronounced from 1903 onwards, flashes of the dispute were already visible in early 1899 – a time during which agriculture was undergoing painful transformation.[44] Periodicals such as the monthly *Matabeleland Times and Mining Journal* and the *Rhodesian Mining Journal*, though overtly committed to covering mining issues, as the titles suggest, could hardly ignore the initial farmers' rumblings about the unfairness of the mining laws. In February 1899, the Matabeleland Printing and Publishing Company, proprietors of the *Matabeleland Times and Mining Journal* (MTMJ), launched a weekly called the *Rhodesian Mining Journal* (RMJ). In courting its readership, the proprietors stressed where the paper's biases lay from the outset. The MTMJ of 25 February 1899 stated in unambiguous terms that Southern Rhodesia was a state that was to be built on the foundations of its mineral resources. However, mindful of not offending the sensibilities of the increasingly restless farmers, the owners of the print media were quick to assuage the fears of the farming sector. The MTMJ immediately censured the Mining Law as farcical, 'unintelligible' and requiring simplification. The proprietors' piercing comments were undoubtedly directed at the BSAC, which doggedly supported the mining industry, however offensive to farmers. Thus, the owners of the new weekly urged that the farmer–miner relations required application of common sense and probity.[45] Launching the *Weekly* on 4 March 1899, the editorship vowed to promote discussion and debate by the public in the paper's columns, of the 'many disputes' which 'have arisen, and more are likely to arise in the near future between the Chartered Company (BSAC – sic) and the claimholder and the owner of the land, and also between the claimholder and the landowner. There are already all sorts of vexed questions with regard to the rights to wood and water'.[46]

The monthly *Matabele Times* also maintained a sharp line of criticism of the 'unintelligible' Mining Law. The editorial commentary on one of the first few recorded litigation cases of 'farmer versus miner' is instructive. In 1898, Hartopp and Company brought a case against Dollar and the Selukwe Gold Mining Company Company, Ltd to the Bulawayo High Court. The plaintiffs were claimholders and sued for damages relating to timber cut from the claims by the Dollars to provide fuel to the Selukwe Company. The Judge settled the case in favour of the contractor who had cut the timber after paying the legalised compensation of £170. The journal was quick to launch a thinly veiled attack on the outcome of the case and charged that it was a travesty of justice. It highlighted some of the problems that the farmer had to contend with and went on to question Clause 60 of the Mines and Minerals Ordinance (1898), which 'gives to the claimholder the exclusive right to all surface within the boundaries of his location'.[47] It also raised a couple of issues that would pervade the farmer–miner controversy for the entire four decades under study. First, the *Matabele Times* wanted to know whether this 'exclusive right to the surface' necessarily transferred 'the ownership in the timber from the farmer to the claimholder'. The paper attacked the principle whereby a prospector could register a block of claims on an already occupied private farm and from that point, the timber on the specific farm ceased to belong to the farmer even though he had occupied it well before the arrival of the prospector.[48] Thus, in these journals, the farmers had found an ally which buoyed up their claims for a revision of the Mining Law to protect them as well.

It is beyond question that farmers and miners equally understood the importance of timber, water and grazing to any farming operation. Without these essentials, Southern Rhodesia would have been less attractive to the immigrant farmer. Before the use of coal and electricity for fuel purposes became widespread, producers of flue-cured tobacco in Marandellas required wood for processing tobacco, and timber for the construction of barns for processing and grading the 'gold leaf'.[49] Similarly, arable farmers and stockbreeders required large tracts of land for grazing their cattle. Thus, with agriculture fast becoming one of the dominant sectors of colonial Zimbabwe's economy, depriving farmers of these crucial resources, yet safeguarding them for one economic group, was bound to generate an aversion to the Mining Law and engender a relentless struggle for the equitable allocation of these scarce resources.

The rapid pace at which mining was consuming wood and timber may well have heightened farmers' calls to the BSAC to revise the legislation, as they were losing out in the grabbing of the resources. In the first few years that farming was beginning to establish itself as a formidable and indispensable sector of the Southern Rhodesian economy, the Secretary for Agriculture was already reporting that 'an enormous quantity of indigenous timber was felled for mining, building, fencing, and fuel purposes during the year 1903–4'.[50] He warned that 'this deforestation is continuing, and must necessarily increase, large tracts of country will, in a few year's time, be quite denuded of timber, and the disastrous results which have followed deforestation elsewhere must be experienced here'.[51] The pulverisation of forests was so widespread that Gann also concluded that '[t]he native woodland at first suffered considerable destruction through the operations of European miners in search of fuel'.[52]

However, as much as miners were engaged in the exploitation of the landscape, farmers were not innocent. As the number of farmers working the land, growing tobacco, maize and cotton increased between 1903 and 1914, the landscape was progressively transformed when bush and forests were destroyed to make way for arable land, as Gann once again observed: 'In time the Rhodesian countryside began to change its appearance as immigrants altered the very landscape'.[53]

The much-vaunted visit of the Directors of the BSAC from London in 1907[54] to 'clear up' some of the deepening differences between miner and farmer with regard to questions over timber, grazing and water rights, hardly made any significant inroads into solving the dispute. If anything, the Directors temporarily assuaged the farmers' increasing dissatisfaction without necessarily introducing any far-reaching amendments in the Mines and Mining Ordinance of 1903, for the dispute lingered on after they had left. Once duly constituted into a formal organisation, the Rhodesian Agricultural Union (RAU) – created from nine separate farmers' organisations in 1904 – attacked the existing mining legislation and criticised all policies and practices the organisation believed hamstrung farmers' operations. Refusing to be taken for granted, in 1905 the RAU ridiculed the BSAC's miserly concession over the terms of compensation offered to farmers for water used by miners.[55] The 1906 RAU Congress roundly declared: 'The existing titles to land and water are worthless'.[56] Acting on a BSAC promise for a land titles reform based on an accord between the two industries after the Directors' visit, the RAU arranged a special meeting to discuss the terms and Gold Belt title alterations desired by farmers. Farmers demanded a uniform title to land eliminating the reservations with respect to wood, water and grazing.[57] The RAU resolutions became the basis for discussion at an inter-industry conference held in Bulawayo in March 1908. But small workers, 'for whom unfettered access to water, wood and grazing could make the difference between profit and loss',[58] vehemently challenged all the concessions to farmers. Consequently the conference was concluded with a compromise between the two antagonistic industries. The conference passed a resolution affirming its conviction in the principle that the ownership of wood, water, and grazing was intrinsic in all titles related to land issues. Thus, the amendments of the Mines and Minerals Ordinance, No. 15 of 1908, were expressly made to render effect to the pledges of the Directors and the resolutions of the conference. But this amendment did not reflect what was expected; instead 'existing mining claims were reaffirmed in their rights', contrary to the new form of Permit of Occupation which was expected to cede to the landowner rights to timber, grazing and water. However, miners, with the support of the Mines Department, and regardless of the views of the Directors and resolutions of the conference, got an exceedingly better deal in response to their demands as hundreds of the old forms of title were issued. This was also in total disregard of the Government's intention that land applications would only be approved and issued with complete title where 'the Mines Department considered that ownership would not interfere with mining'.[59]

Farmers were incensed by the outcome of this conference so that the farmer–miner dispute continued and spilled over into the public domain.[60] Between 1910 and 1911 the question of the respective rights of farmers and miners dominated the discussion columns of the *Rhodesia Herald* and *Bulawayo Chronicle*. The rapid development of the gold mining industry in 1908 and 1909 as well as the large areas that had been pegged off by the 'genuine prospector' and the 'speculative pegger' aroused farmers to the need to protect their interests as landowners.[61]

In 1910, suggestions that yet another conference was going to be held in order to resolve differences between miners, and address the problems that made farmers operations difficult, generated a lot of discussions between the two sectors as well as their sympathisers. A pervading criticism in the newspapers was that the Mining Laws were too anachronistic to be of any value to the emerging agricultural industry. It was the archaic nature of the laws that engendered the farmer–miner hostilities. Those who sympathised with farmers charged that the mining laws of Southern Rhodesia were written when agriculture was virtually non-existent and hence the requirements of the farming population were hardly taken into account. Agriculture had advanced to a stage where farmers' needs had to be re-aligned with those of the mining sector rather than maintaining the 'immensely preponderating rights that now, when the country is really being settled by genuine farmers who desire to live out of the land, many of them are placed in the greatest difficulty', argued the *Herald*'s agricultural correspondent.[62]

Just what was the state of public opinion on the farmer–miner debate? A few letters to the editor of the *Herald* indicate ordinary white settlers taking sides with either the farmer or the prospector, often emphasising the importance of one over the other. Interestingly, some took a non-partisan line often leaning towards the suggestion that both industries were of crucial importance to the colonial economy. The ambivalence of one 'anonymous' writer is symptomatic of this view. Although he was perceptibly on the prospectors' side, as the tone of his letter seemed to suggest, he was also sympathetic to the farmers' predicament. He wrote: 'Personally, I know nothing of farming and in any question between farmers and prospectors, I will confess my predilections are in favour of the latter, but being, I hope, a fair-minded man I recognise that in certain cases the farmers have a real grievance, although I think this has been exaggerated when applied to farming as a whole'.[63] The writer took on the *Herald*'s agricultural correspondent for suggesting that the 'genuine seeker' for gold reefs was a 'rare type', but went on to state that it was this class against whom farmers had a real grievance. He was at odds with the correspondent's view, which seemed to intimate that restrictions had to be imposed on 'this rare genuine type of prospector', on whom the survival of the mining industry of Southern Rhodesia hinged. Without prevarication, the same writer unflinchingly asserted that where the speculative prospector was concerned he fully sympathised with the farmer. But, he reminded the correspondent that 'there is also such a person as the speculative landowner, as instanced by the scores of farms that have been located in the past, solely, or principally, for the native timber that was on them, and also farmers who have ploughed land bearing abandoned workings in order to prevent prospecting'.[64] It would seem that this was the kind of insight, of bringing the two industries on an equal footing or finding ways to make them complement each other that the colonial state lacked. What then was the state's stance on this dispute?

THE STATE ATTITUDE TO IMPLICATIONS OF DISPUTE ON SETTLE-MENT IN RHODESIA

There were fears within the colonial administration that the conflict between farmers and miners could scare away investment as well as potential immigrants to the colony at a time when it needed to attract more settlers to occupy various economic and administrative positions. Perhaps much to the chagrin of the colonial state, there were people outside the colony watching closely the treatment meted out to farmers by the BSAC and wishing to make this known to would-be emigrants. A case in point is that of R. Cross, a prominent farmer and stock-raiser of the Queenstown District of the Cape Colony, who in April 1907 attended the Agricultural Show in Bulawayo as a judge. Upon his return to Queenstown, Cross launched a diatribe published in the *Queenstown Daily Press* and headlined *Straight Talk from a Leading Expert*, against how farming operations and the whole question of land tenure were hemmed in by restrictive legislation privileging the prospector. Cross dissuaded fellow South African farmers from going up north across the Limpopo River, 'Do not think of it under the present conditions', warned Cross.[65] Cross rehashed the well-known 'disabilities' that farmers were exposed to on their farms, such as water, grazing and timber rights. He emphasised just how the farmer could easily be 'treated as a tresspasser' by a prospector or owners of mineral claims: 'I have heard of a case within ten miles of Bulawayo where the bulk of a farmer's land was pegged in claims and he not only got nothing, but was debarred from the use of the pegged part in any shape or form.'[66] Cross ended his 'Straight Talk' memorandum with some salutary and yet sober advice for the BSAC:

> I am not in any way interested in Rhodesia or likely to be, but if the Chartered Company does not take steps to give more favourable and secure titles to the land, which will induce people to take up the vast stretches of unoccupied land, Rhodesia will remain a wilderness. To make a country you must have population, and the soil worked, and this too, before gold mining ... As the Directors of the Chartered Company are about to visit Rhodesia, I should strongly advise them to give a hearing to the farmer's grievances, and give them more secure titles, and so place them on a firm foundation in order to improve and develop the land'.[67]

It was precisely the potentially damaging effects of such 'straight talks' on the colony's drive for immigration and settlement that the BSAC government sought to counteract. Despite the gravity of the farmer–miner controversy, the sixth edition of the *Handbook for the use of Prospective Settlers on the Land* (1910?) trivialised the conflict by underrating its possible impact on those contemplating to migrate to the colony. The language in the handbook, issued ironically by the Minister of Agriculture and Lands, was couched in temperate terms that underscored the importance of mines to farms:

> What might at first appear to be a disparagement of the land rights of the owner is not in practice found to be any very great hardship, as the law prohibits prospecting and pegging on cultivated lands or near homesteads and buildings. It must be remembered that the establishment of a progressive mine in the neighbourhood of one's farm tends to enhance the value of the remaining land, and provides a market at the farmer's door for a good deal of his produce. Again, the provisions of the present mining law have been in operation now for nearly 20 years, and however strange they may read to those acquainted with a different state of affairs, the progress of the farming industry in Southern Rhodesia proves that they are not intolerable in practice.[68]

TOWARDS A RESOLUTION OF THE FARMER–MINER CONTROVERSY

A conference held in 1914 between the visiting directors of the BSAC and various representatives of the mining and agricultural sectors ended in some concessions to demands made by farmers. These minor concessions did not dent large mining companies profoundly as they did small workers, who became the most vocal against the concessions as they threatened to knock their narrow profit margins.[69] In the Legislative Council Mr Begbie, a representative of the small workers, charged that farmers were 'asking for everything' and advised against impediments being instituted in the way of prospectors on whom the colony relied for new mineral discoveries.[70] Small workers persistently defended their rights against farmers. In 1924 Gold Belt titles, as well as related rights to wood and water, fell under the government's spotlight. The Rhodesian Small Workers' and Tributors Association vehemently protested to the Legislative Assembly against any amendment affecting Gold Belt Titles. In this protest small workers and tributors could depend on the support of large mining capital. Thus, an alliance of 'mine owners, small workers, miners and persons directly interested in the mining industry', collectively petitioned Sir Charles Coghlan, the Premier of Southern Rhodesia, against loss of protection accorded to them under the old Gold Belt title:

> We cannot do otherwise than enter a strong protest against the elimination or weakening of the clauses in Gold Belt title reserving wood and water for mining purposes and reserving trading rights, and we do this solely because we feel that this change has been made without sufficient consideration of the welfare and security for the mining industry, without that legislative sanction which we think such a complete change of old-established policy calls for and without giving to

our representatives the opportunity of discussing the change before it had been finally decided upon.[71]

Even though the Chamber of Mines' vested interests were not in any way under threat, it rushed to the aid of the small workers and responded sharply to proposals by the RAU to alter the Gold Belt title, and the RAU's demand for the 'deletion of Clause 15 which reserves indigenous timber to the miner'.[72] In support of the small workers' and tributors' protests, the Chamber produced a pamphlet emphasising the *Importance of the Mines to Farmers*.[73] In collusion with the Ministry of Mines, the Chamber of Mines also strategically issued the same pamphlet as part of a psychological ambush intended to undermine the increasingly strengthening position of farmers as the date of the proposed main conference (April 1925) to resolve the farmer–miner differences was approaching: 'In view of the forthcoming miners-farmers conference … it is felt by the Mining Industry that farmers perhaps do not fully realise to what extent they are dependent upon the premier industry, and how its careful preservation will ensure for many years to come the circulation of large sums of money throughout the country, and maintenance of the internal produce and cattle markets'.[74] But, as Phimister has argued, the colonial state was not 'neutral in the various conflicts within the mining industry and between it and commercial agriculture'.[75]

For the period 1890–1922, it was BSAC policy to favour large capital, albeit with certain reservations. However, the most-favoured-status for large capital was to change with the granting of responsible government to settlers in 1923. As Palmer made clear, 'the new settler government, which took over from the BSA Company in 1923, did all it could to encourage the further development of European agriculture'.[76] Responsible government proactively put agriculture on a firmer footing by providing it with incentives and loans, cutting out African competition by ushering in what has been termed the 'era of commodity control',[77] but for most of the time it stayed clear of the farmer–miner controversy. The state went only as far backing the farmer–miner to resolve the friction between them, hence the conference held in April 1925.

THE FARMER–MINER CONFERENCE OF 1925

The farmer–miner conference was held against the backdrop of a period, which Palmer has called 'the economic triumph of European agriculture, 1915–1925'.[78] The farmer hitherto regarded as a second-class citizen by the miner, swiftly had so much economic clout that he demanded more recognition and more privileges previously the preserve of the miner alone. Held in Salisbury, the conference was meant to discuss certain amendments to the Mining Law which had been suggested by the RAU.[79] Presided over by Sir Morris Carter, the conference was attended by a motley mix ranging from mining and agricultural ministers;

representatives of the Rhodesia and Salisbury Chamber of Mines; small workers; RAU; a government agriculturalist; to the General Manager of the BSAC.

The farmers' representatives went straight to the crux of the matter that had dogged them for over two decades. They contended that the 'disabilities' that affected them were inherited from the past whereby the BSAC regarded Southern Rhodesia as the Second Rand. RAU members insisted that, as a result of this fixation, 'the value of the land and the possibility of its settlement did not enter into the calculations of the early settlers'.[80] The farmers' representatives further challenged the BSAC's preferential treatment of mining by holding land and regarding it as its private property. The farmers challenged the BSAC's monopolistic hold on land and reminded the Company that it was unjust to maintain such land as its private property while simultaneously using it and its products for the assistance and development of one sector only – the mining industry. Such lop-side assistance, argued RAU, was palpably evident in the passage of the first consolidated Mining Law in 1903 and its protection of prospectors' rights. This was made possible by a 'large majority' of elected members of the Legislative Council representing mining interests, which, in the eyes of the farming sector, gave the BSAC (and mining) an unfair advantage over farmers. In reference to the 1914 amendment of the Mines and Minerals Ordinance, the last time it was revised and secured minor concessions to the farmers, RAU representatives stressed that not much came their way because the majority of the Legislative Council still favoured mining interests'.[81]

However, with the 'entirely new position' formed by the 1918 Privy Council decision that land ownership was vested in the Crown, the farmers' representatives requested the BSAC to act accordingly by revising the Ordinance so that it could reflect the new status quo. The position obtaining under the Mining Ordinance was, according to Mr. Huntley, a representative of the farming community, 'little short of intolerable. Not only does it seriously affect the position of the settlers already domiciled in the colony, but it tends to make land settlement on any appreciable scale impossible.' Huntley warned further that this Ordinance was having negative effects to the 'very industry it was created to assist', given available evidence demonstrating that 'under the existing Mining Ordinance discoveries of gold by owners of farms are sometimes concealed for fear of the damage which their development under existing conditions would entail'. Stung by such criticism, miners' representatives, especially small workers, hit back at the farmers reminding them that 'for many years the country was carried on entirely through the efforts and the results of the mining industry, and …but for the mining industry the farmers would be paying a good deal more than 75% of the income tax today'.[82] Thus, small workers steadfastly defended their rights at this conference for they had a lot to lose should farmers' agitations result in some concessions.

Four principal issues dominated the debates between farmers and miners at the conference, namely wood, water, grazing rights, and the destruction of the land surface left in the wake of open mine workings. The destructive effects of base metal mining operations on land belonging to farmers were one of the prickly issues between farmers and miners. Farmers specifically singled out the problem of dyke chrome. The mining of chrome ore, unlike gold for example, resulted in splitting the surface of the land. Although representing the small workers in Gwelo (Gweru), L.I. Davies concurred that this method undoubtedly caused destruction of the surface 'to such an extent that you lose whole farms'.[83] To demonstrate how the operations of base metal mining despoiled the landscape privately owned by farmers, G.A. Dobbin (RAU) gave the following vivid description of a transport rider with 30 wagons and 90 oxen spans who had been transporting chrome to Banket from the western side of Umvukwes 'riding all through the wet season … and they have destroyed terrific tracts of country. They have two tons on a wagon and a double span and they have been pulling through the mud, and they have practically destroyed farms with it.' According to Dobbin, much of the destruction of the surface was due to the failure by transport riders to stick to a single track: 'When they have made one track, they make another the next day, when it could easily have been got over by timbering the road, and they would have been able to carry double the amount of chrome, and it would have been cheaper'.[84] Thus, for the farmer, as J.A. Edmonds (RAU) explained, 'the source of irritation over this matter is caused by making a new road every time you shift your chrome'. So damaging and widespread was chrome mining that a government official was later to report that 'on some farms surface workings have been opened out many miles in length, and the damage done to the land has indeed been serious, Camp fences have been undermined, and many areas ruined for pastoral purposes'.[85]

The question over who had prior right of grazing between the prospector and the farmer on his own land was a huge source of dispute between the two industries. This question was intertwined with the question of mining base metal claims and the hauling of the ore to the railhead over occupied land. The later activity entailed the grazing of large numbers of cattle, and in some instances entire farms were reportedly cleared of grazing as a result. Thus farmers argued against the deprivation of grazing land by the miner under the existing law: 'At the present time, the only thing the farmer appears to own is the air; he does not own the grass because the miner can come onto his farm and if there is not sufficient food for the miner's cattle, the farmer has to sell his cattle or buy food for them. That is an impossibility! That case has actually occurred within 20 miles of Salisbury'.[86] The ownership of grass or lack thereof cut to the core of the very economic survival of the farmer. In this regard, the RAU representatives rejected the idea of giving the farmer open access as that would also undermine their very existence as miners. Some farmers dismissed the oft-repeated benefits that would supposedly accrue to them should they have mines on farms. For example, farmer Huntley deprecatingly stated that owing to the fact that prospectors are 'so hard up', farmers end up having to 'give away what we should be bleeding them for as a matter of fact, if you were unjust, but in trying to be just and encouraging them to go ahead, you lose'.[87] Thus, farmers viewed grass as a critical resource and grazing as an activity indispensable to the farming enterprise. Hence, their pursuit of legal protection against the miner, and demand for unfettered control over the resources.

Apart from grass, farmers considered wood or timber as equally important to the farming enterprise. Under the mining law, the miner was obliged to cut and transport timber from all land under the Gold Belt title for *bona fide* mining purposes without payment. The miner could also cut and take away timber from land which fell under non-Gold Belt title, but he had to pay to the farmer a fixed amount, varying from 5 shillings to 15 shillings per cord depending on the state of the land.[88] The farmers liked to perceive and define themselves in relation to their environment as people who were inextricably linked to their land in a state of permanence and communed with the landscape better than miners. Huntley, once again, strikingly expressed this view at the 1925 conference, albeit in terms somewhat disparaging to the miner:

> We go to a farm and live there. My farm is my home; I know every tree and valley on it. It is my home, but I might get up one morning and find roads cut across it and a man desecrating my trees and valleys, and when you think that these things are done on a man's home, it is that which makes me so sensitive and serious on the point.[89]

What this declaration demonstrates is that environmental sensitivity during the period under study was largely but not exclusively the preserve of the more highly capitalised sections of settler agriculture. There were a few farmers of independent means who were also at the forefront of 'progressive', agriculture, as Phimister has shown.[90] However, these paled into insignificance when one considers the extent to which farmers like Charles Southey, a founder member of the Mazoe Farmers' Co-operative Society, cultivated their land for years on end without even a semblance of conservation practices, resulting in most lands being wrecked in the process.[91]

Concrete evidence on the extent to which farmers were affected by the actual cutting of timber by miners is not readily available. However, the Assistant Director of Lands' observations indicate that some of this loss was probably only measurable in terms of being deprived of the opportunity of 'selling wood on more advantageous terms to others'.[92] He doubted the existence of several cases in which farmers my have been dispossessed of any timber which they required for their own agricultural ventures. He, however, conceded that the cutting and hauling of timber by miners or their contractors damaged farm roads and caused 'general annoyance to the farmer' as well as creating conditions that encouraged erosion and desiccation of the countryside. He further confirmed the existence of farms on Crown land which had been completely stripped of timber to the extent that they could hardly be set aside for settlement purposes. Large landowners who possessed about 3000 acres of land and above were not likely to have been acutely affected by the felling of timber by miners, although this position was thought to change as the holdings became smaller, and became even more serious with a corresponding increase in the production of flue-cured tobacco.[93]

Demands by farmers that prospectors and miners should compensate the former for the damage done to land surfaces, for failure to close up open shafts in disused mines, for wood cut on their farms and grass fed to their draught animals were flatly rejected by the miners. In one of the instances where the Rhodesia Chamber of Mines spoke in support of the small worker, D.V. Burnett referred to the paradox inherent in the minimisation of hardships for farmers, for this meant automatically creating difficulties for the small worker. Regarding the question of wood, Burnett argued that asking the small worker to pay for the wood cut from farms would squeeze them out of mining operations: 'The small worker requires about 200 or 300 cords of wood for his mine, for his two or five stamp battery, and this would amount to approximately £50 or £70 per month. I should like to ask how many of those small workers are making that amount?'[94] Burnett further warned that farmers were expecting too much from small workers as the majority of them failed to even break even. Burnett was strongly supported by J.G. McDonald, also a representative of large mining capital. McDonald reminded farmers that they knew from the beginning that there were restrictions when they took up the titles on the Gold Belt. He argued that it was only after the farmers had pressurised the Government to give them the title that it gave in but made it clear, through mining legislation, that timber was to be reserved to miners. He therefore urged farmers not to insist on their request for compensation for 'the small worker is in a precarious position indeed, while the farming industry today is going on in leaps and bounds. Let it be generous to the small worker of whom 75 per cent are on the verge of ruin.'[95]

Like the mini-conferences that came before it, the April 1925 conference was equally abortive. Despite a thorough discussion of the diverse differences between the two industries as well as 'several concessions to the farmers' being 'acceded to' by the mining industry, the farmers were not satisfied with the result of the deliberations at the conference.[96] In fact, no sooner was the conference poised to begin than a government official expressed doubt as to whether 'any satisfactory understanding will ever be reached', as 'each party tends to magnify the needs and importance of its own position and neither will recede there-from'.[97]

Another conference in 1926 held at the Legislative Assembly Chamber on water, grazing, wood, protection of shafts and trenches, and fencing failed to yield any positive results for the farmers.[98] Farmers resolved to approach the Government with a proposal to appoint a royal commission of inquiry to 'settle the outstanding differences between the mining and farming industries' and address the 'great disabilities landowners are suffering under owing to the incidence of the mining laws of Southern Rhodesia'.[99] But, as always, miners remained tenaciously stuck to their positions and refused to accede to any further demands made by farmers. What irked the Rhodesia Chamber of Mines most was that these demands were being made at a critical time when an economic recession was setting in and 'when gold mining in this Colony, is showing somewhat disquieting signs of decline'.[100] The Chamber of Mines declared that what miners needed at this point was not hurdles in the path of mining operations but assistance in the form of withdrawing the demands. Miners also blamed the farmers for being less than candid:

> The mining industry has always been willing to meet the farmers and discuss differences with them, and to make reasonable concessions to them where possible, but it can hardly be expected to give up rights which it has possessed for so long and which are necessary to enable it to carry on its operations in opening up and developing the mineral resources of the country.[101]

To the extent that farmers and miners held different assumptions about development and/or improvement, these rarely entered public debate in the period under study, beyond the point that both sides were competing for resources (water and timber). Indeed, the methods of some tobacco farmers were almost indistinguishable from miners in the 1920s and '30s, as indicated at the beginning. Nevertheless, efforts to resolve the farmer–miner antagonism culminated in a Government-sponsored conference in October 1933.

THE MINERS–FARMERS CONFERENCE, 1933

Representatives from across the agricultural and mining spectrum, and the Government, attended the conference. Unlike at the 1925 conference, the question of timber rights mostly dominated the 1933 conference. Presenting arguments reminiscent of those advanced at the 1925 conference, miners' representatives unanimously agreed that the wood rights, which included free wood rights on crown lands, free rights on land held under Gold Belt title and compensation to farm owners, were indispensable to their businesses, and argued for the retention of the timber rights in the existing law. Sir Ernest Montagu, representative of the Salisbury Chamber of Mines, argued that of all issues they had discussed in the past with farmers, wood rights were perhaps the most essential especially to the small worker. The small worker 'could not get on without wood' and if 'the small men were stopped from getting wood it would bring their industry to a close'.[102] Sir Montagu raised the vexed question of compensation for wood cut by the small worker and contended that most of these miners lacked the financial wherewithal to pay for the wood as they 'worked on a small margin [of profit – *sic*]'.[103]

Obviously, miners' perceptions about the significance of wood rights were entirely different from those of the farmers. As far as farmers were concerned, the root of the dispute over timber rights was not the question of payment. What gnawed their minds was the question about who was benefiting from the very process of cutting wood, and the damage done to land surfaces. H.M. Hutchinson, a RAU representative, raised two issues related to wood-cutting that irked the farmer. First was the discriminatory practice whereupon some farmers were 'absolutely cut out of wood on their farms, while others were not cut out at all'. Second, Hutchinson raised the point about the denudation of farms surrounded by mines where these had 'cut out all the wood' and in the process those contracted to cut wood had created roads 'all over the farms'. He opined: 'They did not work so much in the dry weather when there was less food for their oxen but in wet weather when they did great damage and caused soil erosion'.[104] In demanding that the whole concept of Gold Belt titles be terminated as they hardly benefited from the resources, RAU farmers also sought to have included in the amended Mines and Minerals Act, a clause that would permit the landowner to 'cut, sell, or use for any purpose he desired such wood as was reserved for his own use'. However much the farmers denied the imputation that they were more interested in monetary compensation from the small worker than anything else, it is inconceivable that their demands for payment were not driven by their entrepreneurial instincts. After all, as we saw at the beginning, early settlers-turned-farmers had survived off the earnings wrung from wood-cutting and selling. Another concession farmers wanted to be added to the amended Act, was a clause which invested powers in the landowner or his agent to 'divide all standing timber equally' on farms where timber was needed for mining purposes. Farmers wanted 50 per cent of the timber reserved for their exclusive use as they deemed it necessary, and the remaining half 'cut by the miner on payment'.[105] As with previous conferences, nothing was resolved. The state, which was better positioned to be the ultimate arbiter, prevaricated for most of the time and left the miners and farmers to sort out their differences. In all the conferences held, the government was content to have an observer status (and accordingly represented by no lesser officials than the mining and agricultural ministers) and did not make arguments for or against the farmer or the miner.

A year after the 1933 farmer–miner conference, the *Report of the Committee of Enquiry into the Economic Position of the Agricultural Industry* (chaired by Mr Max Danziger) was released.[106] The Danziger Report noted that 'the farming community is facing a crisis'. Surprisingly, it steered clear of the farmer–miner dispute and thus did not ascribe part of this 'farming crisis' to the issues at the centre of the controversy. However, it gave the agricultural industry a new lease on life by arguing that while the mining industry was still of 'very great importance', a 'white agricultural population must be the basis on which to build a white Colony', rather than 'establishing a white Colony on mining and on secondary industries'.[107] Throughout the farmer–miner dispute, the farmer repeatedly faulted the miner for cutting and hauling wood and causing soil erosion. While the report did not spare African peasants from liability regarding soil erosion, the Report, somewhat refreshingly, blamed white farmers for similarly causing soil erosion and for carrying on their 'obligations in an extravagant manner'.[108] It brusquely stated: 'In practically all cases farmers have endeavoured to develop their farms too quickly and tried to do in a few years what it has taken generations to do in other countries'.[109]

The impulsion to formally institute a conservation regime came about only when colonial authorities became startled at the rapidity of environmental degradation. Moves towards conservation began in earnest with the appointment, in 1939, of the *Commission to Enquire into the Deterioration and Preservation of the Natural Resources of Southern Rhodesia* (chaired by Mr. Justice McIlwaine).[110] The McIlwaine Commission, just as the Danziger Committee before it, was equally scathing in its accusation of the white farmer's role in causing soil erosion. In its report, the Commission noted that in spite of the steady expansion of the mining industry, and increasing access to export markets, particularly for maize, tobacco and cattle, all of which enhanced agriculture, in general 'for a considerable time questions of erosion and other abuse of the land do not seem to have received much attention'. The Commission also singled out farmers' increasing use of the plough, egged on by a lucrative market for maize, for precipitating soil erosion, as 'ever-increasing areas of rich, virgin land' were subjected to this new technology, with deleterious environmental consequences: 'In an effort to maintain or increase their output, some farmers ploughed up natural hollows, hillsides and narrow valleys between hills and, in a very short time, many acres of valuable pasture and woodland were converted into a donga-scarred waste'.[111]

The white tobacco farmer was as ruinous to the soil as his maize growing counterpart. The commission rapped tobacco farmers' lethargic stance on anti-erosion measures, ascribed by witnesses to the growers' practice of rotation after two successive crops. Moreover, the tobacco growers' 'ridge and furrow system' was singled out as one that increased the collection of water in depressions, whereupon a downhill gradient had to be made in order to promote drainage. The Commission concluded with a loaded commentary on how Southern Rhodesian settler farmers in general had actively caused 'much destruction and dissipation of the Colony's most important asset – the soil' all in the 'comparatively short period of European occupation and owing to the conditions created thereby'.[112]

Concerning the respective rights of farmers and miners the Commission argued, if hypothetically, that the farmer – even presuming that he enjoyed 'absolute' rights to wood and timber – had much the same potential to damage the environment as the miner, for 'he might sell it to the miner or exploit it in other ways with the same results as exist at present'.[113] Attempting to strike a balanced view in its observations, the Commission also focused on the destructive effects of mining activities on the colony's forests as they had done with farmers. Non-partisan to the end, and maintaining a discreet approach so as not to be seen to be judgemental on the merits of the farmer–miner case, the Commission noted that notwithstanding the importance of the mining industry to the colony, the time had arrived when the rights enjoyed by miners had to be changed in ways that did not just favour them but had to be consonant with 'the interests of the whole community'.[114] But the Commission was keen on demonstrating how much wood and timber the mining industry was consuming. Based on figures from the Conservator of Forests, the Commission estimated the quantity of wood consumed by the mining industry to be about 200,000 cords a year for generating power.[115] This figure was estimated to represent the destruction of trees on land corresponding to 40,000 acres per annum. Pointing to the difficult regenerative capacity of indigenous forests, the Conservator estimated that the trees cut would take approximately 50 to 60 years to return to their former grandeur. Furthermore, the absence of a substitute for wood for roasting plants at some mines caused increased cutting of trees. To this end, the Commission expressed its misgivings when it noted that 'in one locality on the railway line, farm after farm is being stripped of trees for lime burning'.[116]

The Commission also censured the practice where miners resorted to the 'wasteful method' of chopping down large trees with the objective of making use of only the branches and abandoning the trunks, thus rendering them useless. Further to this, in arranging contracts for wood and timber supplies miners were also in the habit of specifying the 'sizes, shapes, hardness, etc.', resulting in so much wastage after the stipulated standard portions had been extracted.[117]

The impact and magnitude of both farmers' and miners' activities on the colony's resources saw the Commission passing a decisive recommendation which resulted in the passage of the Natural Resources Act in 1941. This was followed by the appointment of a Natural Resources Board which had powers to ensure the conservation and 'wise utilisation' of the colony's natural resources such as soil, water, minerals, trees, grasses, vegetation, etc.[118] Thus, unlike in other settler societies, it took the colonial state in Southern Rhodesia almost forty years to respond seriously to demands for formal conservation of natural resources, despite ample evidence, in the first decade of the twentieth century, of an ongoing unsustainable environmental despoliation at the hands of both farmers and miners. As Anderson has shown in Kenya,[119] settler worries about conservation in colonial Zimbabwe were not entirely driven by a profound environmental consciousness, at least during the period under discussion. The majority were goaded into taking up conservation measures because of their apprehension about the economic crisis that would ensue as a result of environmental degradation.

CONCLUSION

This paper has demonstrated that at the heart of the miner–farmer dispute was a struggle by each class to define its relationship to the environment by ensuring unfettered access to natural resources as well as the space that was the Gold Belt. It has demonstrated that both farmers' and miners' concerns about the distribution of the resources within the Gold Belt were not motivated by a benign regard for the protection of the environment. The concerns were largely about satisfying their production and consumption needs, hence the repertoire of claims, counterclaims and contestation, all waged to fortify their conflicting positions. This study provides an alternative way of understanding the making of white settler identities and represents an ecological narrative of settler colonisation and entrenchment in colonial Zimbabwe. Above all, the controversy and the actual degradation of forests and the land at the hands of both miners and farmers ultimately laid the basis for a state-sanctioned conservation regime in colonial Zimbabwe. This historical case does indeed support the idea that farmers and miners had conflicting logics of appropriation. But what this article has demonstrated is that such incompatibility, while it was never resolved, was contained through fluctuating political struggles and balance shifts in Southern Rhodesia, from miners to farmers in the 1920s, but back in the 1930s, as a result of the Great Depression, to miners.

NOTES

This article is based on research which I conducted as a British Academy Visiting Fellow in the Department of History, University of Sheffield, June–August 2006. I am very grateful to the British Academy for funding this research project. I would like to thank Prof. Ian Phimister (i) for hosting me in the Department and (ii) for his critical insights into this topic as well as generously sharing his personal collections with me. Last but not least, my sincere gratitude to Prof. Alois S. Mlambo, Department of Historical Studies and Heritage, University of Pretoria, and Prof. Jane Carruthers, Department of History, University of South Africa, for their helpful comments. Lastly, I wish to thank the anonymous reviewers of my article for their helpful comments and suggestions.

1.   For historical writings on 'disputed territories', see, for example, Jane Carruthers, 'Contesting Cultural Landscapes in South Africa and Australia: Comparing the Significance of the Kalahari

Gemsbok and Uluru – Kata Tjuta National Parks', in *Disputed Territories: Land, Culture and Identity in Settler Societies*, ed. David S. Trigger and Gareth Griffiths (Hong Kong: Hong Kong University Press, 2003), 233–268; Terence Ranger, *Voices from the Rocks: Nature, Culture and History in the Matopos Hills of Zimbabwe* (Harare: Baobab Books, 1999); Norman Etherington, 'Genocide by Cartography: Secrets and Lies in Maps of the South-eastern African Interior', 1830–1850', in *Disputed Territories*, 207–232; See also chapters on the theme: 'Settlers and Africans: Culture and Nature' in the collection *Social History and African Environments*, ed. William Beinart and JoAnn. McGregor (Oxford, James Currey, 2003), 197–266.

2. Carruthers, 'Contesting Cultural Landscapes'; Ranger, *Voices from the Rocks*.

3. Katherine G. Aiken, '"Not Long Ago a Smoking Chimney was a Sign of Prosperity": Corporate and Community Response to Pollution at the Bunker Hill Smelter in Kellog, Idaho', *Environmental History Review* 18 (Summer 1994): 67–86; Robert Kelley, 'Mining on Trial', in *Green Versus Gold: Sources in California's Environmental History*, ed. C. Merchant (Washington D.C., Island Press, 1998), 120–125; John Wirth, *Smelter Smoke in North America: The Politics of Transborder Pollution* (Lawrence: University Press of Kansas, 2000).

4. David Lowenthal, 'Empires and Ecologies: Reflections on Environmental History', in *Ecology and Empire: Environmental History of Settler Societies*, ed. Tom Griffiths and Libby Robin (Pietermaritzburg: University of Natal Press, 1997), 232.

5. Lewis. H. Gann, *A History of Southern Rhodesia: Early Days to 1934* (London: Chatto & Windus, 1965), 162.

6. Richard Hodder-Williams, *White Farmers in Rhodesia, 1890–1965* (London: The MacMillan Press, Ltd, 1983), 56.

7. Frank Clements and Edward Harben, *Leaf of Gold: The Story of Rhodesian Tobacco* (London: Methuen & Co. Ltd, 1962), 68. Also cited in Hodder-Williams, *White Farmers in Rhodesia*, 56.

8. Ian Phimister, *An Economic and Social History of Zimbabwe, 1890–1948: Capital Accumulation and Class Struggle* (London: Longman, 1988), 130. See also, JoAnn McGregor, 'Conservation, Control and Ecological Change: The Politics and Ecology of Colonial Conservation in Shurugwi, Zimbabwe', *Environment and History* 1 (1995): 257–260.

9. Percy. F. Hone, *Southern Rhodesia* (New York: Negro Universities Press, 1969), 293–297; Murray, *The Governmental System in Southern Rhodesia*, 126; Ian. R. Phimister, 'History of Mining in Southern Rhodesia to 1953' (Ph.D. diss., University of Rhodesia, 1975), 199–202, and Phimister, *An Economic and Social History of Zimbabwe*, 61–62 and 96; M.E. Lee, 'Politics and Pressure Groups in Southern Rhodesia, 1898–1923' (Ph.D. diss, University of London, 1974), see particularly Chapter 4, 'Changing Conflict: The Growth of Agriculture', 79–102.

10. Lee's thesis discusses 'the political development of the settler population in Southern Rhodesia from 1898 to 1923, and demonstrates the influence of local interest groups on that development', 2.

11. Vimbai C. Kwashirai, 'Dilemmas in Conservationism in Colonial Zimbabwe, 1890–1930', *Conservation and Society* 4 (2006): 541–561.

12. Kelley, 'Mining on Trial'.

13. J. Wirth, *Smelter Smoke in North America*. On the same subject see also, J. E. Read, 'The Trail Smelter Dispute', *The Canadian Yearbook of International Law* 1 (1963): 213–229; and D. H. Dinwoodie, 'The Politics of International Pollution Control: The Trail Smelter Case', *International Journal* 27, 2 (Spring 1972): 219–235.

14. Aiken, '"Not Long Ago a Smoking Chimney was a Sign of Prosperity"'; See also, Randall Rohe, 'Mining's Impact on the Land', in *Green Versus Gold*, 125–135.

15. K. Morse, *The Nature of Gold: An Environmental History of the Klondike Gold Rush* (Seattle: University of Washington Press, 2003).

16. Chad Montrie, *To Save the Land and People: A History of Opposition to Surface Coal Mining in Appalachia* (Chapel Hill: University of North Carolina Press, 2003).

17. *Report of the Commission to Enquire into the Preservation of Natural Resources of the Colony, 1939* [chaired by Sir Robert McIlwaine] (CSR – 40 – 1939], Hereafter referred to as the McIlwaine Report).

18. Murray, *The Governmental System in Southern Rhodesia*, 121.

19. S.J. Du Toit, *Rhodesia: Past and Present* (London: William Heinemann, 1897), 123–124.

20. NAZ: S1215/1081/4: Notes of Conference between Miners and Farmers, 18 Oct. 1933; and the *Bulawayo Chronicle*, 30 Mar. 1935.

21. Arthur. Keppel-Jones, *Rhodes and Rhodesia: The White Conquest of Zimbabwe, 1884–1902* (Pietermaritzburg: University of Natal Press, 1983), 363.

22. Henri Rolin, *Les Lois et l'Administration de la Rhodésie* cited in Philip. Mason, *The Birth of a Dilemma: The Conquest and Settlement of Rhodesia* (London: Oxford University Press, 1958), 270.

23. NAZ: S480/59C: Conference of Miners and Farmers, 1925, 7

24. [National Archives, UK] DO 35/841/95/3: Mining, House of Commons, 8 Jun. 1939.

25. Mason, *The Birth of a Dilemma*, 253.

26. *Chimurenga/Umvukela* are respectively Shona/Ndebele terms meaning 'uprising'.

27. Robin Palmer, *Land and Racial Domination in Rhodesia* (London: Heinemann, 1977).

28. Henry. V. Moyana, *The Political Economy of Land in Zimbabwe* (Gweru: Mambo Press, 1984), 70.

29. For a detailed outline of the administrative and political structures of the mining industry, see Murray, *The Governmental System in Southern Rhodesia*, Chapter 5, 'The Mining Sector', 118–161.

30. Keppel-Jones, *Rhodes and Rhodesia*, 581.

31. Gann, *A History of Southern Rhodesia*, 161.

32. Keppel-Jones, *Rhodes and Rhodesia*, 370.

33. Gann, *A History of Southern Rhodesia*, 162; Phimister, 'History of Mining in Southern Rhodesia to 1953', 58.

34. Gann, *A History of Southern Rhodesia*, 162.

35. Phimister, *An Economic and Social History of Zimbabwe*, 58.

36. Phimister, 'History of Mining in Southern Rhodesia to 1953', 99.

37. Henri Rolin, *Rolin's Rhodesia*, (First published in Brussels, 1913), Reprinted, Volume 21 (Bulawayo: Books of Rhodesia, 1978), 244.

38. Phimister, 'History of Mining in Southern Rhodesia to 1953', 200.

39. Murray, *The Governmental System in Southern Rhodesia*, 120.

40. Phimister, *An Economic and Social History of Zimbabwe*, 61–62.

41. National Archives of Zimbabwe [hereafter NAZ] S.1246, General, 1917–45: Memorandum on Titles to Land on the 'Gold Belt' of Southern Rhodesia: The Rights of Miner and Farmer. Issued by the Department of Lands, 1928?.

42. *Southern Rhodesia: Handbook for the use of Prospective Settlers on the Land*, (Issued by Direction of the Hon. Minister for Agriculture and Lands), Sixth Edition (No Date), 27.

43. Phimister, 'History of Mining in Southern Rhodesia to 1953', 62.

44. Phimister, *An Economic and Social History of Zimbabwe*, 58.

45. *The Matabele Times and Mining Journal*, 25 Feb. 1899.

46. *The Matabele Times and Mining Journal*, 4 Mar. 1899.

47. *The Matabele Times and Mining Journal*, 18 Feb. 1899.

48. *The Matabele Times and Mining Journal*, 18 Feb. 1899.

49. Hodder-Williams, *White Farmers in Rhodesia*, 53.

50. *Southern Rhodesia: Report of the Department of Agriculture for the Year ended 31 Mar. 1905*, presented to the Legislative Council, 13.

51. *Report of the Department of Agriculture for the Year ended 31 Mar. 1905*, 13.

52. Gann, *A History of Southern Rhodesia*, 171.

53. Gann, *A History of Southern Rhodesia*, 171.

54. Gann, *A History of Southern Rhodesia*, 167.

55. Phimister, *An Economic and Social History of Zimbabwe*, 62.

56. Phimister, *An Economic and Social History of Zimbabwe*, 62; also cited in Lee, 'Politics and Pressure Groups', 84. The direct quotation derived from the *Proceedings of the Rhodesia Agricultural Union Congress, 1906*, 102.

57. The Gold Belt title was applicable to Mashonaland only because most of Matabeleland was alienated under the Victoria Agreement Title. Here miners did not enjoy any special privileges in the form of reservations. See NAZ S.1246: Memorandum on Titles to Land on the 'Gold Belt' of Southern Rhodesia.

58. Phimister, *An Economic and Social History of Zimbabwe*, 96.

59. NAZ S.1246: Memorandum on Titles to Land on the 'Gold Belt' of Southern Rhodesia.

60. *An Agricultural Survey of Southern Rhodesia: The Agro-Economic Survey*, Part 11 (Salisbury: The Government Printer), 15–16.

61. *Rhodesia Herald*, 19 Aug. 1910.

62. *Rhodesia Herald*, 9 Sept. 1910.

63. *Rhodesia Herald*, 14 Oct. 1910.

64. *Rhodesia Herald*, 14 Oct. 1910.

65. Cross's lengthy observations were reproduced verbatim in the *Bulawayo Chronicle*, 10 Aug. 1907.

66. *Bulawayo Chronicle*, 10 Aug. 1907.

67. *Bulawayo Chronicle*, 10 Aug. 1907.

68. *Southern Rhodesia: Handbook for the use of Prospective Settlers on the Land*, (Issued by Direction of the Hon. Minister for Agriculture and Lands), Sixth Edition (No Date), 25–26.

69. Lee, 'Politics and Pressure Groups in Southern Rhodesia', 141.

70. *Southern Rhodesia Legislative Council Debates*, First Session, Sixth Council, Volume 4, Number 1, 1914, columns, 86 and 84. Also cited in Phimister, 'History of Mining in Southern Rhodesia', 201.

71. NAZ S480/92: Petitions and Protests to Legislative Assembly against Amendment affecting Gold Belt Titles 1924: Petition from the Rhodesian Small Workers' and Tributor's Association to Sir Charles Coghlan, Premier of the Colony of Southern Rhodesia, 1924.

72. NAZ S480/92: Petitions and Protests to Legislative Assembly against Amendment affecting Gold Belt Titles 1924: Minutes of the Rhodesia Chamber of Mines.

73. *The Importance of the Mines to the Farmers*, issued by the Chamber of Mines, the Chamber of Mines, Salisbury, (Bulawayo, May 1924).

74. *The Importance of the Mines to the Farmers*.

75. Phimister, 'History of Mining in Southern Rhodesia', 202.

76. Palmer, 'Agricultural History of Rhodesia', 236.

77. *An Agricultural Survey of Southern Rhodesia: The Agro-Economic Survey* (Salisbury, Government Printer, Part II, no year of publication), p. 17. See also Palmer, 'The Agricultural History of Rhodesia'. Between the 1930s and 1950s the state introduced statutory Marketing Boards to oversee the marketing of the main agricultural products such as maize, cotton, tobacco, beef and

dairy products, either solely by those boards or in conjunction with a European co-operative movement. This was done to undercut competition from African farmers, among other things.

78. Palmer, 'Agricultural History of Rhodesia', 235.

79. *Rhodesian Manual: African Manual on Mining, Industry, and Agriculture* (Incorporating Rhodesia and Portuguese East Africa (Johannesburg: Mining and Industrial Publications of Africa, Ltd., 1927), 93.

80. NAZ: S480/59C: Miner versus Farmer: Conference between representatives of the Mining and Farming industries held at Salisbury, Southern Rhodesia, 24 and 25 April 1925.

81. NAZ: S480/59C: Miner versus Farmer: Conference.

82. NAZ: S480/59C: Miner versus Farmer: Conference.

83. NAZ: S480/59C: Miner versus Farmer: Conference, 8

84. NAZ: S480/59C: Miner versus Farmer: Conference, 61.

85. NAZ: S480/59C, Miner versus Farmer: Memorandum prepared by Acting Assistant Director of Lands, Department of Lands, 23 Feb. 1927.

86. NAZ: S480/59C, Miner versus Farmer: Memorandum prepared by Acting Assistant Director of Lands.

87. NAZ: S480/59C: Miner versus Farmer: Conference, 127.

88. NAZ: S480/59C, Miner versus Farmer: Memorandum prepared by Acting Assistant Director of Lands.

89. NAZ: S480/59C: Miner versus Farmer: Conference, 6.

90. For more on this see Phimister, *An Economic and Social History of Zimbabwe*, 227–229.

91. Phimister, *An Economic and Social History of Zimbabwe*, 228.

92. NAZ: S480/59C, Miner versus Farmer: Memorandum prepared by Acting Assistant Director of Lands.

93. NAZ: S480/59C, Miner versus Farmer: Memorandum prepared by Acting Assistant Director of Lands.

94. NAZ: S480/59C: Miner versus Farmer: Conference, 26.

95. NAZ: S480/59C: Miner versus Farmer: Conference, 26.

96. *Rhodesian Manual*, 94.

97. NAZ: S480/59C, Miner versus Farmer: Memorandum prepared by Acting Assistant Director of Lands.

98. NAZ: S480/59C: Miner Versus Farmer: Minutes of a meeting held in the Legislative Assembly Chamber, Salisbury, 12 & 13 Feb. 1926.

99. NAZ: S480/59A: Farmers versus Miners: Copy of Letter from the Secretary of the Rhodesian Agricultural Union, Salisbury, to the Secretary for Mines and Works, Salisbury, 23 Aug. 1927.

100. *Rhodesia Manual*, 94.

101. *Rhodesia Manual*, 94.

102. NAZ: S1215/1081/4: Conference between Miners and Farmers, 1933: Notes of Conference between Miners and Farmers held at the Legislative Buildings, Salisbury, 18 Oct. 1933.

103. NAZ: S1215/1081/4: Conference between Miners and Farmers, 1933.

104. NAZ: S1215/1081/4: Conference between Miners and Farmers, 1933.

105. NAZ: S1215/1081/4: Conference between Miners and Farmers, 1933.

106. *Report of the Committee of Enquiry into the Economic Position of the Agriculture Industry*, (chaired by Mr. Max Danziger) (CSR –16 – 1934),1. (Hereafter referred to as the Danziger Report). The Committee identified about eighteen causes of the poor state of the farming industry among which were a heavy slump in the value of farming commodities on export as well as local markets, redundant creameries, unrestricted imports of agricultural commodities, etc. For more details on the causes see pages 2–3 of the Report.

107. The Danziger Report, 1.

108. The Danziger Report, 29.

109. The Danziger Report, 29. Colonial officials had, in the past, a field day blaming African agricultural methods for being unscientific and hence causing wanton environmental degradation while at the same time they paid scant attention to the European farmers' so-called scientific methods which equally caused soil erosion – as the Enquiry into the Economic Position of the Agriculture Industry was to reveal. For further discussion on this see: McGregor, 'Conservation, Control and Ecological Change', 261–266; Terence Ranger, *Peasant Consciousness and Guerrilla War in Zimbabwe* (London: James Currey, 1985), 68–69.

110. The McIlwaine Report, 15–16.

111. The McIlwaine Report, 16.

112. The McIlwaine Report, 19.

113. The McIlwaine Report, 36.

114. The McIlwaine Report, 34.

115. A cord was assumed to represent a 'pile of wood eight feet long, four high and four wide', definition given in McIlwaine Report, 34.

116. The McIlwaine Report, 36.

117. The McIlwaine Report, 37.

118. C.H. Thompson and H.W. Woodruff, *Economic Development in Rhodesia and Nyasaland* (London: Dennis Dobson Limited, 1953), 50.

119. D. Anderson, 'Depression, Dust Bowl, Demography, and Drought: The Colonial State and Soil Conservation in East Africa During the 1930s,' *African Affairs* 83 (1984): 321–344, p. 324.

Empirical Knowledge, Scientific Authority, and Native Development: The Controversy over Sugar/Rice Ecology in the Netherlands East Indies, 1905–1914

Suzanne Moon

INTRODUCTION

In the Madiun[1] region of central Java in 1903, the *Binnenlands Bestuur*, the colonial civil service, reported complaints from indigenous farmers who regularly worked their land in rotation with European sugar planters. Farmers claimed that sugar growers were ruining the land by using chemical fertilisers on the cane crop.[2] The Resident, the head of the *Binnenlands Bestuur* for the region, had been sufficiently alarmed by these complaints in light of recent declines in rice and peanut production that he contacted Melchior Treub, the highly respected head of the Royal Botanical Garden in Buitenzorg.[3] Treub acknowledged the possibility that chemical fertilisers could have created some problems in soil fertility as such effects were well documented in Europe. If declines in rice and peanut production had occurred, then it was certainly possible that the chemical fertiliser favoured by sugar planters, sulphate of ammonia, was to blame. The Resident accordingly suggested that Madiun planters immediately forego the use of chemical fertilisers and switch to natural alternatives.[4] J.D. Kobus, a prominent sugar scientist, denied that fertilisation practices had produced negative effects. He claimed instead that rice yields usually improved as a consequence of rotation with sugar, and recommended that sugar planters continue as usual.[5]

For ten years, the ecological relationship between sugar and rice crops, when these were grown in rotation, remained a controversial subject, pitting scientists from the East Java Experiment Station of the General Syndicate of Sugar Manufacturers[6] against scientists from the Netherlands East Indies Department of Agriculture. What made the controversy particularly difficult to resolve was both the complex and contradictory nature of the data, and accusations of bias on both sides that hampered any consensus-building on interpreting that data. Some critics in the colony and in the Netherlands had long accused sugar planters of practices that impoverished local farmers; these critics believed that any result coming from scientists working for the sugar industry would be biased conveniently to overlook harm to local communities.[7] Supporters of the sugar industry, on the other hand, felt that such criticisms were unfair and took an unduly skewed view of the best interests of farmers, ignoring the positive

Suzanne Moon

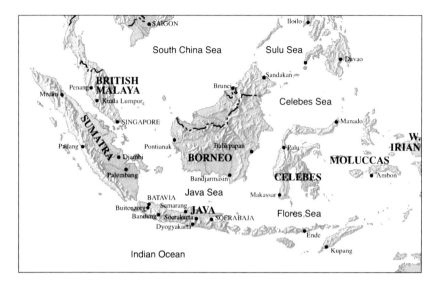

MAP 1. The Netherlands East Indies

economic contributions of sugar plantations to the regions. In the view of the sugar industry, narrow-minded critics would unduly weigh the value of isolated negative cases and ignore the places where rice crops had remained strong in rotation with sugar.

At issue was not merely the ecological effects of sugar cane on subsequent rice crops, but the question of who had the credibility to interpret this relationship reliably. For those who were suspicious of the sugar industry's motives, the Department of Agriculture, which had been formed with the explicit mission to improve indigenous agriculture, seemed most likely to remain impartial.[8] On the other hand, the sugar industry had built a respected scientific institution in East Java, and sugar scientists bolstered their interpretive credibility by arguing that they had a superior understanding of the day-to-day conditions of cultivation on Java. The rotation of sugar and rice on the same land blurred the boundaries between European and indigenous production. In the same way, the controversy over the ecological relationship between rice and sugar called into question the seemingly clearly demarcated boundaries of scientific authority in the colony.

The perception of political or economic interest biasing interpretation made scientific credibility the important underlying issue of this debate. Claims to superior empirical knowledge of conditions on Java became the favoured strategy for asserting the credibility of scientific claims and claimants. This paper explores how and why superior empirical knowledge became a critical tool for combating accusations of bias and for conferring interpretive authority in an otherwise intractable debate.

SUGAR, RICE AND THE ETHICAL POLICIES:
BACKGROUND TO THE CONTROVERSY

The system of rotating rice and sugar on the same land reached back to the mid-nineteenth century, when a system of production called the Cultivation System was the colonial government's main source of revenue. Under the Cultivation System, villages were required to grow certain export crops mandated by the colonial government in lieu of taxes. The government auctioned off these crops, making a healthy profit.[9] Under this mandate, villages in regions where sugar grew best added sugar to their customary rotation of rice and other dry season crops. By 1870, the advent of more liberal thinking in the Dutch government motivated a change in policy, and reformers passed a number of reforms that were meant to remove the government from export production. Dutch liberal reformers argued that the Cultivation System was outdated because the government interfered too much with trade that would operate more efficiently in the private sphere. The Sugar Law of 1870 mandated that the government would gradually withdraw from sugar production, opening Java to private investment in 1878.[10] At the same time however, reformers wished to safeguard Native peasants from predatory practices of wealthy European or Chinese investors.[11] The Agrarian Law of 1870 guaranteed that Natives in possession of land would maintain their customary rights, and that they could, if they chose, obtain the right to private ownership.[12] A further ordinance in 1875 formally mandated that Native lands could not be alienated to non-Natives.[13] Forbidden from owning land, European investors had to lease land from individuals or villages in order to grow sugar crops on Java, and could not require that land be taken out of rice production for too long. Private sugar planters therefore had to accommodate their needs to those of the villages, and their cane planting to rotation with rice, which had become, under the Cultivation System, ordinary practice.[14]

A sugar manufacturer would negotiate leases with a number of villages in order to gain access to a large contiguous block of land. The lease ran for 21½ years, and during that time a typical rotation was sugar cane for 15–18 months, followed by rice, a dry season crop like peanuts or maize, and a second rice crop.[15] While not allowing sugar planters continuous and sole access to (and control over) the land, the system did give them access they might not otherwise have had to land that was ideally suited for sugar cultivation. Both rice and sugar thrive when there is good and well-controlled access to water, so the flat, rich, and well-watered valleys where wet rice, or *sawah*, agriculture was most successful were the same places that made the best locations for sugar planting.

Because of the similar needs of the two crops, colonial agricultural experts tended to view rice and sugar as complementary, at least ecologically. Sugar planters frequently ploughed more deeply than did indigenous farmers, and on some soils this provided an improving effect for subsequent crops as well as

Suzanne Moon

FIGURE 1. Indigenous labourers prepare cane for planting.(J. Sibinga Mulder, *De Rietsuikerindustrie op Java*, N/V H.D. Tjeenk Willink & Zoon, 1929, p. 23.)

for the sugar itself.[16] The theory of rice/sugar complementarity was, however, primarily grounded on the basic assumption that greater access to water improved rice harvests.[17] Colonial officials tended to view new irrigation works, which were promoted and sometimes partially funded by sugar manufacturers, as equally beneficial for indigenous farmers.[18] Deeper, scientific studies of the effects of sugar on rice harvests were rare before the turn of the twentieth century.[19]

While many Europeans assumed that the ecological relationship between rice and sugar was harmonious, they would not make the same claim about the social and economic relationship between rice and sugar growers.[20] Indeed from the onset of private sugar planting, the relationship had been contentious and the object of much criticism among reformers in the European community. Despite the assumption that irrigation works would benefit everyone, rice farmers and sugar planters often contested the distribution of irrigation water. Towards the end of the nineteenth century, sugar planters adopted new sugar processing technologies that used more water, exacerbating these conflicts.[21] Government regulations that partitioned water by day- or night-use eased, but did not resolve, the conflicts.[22] More contentious over the long term were the linked questions of rent paid by sugar planters to farmers for use of their land, and the wages paid to field workers who provided seasonal labour while land was planted to sugar.[23] The amount of land rent was based on the estimated value of the rice harvest that farmers would forego. Because sugar was a much more valuable crop than rice, sugar planters stood to make good use of their investment in land rent, despite their complaints about the system. Indigenous farmers, on the other hand, frequently entered a cycle of debt, in which they spent their rent payments early in the three-year crop rotation, leaving them vulnerable to loan sharks and pawn shop operators for later shortfalls. Sugar planters also came under fire for paying low wages to workers – most of whom needed hard cash to pay their tax obligations. Sugar planters emphasised the good they did in bringing wage labour to the indigenous people. Opponents of the sugar industry accused them of exploiting a population who had become caught in a cycle of debt. Sugar planters were known to bribe the village leaders with whom they negotiated to get more favourable terms for themselves, a practice which left ordinary people at the mercy of both the sugar planters and their own leaders.[24]

The problems in the sugar/rice regions gained increased political visibility with the formal advent of what became known as the Ethical Policies in 1901. These policies, which bore a family resemblance to civilising missions elsewhere in the colonial world, declared that the government of the Netherlands East Indies was obliged to work actively to improve the welfare of the indigenous peoples.[25] While the Ethical Policies advocated far-reaching reforms in many aspects of public life (for example greater press freedoms, freedom of political organisation for indigenous people, and more widely available education), they contained a significant technological component as well. The earliest Ethical planners argued that the lives of indigenous peoples could be improved if their technologies of production (primarily agricultural, as most indigenous people worked in agriculture) were made more efficient. The slogan 'irrigation, education, and emigration' captures the initial technological thrust of Ethical thinking: adding new irrigation works to improve harvests of indigenous crops (especially rice), educating mid-range officials, farmers and trades people in new, more productive ways of working, and encouraging farmers to emigrate from the crowded and subdivided lands of Java to the Outer Possessions where they would establish prosperous, rice-farming communities.[26] The Department of Agriculture was established to address shortcomings in indigenous production. Like civilising mission projects elsewhere in the world, the government took the responsibility of both identifying the main problems of indigenous society, and devising solutions, a process that came to be called 'the development of the Native peoples'. The existing conflicts in the rice/sugar lands gained new political visibility with the advent of the Ethical policies.

Suzanne Moon

SUGAR YIELDS, RICE YIELDS AND THE QUESTION OF CHEMICAL FERTILISERS

While the ecological relationship between rice and sugar had not played much role in the earlier discussions about the effects of the sugar industry on indigenous farmers, in 1905 Zeno Kamerling brought the subject into public debate in a thirteen-part series of articles published in the *Soerabaiasch Handelsblad*, a general interest newspaper published in Surabaya, East Java.[27] Kamerling was a former employee of the sugar industry, working at the Kagok sugar experiment station on Java. During his tenure there, he had argued vehemently against the sugar industry's use of sulphate of ammonia fertilisers, claiming that these fertilisers degraded the soil and that natural manures were superior.[28] Although he had left the sugar industry to take up a position teaching natural science at a European academy in Batavia, he continued his campaign against standard fertiliser practices in the sugar industry with the articles in the *Soerabaiasch Handelsblad*. Kamerling asserted that the use of sulphate of ammonia fertiliser on sugar cane caused nitrogen depletion and structural damage to the soil. He further posited that this fertiliser greatly increased the cane crops' uptake of nitrogen, providing short-term yield improvements for sugar, but at the cost of long-term nitrogen depletion of the soil.[29] As evidence he pointed to the declines in the rice and peanut yields of indigenous farmers in the region of Madiun.

In Kamerling's earlier work on the topic (in 1900), he argued that protecting the soil (by not using sulphate of ammonia) was in the best interests of the sugar industry. He had little to say about the effects on indigenous farmers.[30] Perhaps motivated by the debates over the Ethical policies, Kamerling changed focus in his 1905 newspaper articles by putting the needs of the indigenous people on equal footing with the long-term health of the sugar industry, and highlighting the consequences of soil depletion for the yields of indigenous rice fields. Despite its specialised and technical content, the editor found the topic of enough general interest to put each instalment of the article on the front page of the newspaper. Kamerling therefore had a relatively large audience, as he subtly transformed the earlier discussion about the technological practice of fertiliser use in sugar growing into a broader discussion about the ecological relationship between rice and sugar production.[31] Kamerling scrutinised the technology of sugar production as an issue with Ethical implications, in effect opening up the 'black box' of sugar manufacturing for the general public (where a black box is a technology not widely understood by non-experts).[32] Rice agriculture became both key evidence of soil degradation, and the central motivation for immediate change. For Kamerling, rice agriculture needed to be understood as the canary in the coalmine. Sugar cultivation had changed over the years, with improved methods masking the degraded fertility of the soil, he claimed, while indigenous rice agriculture was being practised as it always had. Therefore, rice agriculture ought to provide the earliest evidence of decreased soil fertility, and Kamerling claimed the situation in Madiun proved just this.[33] The consequences for rice agriculture were serious, Kamerling argued, because the very difficulty of changing indigenous practices meant that decreases in fertility would devastate indigenous agriculture quickly, and possibly irreversibly. Solving the problem could not wait for the declines to show up in sugar production.

Kamerling's strenuous insistence for immediate action was hampered by the weakness of his evidence. Indeed, Kamerling admitted that direct evidence of a decline in fertility over the previous thirty years was lacking. Many sugar producers showed a steady or even slightly increasing production of cane sugar during that time, which Kamerling attributed to factors that he argued masked the slow decline in soil fertility: more intensive and deeper cultivation of the soil, cultivation of larger areas, heavier use of fertilisers, and new canes.[34] Lacking evidence from the sugar industry, Kamerling marshalled indirect evidence to make his case. In addition to the declining rice yields mentioned earlier, he also pointed to European examples in which the sole use of sulphate of ammonia for other crops, specifically barley, had led to serious soil degradation.[35] The European example was of special importance to Kamerling because it resulted from forty years of experiment and observation, and therefore showed the decline even though it was quite gradual. To make the link to sugar he cited studies of sugar production in the West Indies, made by F.A.F.C. Went, a well-known and respected Dutch scientist. In Surinam especially, Went concluded that a steady decline in soil fertility occurred after about seven years of continuous cultivation of cane.[36] These cases served not only as evidence against sulphate of ammonia, but as cautionary tales – by the time undeniable direct evidence was available, it might be too late to prevent ecological and economic disaster on Java.

Not satisfied simply with showing soil degradation, however, Kamerling asserted that rice was a soil improving crop, which made the Ethical implications of the rice/sugar relationship even more clear. While sugar cane, with the help of sulphate of ammonia, would remove nitrogen from the soil, the silt from the irrigation water used in flooded rice agriculture added nitrogen back to the soil.[37] By the time the land recovered to some degree from nitrogen depletion, it would be time to rotate back to sugar production. In essence Kamerling was demonstrating that not only did wealthy sugar planters harm the yields of poor, indigenous rice farmers, but the rice farmers were in fact ecologically subsidising sugar production.

Suzanne Moon

FINDING THE TRUTH: SCIENTIFIC BIAS, THE SUGAR INDUSTRY,
AND THE DEPARTMENT OF AGRICULTURE

As this was not the first time Kamerling had raised this issue, the sugar industry
had already proposed a way of resolving the question. J.D. Kobus, the prominent
sugar scientist who would spend the next several years debating the issue with
Kamerling, had proposed that the sugar experiment stations, with witnesses from
the *Binnenlands Bestuur*, run a series of experiments to determine whether or
not sulphate of ammonia was indeed creating problems in soil fertility. Kamerling took issue with this proposal in his article, bluntly claiming that the sugar
industry was too biased to be trusted with such important work.[38] He argued that
only the Department of Agriculture (despite its youth, having only been established in January of that year) could put this research on an unbiased footing,
and then only if they committed to repeating the experiment over many years,
as had been done in Europe. Kamerling doubted the ability and desire of the
Binnenlands Bestuur and the sugar industry to interpret the evidence properly:

> In the best case they might continue for four years or so, (while as the research
> in Europe teaches, such experiments must continue for thirty or forty years to
> get convincing results) and then they should conclude, on the basis of four years
> of negative results, and contrary to all practical experience in all other countries,
> that the exclusive use of chemical fertilisers does no harm.[39]

Such an accusation of bias was certainly not surprising. Even Kobus acknowledged
the sugar industry's appearance of interest in the outcome when he suggested
that the fertiliser experiments be witnessed by the *Binnenlands Bestuur*.[40] The
naming of the Department of Agriculture as an unbiased scientific authority did
not go unchallenged by the sugar industry, however, whose members did not
regard the government as disinterested, but rather as carrying their own biases
which might shape their interpretation of the data.

The young Department of Agriculture, while created by the government to
serve the interests of the entire colony, had yet to earn unanimous recognition
of its own authority and credibility. One critic complained to the Seventh Sugar
Congress in 1905 that sugar planters knew their own regions far better than representatives of the civil service, whose officers came and went, and who relied
naively on informants without knowing the real situations within the villages.
The same criticism was easily levelled against the Department of Agriculture.
For this critic and others, the closeness of sugar planters to the situation in the
villages was an advantage that government departments did not have. In this
view, their closeness allowed them to interpret local complaints critically; planters
clearly viewed themselves as best placed to analyse and correct the complicated
social and economic problems of the sugar/rice villages.[41]

The primary reason that sugar planters accused the government of bias was the government's particular attachment to the notion of Native development through improvements to rice farming. For many critics, improving the social welfare of the Native people needed to be understood more broadly than simply improving Native agricultural practices.[42] The Department of Agriculture had focused their efforts (and pinned most of their hopes) on improving the productivity of rice agriculture, the most widespread economic activity on the island of Java.[43] Opponents claimed that improving welfare through rice farming would never work, because rice farming simply would never pay.[44] In this view, the sugar industry, by offering paid employment, as well as opportunities to make money through support activities like transportation, was essential to the welfare of the farmer, who would never make ends meet on rice alone. The implicit critique from the sugar industry was that the colonial government was biased towards keeping rice farmers operating as rice farmers. The sugar industry feared that anything that was perceived as hurting rice farmers would not get a fair hearing.

The fear of government bias was exacerbated by the publication in 1907 of the first volumes of the *Investigation into the Declining Welfare of the Native Peoples*. In the section of the report that looked at sugar villages, the authors noted the widespread belief (among indigenous farmers) that sugar planting caused a decline in rice yields.[45] Some disputed the truth of the *Investigations*. In the November issue of the *Indische Gids*, one critic questioned the veracity of the statistics used in the *Investigations*, suggesting that they were 'taken out of the air' and that they demonstrated the subjective beliefs of the reporters while maintaining a veneer of objectivity.[46] Further, the author stated that H.E. Steinmetz, the chair of the committee that produced the report was 'a long-time, known opponent of the sugar industry, whose views come to the fore in the list of questions [asked of village heads]'.[47] The author goes on to wonder how many of the reporters came under the chair's influence, and as a result perhaps even generated fictional numbers. Despite this sort of criticism, the Minister of the Colonies in the Netherlands strongly backed the reliability of the *Investigations*.[48] Supporters of the sugar industry maintained, however, that government officials might have their own anti-sugar agenda that coloured their interpretations, or perhaps even led to falsification of the data. In the matter of the relationship between sugar and rice farming, neither side in the debate believed that the facts of the matter could be reliably determined by those on the other side due to economic or political biases. To resolve the question, investigators had to find some way to demonstrate that their evidence was unbiased. Empirical knowledge became the most important weapon in this dispute.

Suzanne Moon

LOCAL KNOWLEDGE AND EXEMPLARY CASES

In a blistering initial response to Kamerling's attack on the sugar industry, Kobus, the scientist who spoke for the sugar industry in this dispute, dismissed nearly every aspect of Kamerling's critique by questioning Kamerling's empirical knowledge of Java.[49] To understand what was happening on Java, Kobus claimed, one needed to understand specifically Javanese circumstances and not use examples from Europe or elsewhere as Kamerling had done. Kobus argued that Kamerling's 'doubtful evidentiary procedures' led Kamerling to see evidence from Europe and the West Indies as properly exemplary for Java, when in Kobus's view it was not:[50] 'The European circumstances which he uses for comparison, diverge so entirely from that which prevails here on Java, that conclusions based on this are entirely worthless for us.'[51]

Kobus's accusation of questionable evidentiary practice was certainly meant to suggest that Kamerling was either incompetent or a biased opponent of the sugar industry (or both), but Kamerling was not in fact doing anything unusual when he claimed that fertilisers would behave in an approximately similar manner anywhere, whether Europe or Java. When establishing matters of fact, scientists must always construct which factors (including place) will be relevant to a particular phenomenon and which will not.[52] Such claims of similarity are, however, frequently called into question during controversies, and are especially easy to question in cases of agricultural experiment or observation, given the normal ecological variability of soil, water, and climatic conditions.[53] Kobus disputed the assumption of similarity by highlighting the differences between Java, Europe and other sugar colonies: the significance of flooded rice agriculture and the sugar/rice rotation for the condition of the soil. He claimed that the periodic flooding of the land with nutrient-rich irrigation water, in combination with a rotation that kept sugar on the land for only eighteen months in three years made the Javanese situation unique and comparable to neither Europe nor the West Indies. Any detrimental effect inherent in sulphate of ammonia alone would be more than compensated for by the periodic benefits of flooded rice agriculture.

Kobus did not stop there however. He dismissed Kamerling's evidence from Java as well, again on the grounds that the evidence chosen was not exemplary – that is, it was incapable of representing the true situation on Java as a whole. For example, Kamerling had used some data from a sugar scientist to show that the all-important silt in irrigation water usually had a deficit in calcium needed by the crops.[54] Kobus criticised that data as being based on too limited a sample size, and being far less valuable than his own data which had been taken daily from five rivers in Java.[55] It is important to note that Kobus did not dispute the importance of empirical evidence in resolving the question. Rather he disputed whether the evidence used could be considered exemplary. He noted, irascibly:

'When someone so rashly judges an entire method of cultivation, he ought to at least be up to speed on the correct investigations.'[56]

Kobus offered evidence to counter Kamerling's claims, evidence that he presented as being truly exemplary of the particular circumstances of the rice/sugar rotation on Java. His strongest counterexample came from tests run in 1904 at a sugar plantation in East Java. The planter involved reported that in the region of his plantation, all the areas that had been planted to sugar had subsequently produced dramatically higher yields of rice than those lands that had never been planted to sugar.[57] Kobus attributed this gain to the positive effects of the deeper ploughing practised by the sugar plantations. That is, when sugar planters prepared the land for the cane crop, they ploughed twice as deeply as did indigenous farmers. When the land was returned to the rice farmers, they would gain benefits from this deeper ploughing. Kobus added evidence from a Native regent in the area of Sidhoarjo, who claimed that yields had nearly doubled in the previous thirty-five years in his region, supporting the contention that the deeper ploughing practised by sugar factories improved the soil, and thereby increased subsequent rice yields.[58]

To what extent did Kobus wish to position these as exemplary cases? He was careful to conclude his critique of Kamerling by stating, 'Establishing the facts through well-done tests is the main thing, and when we are careful not to generalise, we shall slowly reach our goal of improving agricultural knowledge and therefore the Java sugar industry and Native agriculture.'[59] Despite this balanced wording, Kobus uses more definite language earlier in his article saying 'The alleged decline in yields of rice after fertilising cane with sulphate of ammonia was shown … [through the East Java experiments] to be impossible.' In a discussion on the topic that took place in the Seventh Sugar Congress in April 1905, Kobus said confidently: 'With regard to the connection between rice harvests and sugar cultivation, I can supply some statistics, that show that the sugar industry has a favourable influence on rice yields.'[60] His lack of conditional language suggests that Kobus did indeed want to assert his own evidence as essentially 'normal', that is, exemplary for the conditions of the sugar/rice rotation on Java.

Establishing the exemplary value of cases proved to be difficult on both sides of the debate. Kobus did not manage to build a consensus of opinion around his own cases, even within the sugar industry. During the discussion at the Seventh Sugar Congress, one planter who had seen rice yields drop by almost one half in his region, responded to Kobus by arguing: '… The numbers cited by Kobus cannot serve as material for comparison. We must account for the situation as it is, not as we wish it to be.'[61] The claims in the *Investigations into the Declining Welfare* also continued to spur discussion on the subject, but without building any consensus that the cases there were exemplary either. One supporter of the sugar industry, a planter named H.J.W. van Lawick expressed his own frustration at the explosion of contradictory evidence by asking how anyone was supposed to tell which cases represented the normal situation and which did not.[62] Complexity of evidence was certainly a problem, as no easily discernable patterns had emerged in the investigations. But complexity was not the only issue, as the question of interests or bias continued to be discussed. How could one be sure that the scientists who chose the cases were not merely choosing them because they fit preconceived beliefs? Van Lawick's proposed solution pointed to the general concern over the problem of bias. Despite his strong belief that experiments would eventually vindicate the sugar industry, he argued that only the Department of Agriculture could resolve the question, and then only if the experiments were personally overseen by Melchior Treub, perhaps the most famous scientist in the Indies, and a man whose reputation for impartiality on scientific matters was readily accepted by most of the European population.[63] Individual authority, however, would not resolve this debate.

Kobus dismissed the notion that only the Department of Agriculture could make a credible experimental inquiry, and organised his own series of experiments between 1906 and 1908 to study the question.[64] He tested a number of different crop rotations on the same land using small plots with different fertiliser inputs to gauge the results of fertiliser use on yields of various crops. While rice did consistently more poorly after sugar fertilised by sulphate of ammonia, the declines were fairly small, and Kobus argued that they were almost within the range of acceptable error for the test.[65] Perhaps the most decisive result of the three year series of tests was that access to irrigation water seemed to have a much more dramatic effect on all the crops than the particular crop rotation.

OMO MENTEK IN MADIUN: THE DEPARTMENT OF AGRICULTURE ENTERS THE DEBATE

Until 1910, the Department of Agriculture, despite calls to act otherwise, stayed out of the controversy. Events in the region of Madiun in 1910, however, brought the question of the rice/sugar relationship to the fore within the Department. During 1910 the area around Madiun had experienced serious outbreaks of *omo mentek*, a disease whose origins were unknown at the time.[66] Rice plants infected with the disease withered and gave little or no grain.[67] By responding for the Department of Agriculture to the situation in Madiun, a third player entered the debate started by Kamerling and Kobus, a young botanist named P. van der Elst. Because many outbreaks had occurred on land regularly rotated with sugar (as much of the land in Madiun was), van der Elst took a fresh look at the question of the effects of sugar cane on rice agriculture, asking whether cane cultivation might make the rice crop more susceptible to the disease. While van der Elst reported very thoroughly on other aspects of the disease, he gave most attention to rice rotation practices, including the rotation with sugar.[68]

Van der Elst demonstrated a link between nitrogen-poor soil and the occurrence of the disease, and asked whether certain rotations resulted in nitrogen exhaustion in the soil. Citing, and rejecting, the findings of Kobus, van der Elst concluded that nitrogen deficits were common on soils where sugar and rice were worked in rotation. Such deficits were only avoided under conditions of very strong irrigation on particularly good soil.[69] Regardless of fertilisation practices, the soil on which sugar was rotated with rice showed noticeable deficits in nitrogen content even a year after the harvest of the sugar crop, despite the flooding that accompanied the rice crop during that time. *Mentek* appeared much more frequently on the nitrogen deficient soils than it did in areas where irrigation was stronger and silt content higher in that essential nutrient.

For the purposes of this essay, van der Elst's method of resolving the debate is more important than the fact that he did resolve the debate. He managed to assert his claims as both credible and authoritative in a way that the exemplary cases of Kobus and Kamerling had not. Van der Elst repeatedly emphasised the breadth and reliability of his empirical knowledge of the region of Madiun, as compared to the work of either Kobus or Kamerling, and played down the idea of normal or exemplary cases. Instead he foregrounded the interplay of different factors that contributed to the seemingly contradictory situations across Madiun, drawing an interpretation that highlighted rather than erased the complexity of the results.

Van der Elst gave a detailed explanation of the many sources of statistics he used to build his case.[70] The multiplicity of his sources of data, and the role of personal experience in compiling and verifying all of this data figured prominently in his description of his research methods. He consulted the *Binnenlands Bestuur* for reports on the recent harvests, the head of the local irrigation districts for information on the strength of irrigation in the area, the meteorological service for weather data, and village heads and farmers for anecdotal reports of the disease. Not relying on second-hand accounts alone, he visited most of the villages, many more than once, to track the progress of crops and disease during 1910, adding the veracity of a scientific witness to these many accounts. He made sure to let his readers know that he had collected data from every village in Madiun, not just a few, and that he cross-checked all of the data himself with his other sources of information to be sure that it was accurate.

The claim he built by doing so was about coverage, a deep empirical knowledge of the entire region of Madiun, not just a few isolated and possibly opportunistically chosen areas. Unlike others who could be accused of picking and choosing data to suit their interests, by compiling so much data on the entire region of Madiun van der Elst bolstered his own assertions of objectivity. He criticised Kobus's experiments in comparison with his own analysis, accusing

Kobus of drawing incorrect general conclusions about the effects of sugar on rice from too few examples, because those few examples could only match a few local conditions, and could not account for others.[71] For van der Elst, the scope of Kobus's knowledge was inadequate to draw useful conclusions about the complex relationship between land, water, and soil that defined the rice/sugar rotation.

Drawing on the Madiun study, van der Elst published in 1913 a more direct and cutting response to Kobus in a sugar industry publication.[72] Van der Elst again argued that only comprehensive empirical understanding of the varying ecological circumstances in the rice/sugar regions on Java could produce a reliable interpretation of the relationship between rice and sugar growing.[73] In this later essay, van der Elst directly challenged the credibility and authority of Kobus in ways he had only hinted at in the *mentek* report. In so doing he undermined the exemplary value of Kobus's findings. Van der Elst accused Kobus of uncritically using data from one source, a planter who had no scientific credentials to speak of. Van der Elst had checked this data and found it to be at best selectively drawn to highlight the best case for planters, and at worst an outright misrepresentation of the facts.[74] Van der Elst scathingly attacked the work of this planter, and implicitly the reliability of unverified reports from sugar planters in general:

> It is most wondrous of all that the administrator of a sugar factory would know his own planting area so little that he ... would call the advantages of cane agriculture so absolutely visible, and even give the impression that the yield was better the longer there had been cane agriculture, while – so it seems – the influence of cane on rice yield in that area absolutely doesn't exist.[75]

Van der Elst saved most of his scorn for Kobus however:

> On the basis of this single test ... Kobus found himself justified to triumphantly declare that further tests were unnecessary. Had he done this experiment himself, it might have been a better basis on which to build conclusion ... he then made an attempt ... to use the circumstances in Sidhoarjo as evidence to stand in for Madiun. The situations in these areas are entirely different.[76]

Van der Elst rejected Kobus's experiments, not because they were poorly done, but because they did not demonstrate their applicability anywhere but in the region in which they were done. After destroying the credibility of Kobus, van der Elst reinterpreted Kobus's data in light of his own empirical understanding of the region. Much as he had found in the *mentek* report, van der Elst judged that the rice paddies in most parts of Madiun did indeed demonstrate some nitrogen exhaustion, and that irrigation factors were the most critical for determining the positive or negative outcome of the sugar/rice rotation. For good measure, his new calculations demonstrated that sulphate of ammonia when used as a sole fertiliser could indeed produce soil exhaustion, a conclusion made somewhat less than momentous by the fact that many sugar planters had by this time stopped that practice on their own, using a new mix of fertilisers that included superphosphates.[77]

CONCLUSION: THE DEPARTMENT OF AGRICULTURE, AND THE IMPORTANCE OF PLACE

Van der Elst based his authority to speak on his intimate knowledge of place, and the growing understanding of the variations between places that he obtained for the Department of Agiculture. Van der Elst's research in Madiun became the definitive study of *mentek* for many years, and his complicated picture of the relationship between rice, water, and sugar put an end to the scientific controversy.[78] Van der Elst continued to refine his understanding of the relationship between irrigation and rice production in a later set of experiments which he coordinated for the Department of Agriculture.[79] Van der Elst's appreciation for the variations in soil and water even within fairly small regions made him suspicious of trying to apply results obtained from local experiments on a larger scale, because, by his own argument, it was never clear how well the results might apply outside a small area.[80] His knowledge of subtle differences from place to place kept his proposals for improvements small-scale and sensitive to local conditions.

In the controversy over the ecological relationship between rice and sugar, empirical knowledge that could be shown to be both broad (i.e., significantly covering an area of study) and deep proved critical to resolving the debate. Van der Elst's comprehensive knowledge of the rice/sugar villages of Madiun allowed him to illuminate the controversial exemplary cases of others in the wider context of ecological variation in the region, strengthening the legitimacy and authority of his interpretations. While sugar planters might have been able to claim a similar depth of knowledge about the land they leased, they were unable or unwilling to match the Department of Agriculture's scope of knowledge, that is, the wide-ranging and comparative understanding of all the places on Java. It was the scope of their studies that allowed the Department of Agriculture to escape the accusations of interest that plagued the exemplary cases of others, and to cement its authority to speak for, and create knowledge pertaining to, indigenous agriculture.

Van der Elst's use of empirical knowledge and attention to local variability is representative of a larger trend in the Department of Agriculture's approach to 'Native development', especially when it came to efforts by the extension service to improve rice yields.[81] As the extension service increased the number of demonstration and test fields across the island of Java in the 1910s, they gained an appreciation for the diversity of social and ecological conditions that contributed to rice yield, and sought to find rice varieties and techniques that increased production by suiting local conditions and tastes.[82] The Department also ran long-term (i.e., 5–10 year), detailed investigations into the social, economic, and technical organisation of agricultural production of villages across Java.[83] From the sugar/rice controversy it is possible to understand the ways that such an approach could help the Department maintain its position as the central authority to speak about indigenous farming, even when such issues might also affect the powerful groups of European investors and plantation operators.

In terms of the cultivated environment of Java, the Dutch approach to knowledge production and agricultural improvement is as notable for what it did not produce as for what it did.[84] The localised approach meant that there were few wholesale dramatic transformations of the cultivated environment, as the Department of Agriculture advocated for regionally-specific, and often small-scale change. The changes did nevertheless have important implications for the cultivated environment across the island. The best example of this is the role that the Department played in introducing a wider variety of dry-season food crops to indigenous farmers.[85] Changing crop rotations and adding new dry season crops was not only meant to improve the food supply, but also to maintain or improve soil fertility in regions with poor irrigation. The Department encouraged indigenous farmers to grow crops like cassava and soybeans as an alternative to a dry-season rice crop, a practice that depleted soils dramatically where irrigation was poor. Before making such recommendations, however, the Department typically investigated the environmental and social relations of production. The result was a broadening of dry season crops across Java that may have had a more notable effect on the Javanese diet than any increase in rice production.[86] Such changes, however, rarely changed the overall character of farming in a region. In an area where land was rotated with sugar, they did not seek to remove sugar planting, but to find crops like soybeans or green fertilisers that would help replenish the soil. Where peasants grew rice in much the way that had been practised for centuries, the Department might try to introduce new crops to rotate with rice, encourage new rice varieties or change some aspect of rice cultivation to make it more productive, but they did not try to remove rice agriculture altogether, nor completely transform its practice. In all respects, the Department refused to play a revolutionary role. A superficial look at cultivation on Java might suggest that no changes had taken place at all, even though this was not the case.

The subtlety of the Dutch approach, and its lack of dramatic environmental transformation contrasts significantly with other stories of the colonial politics of agricultural and environmental change. In one part of James Scott's illuminating study of 'authoritarian high modernism,' he examines a mindset towards technological improvement of agriculture that de-emphasised the significance of local difference and local practices in agriculture, in order to create a less complex, more easily managed reality in its place.[87] In such cases, scientific theories, or technological assumptions took precedence, at least among high-level officials, over the detailed, empirical understanding of the places and people they wished to transform. Such cases can indeed be found in colonial history,[88] and Scott's insights are important for understanding colonial projects of transformation.

Not all colonial improvement projects, however, chose the path of simplification, as this essay has shown. How do particular scientific approaches, whether theoretical work that simplifies or empirical studies that complicate, gain enough authority to become the basis of transformation? In the Netherlands East Indies, scientists and officials had to work to establish their credibility to speak to problems of indigenous agriculture and to define the ecological realities of the colony. Unlike the cases that Scott describes, the authority of the colonial government was not sufficient in itself to grant credibility to the Department of Agriculture with the wider community of the Netherlands East Indies. Detailed, empirical knowledge of entire regions, not simplifying scientific assumptions, became the vital tool with which the state established its authority and credibility as a producer of scientific knowledge. This approach allowed them to defend their knowledge as the privileged basis for policies of colonial development in the face of accusations of bias. In the Indies, a more complicated view of the colony could serve as vital a political purpose as a more simplified view. At the same time, scientists put this hard won empirical knowledge to use in a restrained, if not necessarily modest way, by rejecting the possibility of wholesale transformation and preferring instead a process of change more in tune with their understanding of the variability of local conditions.

NOTES

1. For place names I use the current standard Indonesian spelling. Names of persons and institutions follow the spelling of the period.

2. Report quoted in Z. Kamerling, 'Achteruitgang in vruchtbaarheid der voor de rietcultuur gebruikte gronden' (Decline in Fertility on Lands Used for Cane Culture), *Soerabaiasch Handelsblad*, 16 February 1905.

3. Het Algemeen Syndicaat van Suikerfabrikanten.

4. Z. Kamerling, 'Achteruitgang in vruchtbaarheid der voor de rietcultuur gebruikte gronden' The Resident's description of the problem, as well as his solution, match Kamerling's so closely, it seems reasonable to assume that Kamerling was his main source of information.

5. Ibid. Kobus was also editor of the *Archief voor de Java-suikerindustrie*, a journal of scientific and technical reports on the sugar industry.

6. Het Proefstation Oost-Java

7. Among the most prominent of the critics of the practices of the sugar industry was H.H. van Kol, a member of the lower house of the Dutch parliament, the Tweede Kamer. See for example his criticisms of the sugar industry voiced in the Tweede Kamer. (Reprinted in the *Archief voor de Java-suikerindustrie*, 1907).

8. On the mission of the Department of Agriculture, see *Regeerings-almanak voor Neder-landsch-Indie*, 1905.

9. For more on the Cultivation System see Cornelis Fasseur, *The Politics of Colonial Exploitation: Java, the Dutch, and the Cultivation System* (Ithaca: Southeast Asia Program, Cornell University, 1992). For the case of sugar production see R.E. Elson, *Javanese Peasants and the Colonial Sugar Industry: Impact and Change in an East Java Residency, 1830–1940* (Oxford: Oxford University Press, 1984).

10. For more on the Sugar Law of 1870 see J.S. Furnivall, *Netherlands Indië* (Cambridge: Cambridge University Press, 1944), p. 165.

11. When I use the terms European, Chinese and Native in this respect, I am referring to the legal category used in the colony at the time, and my use of the capitalisation is meant to be a reminder that these were constructed categories, not natural distinctions.

12. Furnivall, *Netherlands Indië*, pp. 178–9. Also see Elson, *Javanese Peasants and the Colonial Sugar Industry*, pp. 127–8.

13. Furnivall, *Netherlands Indië*, p 180.

14. Ibid., pp. 178–80. The Agrarian Law also regulated which land could be considered Native land, and which 'waste' land that would then fall under state control.

15. Clifford Geertz, *Agricultural Involution: The Processes of Ecological Change in Indonesia*, pp. 87–9. Although frequently (and fairly) criticised for its analysis of labour relations in the sugar lands, *Agricultural Involution* is still a good introduction to the intricacies of the sugar leasing practices on Java. See also Elson, *Javanese Peasants and the Colonial Sugar Industry*. Sugar manufacturers often kept different portions of the land at different stages in the rotation so that there would be a yearly sugar crop.

16. J.D. Kobus, 'Achteruitgang in vruchtbaarheid der voor de rietcultuur gebruikte gronden?' *Archief voor de Java-suikerindustrie*, vol. xiii (1905): 282–3. This was confirmed with some qualifications in P. van der Elst, 'De Padioogstmislukking in de Residentie Madioen in 1910: Een onderzoek naar de oorzaken der Omo Mentek en naar Nawerking van Suikerriet op Padi in die Residentie', *Mededeelingen van het Proefstation voor Rijst c.a.* (Batavia: G. Kolff, 1912), p. 44.

17. See for example H.C.H. de Bie, *De Landbouw der Inlandsche Bevolking op Java* (Batavia: G. Kolff & co., 1902) for a scientific study that gives detailed attention to irrigation practices of various kinds of indigenous rice agriculture.

18. For examples from the time see: 'Irrigatiegrieven tegen de suikercultuur', no author given, *Archief voor de Java-suikerindustrie*, Bijblad, vol. xvi (1908): 613–20; D. van Hinloopen Labberton, 'Invloed van de suikerfabriek op hare omgeving', *Archief voor de Java-suikerindustrie*, Bijblad, vol. xvi (1908): 796–7. Others argued that only careful study of agricultural conditions could guarantee such benefit, see for example: J. Homan van der Heide, *Beschouwingen aangaande de Volkswelvaart en het Irrigatiewezen op Java* (Batavia: G. Kolff, 1899); J. Nuhout van der Veen, *Irrigatie en Landbouw op Java* ('s Gravenhage: G.A. Kottman, 1907). For more recent studies of irrigation in Indonesia see: Wim Ravesteijn, *De Zegenrijke Heeren der Wateren: Irrigatie en Staat op Java,1832–1942* (Delft: Delft University Press, 1997); Petrus van der Eng, *Agricultural Growth in Indonesia since 1880* (Groningen: Universiteitsdrukkerij, Rijksuniversiteit Groningen, 1993), especially pp. 57–60. Van der Eng disputes the contention that irrigation works on Java predominantly benefited sugar planters. For more on the problems of irrigation systems for rice farmers see Anne Booth, 'Irrigation in Indonesia, pt. I', *Bulletin of Indonesian Economic Studies*, vol. 13, no. 1 (March 1977).

19. A survey of the *Archief voor de Java-suikerindustrie*, the journal for scientific sugar research in the Indies show no studies of the effects of sugar growing on subsequent rice or secondary crops before 1910. Later studies of the relationship cite no previous studies. There are one or two studies of the effects of sulphate of ammonia fertiliser on rice, including Kreischer, 'Verslag omtrent in den Westmoesson van 1898/1899 genomen bemestingsproeven met zwavelzure

ammonia op met padi beplante sawahs in de Controle Afdeeling Kota Pasoeroeran en Grati', *Archief voor de Java-suikerindustrie, 1899*. One exchange of articles by H. Prinsen Geerligs ('Iets over de bemesting van het suikerriet', *Archief voor de Java-suikerindustrie*, vol I: 161) and C.J. van Lookeren Campagne ('Een bemestingskwestie', *Archief voor de Java-suikerindustrie*, vol 1: 397) studied the effect of irrigation water used during rice cultivation for the subsequent sugar crop.

20. For more on the problems described in this paragraph, see Elson, *Javanese Peasants and the Colonial Sugar Industry*.

21. Margeret Leidelmeijer, *Van Suikermolen tot Grootbedrijf: Technische Vernieuwing in de Java-suikerindustrie in de Negentiende Eeuw* (Amsterdam: NEHA, 1997).

22. For a description of day–night regulations and their problems written during the colonial era, see 'Het Wadoek Stelsel' (no author given), *Koloniale Studiën*, 1920: 65–91.

23. For discussions of this subject from the time see: D. van Hinloopen Labberton 'Invloed van de suikerfabriek op hare omgeving', *Archief voor de Java-suikerindustrie, Bijblad*, vol. xvi (1908): 749–802; E. H. s'Jacob, *De Economische Betekenis van de Suikerindustrie op Java* (Batavia: G. Kolff, 1903). For opposing views, arguments in the Tweede Kamer are a good resource. See, for example, the debate from the the 1909 colonial budget talks, reprinted as 'Stemmen in de Tweede Kamer over de suikerindustrie op Java', *Archief voor de Java-suikerindustrie, Bijblad* vol. xvi (1908): 1051–72.

24. R.E. Elson, *Javanese Peasants and the Colonial Sugar Industry*; Loekman Soetrisno, The Sugar Industry and Rural Development: The Impact of Cane Cultivation for Export on Rural Java, 1830–1934 (Dissertation, Cornell University, 1980).

25. Speech by Queen Wilhelmina cited in P. Creutzberg, *Het Economische Beleid in Neder-landsch-Indië* (Groningen: H.D. Tjeenk Willink, 1972), vol. 1, p. 173, see footnote 1. For a good overview of the Ethical polices see Merle Ricklefs, *A History of Modern Indonesia since c.1300* (Stanford: Stanford University Press, 1993). Among the most famous of commentaries on the problems of the colonial system for indigenous people, and the obligations of the colonial government is C. Th. van Deventer, 'Een Eereschuld', *De Gids*, vol. 63 (1899). P. Brooshooft applied the term 'Ethical' to these policies retrospectively in his short work, *De Ethische Koers in Koloniale Politiek* (Amsterdam: DeBussy, 1901).

26. The Outer Posessions included all the other islands in the archipelago. Of particular interest for Javanese emigration were Borneo, Sumatra and Sulawesi.

27. Z. Kamerling, 'Achteruitgang in vruchtbaarheid der voor de rietcultuur gebruikte gronden' (Decline in Fertility on Soil Used for Cane Culture), *Soerabaiasch Handelsblad*, 16 Feb. 1905 – 31 Mar. 1905.

28. For details on Kamerling's background see *Onderzoek Naar de Mindere Welvaart der Inland-sche Bevolking op Java en Madoera*, Batavia, vol. 5a (1907): 200. For references to his earlier work see Kamerling, 'Achteruitgang in vruchtbaarheid', 21 March 1905. The introduction of chemical fertilisers to Java came primarily through the agency of sugar producers in the late 1800s. The use of sulphate of ammonia was discussed in the sugar industry publications, especially in *Archief voor de Java-suikerindustrie*. In the Netherlands, the introduction of chemical fertilisers in the mid- to late-nineteenth century had produced dramatic yield improvements. See Jan Bieleman, *Geschiedenis van de Landbouw in Nederland, 1500–1950: Veranderingen en Verschiedenheid* (Amsterdam: Meppel, 1992).

29. Kamerling, 'Achteruitgang in vruchtbaarheid', 16 Feb. 1905.

30. Ibid., 21 March 1905

31. I am using the term ecological here to refer to the effects of the sugar/rice rotation on soil, water, and plants, to distinguish this part of the debate from the broader discussions of land rent and labour usage that focused on the relationship between planters and farmers.

32. The term 'black box' is used frequently in the history of technology and science and technology studies to refer to the status of technologies whose inner workings are not considered important

to wider society. Opening a black box means taking the inner workings of a technology to public scrutiny.

33. Kamerling, 'Achteruitgang in vruchtbaarheid der voor de rietcultuur gebruikte gronden', 16 Feb. 1905.

34. Ibid.

35. Ibid., 21 Feb. 1905. He cites studies by Lawes and Gilbert from Rothamstead, England as reported by the USDA in 1895.

36. Ibid., 16 March 1905. He cites F.A.F.C. Went, 'Waarnemingen and Opmerkingen omtrent de rietsuikerindustrie in West Indië' (no publication info.)

37. Ibid., 8 March 1905. He cites studies by van Lookeren Campagne and Prinsen Geerligs.

38. Ibid.

39. Ibid., 16. Feb. 1905.

40. Ibid. Kamerling quotes Kobus's memorandum addressed to the Resident of Madiun in the first instalment of the series of articles.

41. Ibid.

42. Ibid., p. 321.

43. Suzanne Moon, Constructing 'Native Development': Technological Change and the Politics of Colonization in the Netherlands East Indies, 1905–1930 (dissertation, Cornell University, 2000).

44. E.H. s'Jacob, *De Economische Betekenis van de Suikerindustrie op Java* (Batavia, 1903), pp. 1–26.

45. *Onderzoek naar de Mindere Welvaart*, vol. 5a (Batavia, 1907), pp. 196–201.

46 Article quoted at length in the *Archief voor de Java-suikerindustrie, Bijblad*, vol xvi (1908): 1040–2, from *De Indische Gids*, Nov. 1908.

47. Ibid., p. 1042.

48 'Stemmen in de Tweede Kamer over de suikerindustrie op Java', *Archief voor de Java-suiker-industrie, Bijblad*, vol xvi (1908): 1037–40.

49. J.D. Kobus, 'Achteruitgang in vruchtbaarheid der voor de rietcultuur gebruikte gronden?' *Archief voor de Java-suikerindustrie*, vol. xiii (1905): 281–93.

50. Ibid., p. 283.

51. Ibid.

52 On replication in science, see H.M. Collins, *Changing Order: Replication and Induction in Scientific Practice* (Chicago, 1985); Bruno Later and Steve Woodlark, *Laboratory Life: The Construction of Scientific Facts* (Princeton: Princeton University Press, 1979).

53. On the importance of place in agricultural field experiments see Christopher R. Henke, 'Making a Place for Science: The Field Trial', *Social Studies of Science*, vol. 30, no. 4 (August, 2000).

54. J.D. Kobus, 'Achteruitgang in vruchtbaarheid der voor de rietcultuur gebruikte gronden?', p. 284.

55. Ibid.

56. Ibid., p. 285.

57. Ibid., p. 282. The sugar planter reported a high yield of 70 piculs/hectare as compared to around 20 piculs on the poorer ground.

58. Ibid., pp. 282–3.

59. Ibid., p. 293.

60. E.H. s'Jacob, 'Problemen van de Dessa', discussion, p. 337.

61. Ibid., p. 338.

62. H.J.W. van Lawick, 'Suikercultuur en Inlandsche Landbouw', *Archief voor de Java-suiker-industrie, Bijblad*, vol. xv (1907): 990–1006.

63. Ibid., p. 993.
64. J.A. van Haastert, 'Verslag Omtrent de Resultaten der Bemestingsproeven in den Aanplant van Oogstjaar 1905/1906', *Archief voor de Java-suikerindustrie*, vol xvi (1908); J.D. Kobus, J.A. van Haastert. 'Padiproeven', *Archief voor de Java Suikerindustrie*, vol. xvi (1908): 571–93. It is unclear how the Department of Agriculture or Melchior Treub responded to these periodic calls for their involvement. The department did not organise experiments, nor did they report on the problem in their yearbooks from 1905–1910. They may not have considered it a top priority, or they may had trouble organising such work when they were beset with other organisational and staffing difficulties.
65. Kobus, van Haastert, 'Padiproeven', p. 578.
66. Early works on *omo mentek* include J. van Breda de Haan, 'Omo Mentek, een aaltjesziekte der padi', *Mededeelingen uit 's Land's Plantentuin*, vol. LIII, Batavia, 1902. Van Breda de Haan hypothesised that the disease was caused by an insect. In the 1960s the disease was determined to have been caused by a virus. H. Toxopeus, 'Landbouwkundig onderzoek: het Algemeen Proefstation voor de Landbouw ALP en resultaten in de praktijk', *Landbouwonderwijs, -voorlichting, en -onderzoek voor de kleine boer op Java, 1900–1940*. NEHA-Jaarboek, NEHA, Amsterdam: 1999.
67. P. van der Elst, 'De Padioogstmislukking in de Residentie Madioen in 1910: Een onderzoek naar de oorzaken der Omo Mentek en naar Nawerking van Suikerriet op Padi in die Residentie', *Mededeelingen van het Proefstation voor Rijst c.a.* (Batavia, 1912).
68. The chapter entitled 'Fertility, Irrigation, and Crop Rotation' is more than a quarter of the entire ten chapter text.
69. See discussion in van der Elst, 'De Padioogstmislukking in de Residentie Madioen in 1910', pp. 26–50.
70. Ibid., pp. i–iv.
71. Ibid. See analysis, pp. 40–50 and footnote about Kobus on p. 45.
72. This is the new name for the *Archief voor de Java-suikerindustrie*.
73. P. van der Elst, 'Over de Nawerking van Suikerriet op Padi in de Residentie Madioen', *Archief voor de Suiker-Industrie in Nederlandsch-Indië*, vol. axe, no. 1 (1913).
74. Ibid., pp. 560–70. In order to estimate the tax for rice producers, the government did test cuttings of rice fields across the island, to determine the yields for that year.
75. Ibid., pp. 563–4.
76. Ibid., p. 559.
77. Ibid., p. 575.
78. A survey of literature from both the Department of Agriculture and the Sugar Experiment Station finds no new articles on the sugar/rice relationship questioning van der Elst's results.
79. P. van der Elst, 'Bevloeiingsproeven bij Padicultuur: Opzet der Proeven', *Mededeelingen van het Proefstation voor Rijst c.a.*, no. 2 (Batavia: G. Kolff, 1916).
80. See comments on this in P. van der Elst, 'Bevloeiingsproeven bij Padicultuur', p. 12.
81. Gé H.A. Prince, 'Landbouwvoorlichting en onderwijs als onderdelen van de koloniale welvaartspolitiek', *Landbouwonderwijs, -voorlichting, en -onderzoek voor de kleine boer op Java, 1900–1940*. NEHA-Jaarboek, NEHA, Amsterdam: 1999. Suzanne Moon, Constructing 'Native Development'.
82. Suzanne Moon, Constructing 'Native Development', pp. 50–69.
83. For example see Egbert de Vries, Landbouw en Welvaart in het Regentschap Pasoeroean: Bijdrage tot de Kennis van de Sociale Economie van Java (dissertation, Wageningen, 1931).
84. I am limiting my conclusions to the effects on agriculture. For a telling study of the Dutch approach to forest management that comes to different conclusions, see Nancy Peluso, *Rich*

Suzanne Moon

Forests, Poor People : Resource Control and Resistance in Java (Berkeley: University of California Press, 1992).

85. Peter Boomgaard and J.L. van Zanden (eds), *Changing Economy in Indonesia: A Selection of Statistical Source Material from the Early 19th Century up to 1940*, vol. 10, 'Food Crops and Arable Lands, Java 1815–1942' (Amsterdam: Royal Tropical Institute, 1990).

86. Ibid.

87. James Scott, *Seeing Like a State: Why Some Schemes to Improve the Human Condition Have Failed* (New Haven: Yale University Press, 1998).

88. In addition to the case studies in *Seeing Like a State*, see also Chris Conte, 'Colonial Science and Ecological Change: Tanzania's Mlalo Basin, 1888–1946,' *Environmental History*, vol. 4, no. 2 (April 1999): 220–44; Christophe Bonneuil, 'Development as Experiment: Science and State Building in Late Colonial and Post-colonial Africa, 1930–1970', *Osiris*, vol. 15 (2001): 258–81.

CHANGING

Agrarian Change, Cattle Ranching and Deforestation: Assessing their Linkages in Southern Pará

Pablo Pacheco

1. INTRODUCTION

The Redenção area in southern Pará is one of the oldest frontiers in the Brazilian Amazon, and its origins are largely linked to the opening of the Belém-Brasília highway in the 1960s. In its early stage, the frontier was dominated by large corporations attracted by public incentives. Over time, medium-scale ranches (*fazendas*) have become a feature and the associated ranchers gained influence in local politics. More recently, smallholders supported by agrarian reform have begun to compete for land by contesting the rights of landholders with large holdings of low productivity. Fifty years of occupation and land use renders this region an ideal case to analyse and evaluate the implications of shifting policies and development of markets for beef and dairy production, and their implications for land tenure, agrarian transformation and patterns of land use.

In order to accomplish this goal, I examine both the factors that drive the formation of the agricultural frontier, and the economic, social and political conditions that shape its further development. The emergent path of development strongly depends on the initial conditions, which in the study area were the availability of cheap land and public incentives. These fostered the development of large-scale, extensive systems for livestock production. However, development of markets linked to investments in beef and milk processing capacity, and increasing competition for land, have over time promoted the modernisation of production systems and fragmentation of estates. In this paper, I analyse the interaction of these factors in the Redenção area.

The theoretical foundations of this paper are diverse. I draw heavily on the emerging science of land-use change which aims to understand the human and environmental dynamics influencing change in the type, location and magnitude of land use and vegetation cover.[1] Furthermore, this work relies on considerations of political economy and regional political ecology – broadly interpreted – in order to explore the politico-economic implications of regional development of the frontier and changes to land use.[2]

I have adopted a step-wise approach to interpreting the varied dimensions of frontier development and their evolution over time. The advantage of this approach is that it provides a flexible framework for pulling together a patchwork

Environment and History **15** (2009): 493–520.

of factors in spite of incomplete data coverage. Most primary social, economic and political information was gathered during semi-structured interviews with 50 rural producers and key informants in southern Pará during 2003 and 2004. A land-use change dataset was produced through remote sensing analysis of Landsat TM and ETM+ images for five years between 1986 and 2002. A supervised classification was performed based on training sites identified during fieldwork. A maximum likelihood procedure was performed in IDRISI software. Deforestation data for the period from 2003 to 2006 was updated with the digital information provided by the Brazilian Institute of Spatial Research (INPE)[3].

This paper is organised in six sections including this introduction. In the second section, I provide a theoretical review of the forces shaping frontier development, much of it drawn from literature produced in the Amazon. The third section depicts the history of land occupation in southern Pará which is surprisingly relatively unknown, and has only been analysed in fragmented pieces elsewhere. In the next section, I analyse the interactions between the expansion of cattle production, land tenure, and local power relationships. In the fifth section, I explore the main land-use change trajectories taking place in the study area, assessing the independent contribution of each type of actor to deforestation. Finally, I provide a synthesis and major conclusions.

2. A REVIEW OF THE FACTORS SHAPING FRONTIER DEVELOPMENT

The factors driving frontier development can be classified as exogenous and endogenous. The exogenous factors are primarily public policy decisions about road construction and land distribution, along with some incentive policies. The location, extent and shape of the road network are important features determining the spatial pattern of land occupation by influencing where people and firms migrate and settle.[4] Land policies exert a large influence on who owns land, shaping the structure of land tenures.[5] In turn, incentive policies determine who benefits from institutional rents provided by the state in the form of fiscal incentives, tax holidays, cheap credit, or roads that facilitate access to markets.[6]

After an initial period of occupation, there are several endogenous interactions that operate on the frontier, contributing to their spatial and socio-economic configuration. The main interactions are related to the influence that roads and urban centres exert on the spatial distribution of farms and ranches, the functioning of land markets that promote either land concentration or fragmentation linked to the arrival of investors interested in buying lands or setting up processing plants (slaughterhouses, meat packing plants, or dairy processing plants), and the local power relations that tend to reinforce local social structures. Opportunities to gain land on the frontier contribute to subsequent affluence. Furthermore, improved road access, better market conditions and new investments stimulate the modernisation of production systems, making them more profitable.

In its initial stage, frontier occupation is motivated by the appropriation of nature and its transformation into natural resources to be traded as commodities. The question about who appropriates such resources and the way in which the resulting economic rents are distributed is related to the origins of property – how resources were originally distributed among groups and individuals. Thus, initial occupation and resource appropriation can be seen as a process of primitive accumulation of capital.[7] In the early stage of frontier development, a logic of natural resource mining predominates.[8] Land and forest are exploited in an extractive way by influential social groups that squat on public land or acquire it, often through fraudulent means. By using the land so acquired, individuals in these groups can make a profit which is then enlarged by capturing institutional rents provided by the nation-state.[9]

Nation-states which still have abundant public lands, such as Brazil, are often committed to development of the frontier for both political and economic reasons. The political goal is to guarantee physical control of their territories.[10] The economic goal is to articulate the frontiers into broader circuits of capital, labour and exchange of goods by converting what are often considered idle lands to productive uses[11]. However, this creates complex challenges related mainly to the distribution of benefits within society and the sustainable use of natural resources.[12]

The motivation underlying public policies aimed at developing the frontier tend to change with shifting priorities of governments. In the Amazon, for instance, there has been a transition from conceptions of regional planning towards neo-liberal notions of market-based development. Regional planning for the frontier focused on infrastructure, large-scale projects, encouragement of growth poles, and infusions of population. Under neo-liberal policies of development, in contrast, market dynamics increased their weight in driving development, with a concomitant decline in the role of public policy.[13]

It is a paradox of frontier development that although the nation-state promotes occupation of the frontier, state institutions have little presence there.[14] The state stimulates the arrival of pioneers seeking to enlarge their self-interest. These pioneers benefit from state inaction (for instance in controlling land squatting), but also from incentives granted through public policy. The state often arrives later in an attempt to expand the provision of services and to capture a portion of the rents from private earnings.[15] However, in recent times and in the face of the growing influence of markets, the state has increasingly attempted to regulate the use of natural resources mainly as a way to halt depletion of forests.[16] Yet the paradox persists due to the limited capacity of the state to influence the behaviour of landholders. This favours a 'law of the jungle' situation in which influential social groups enforce their interests.[17]

Pablo Pacheco

The fiscal and credit incentives provided by the state have proved to be relatively efficient mechanisms for promoting occupation of land in the frontiers. Occupation of land may be driven by speculation stimulated by expectations of profit from both agricultural development and state subsidies.[18] However, as the frontier develops, occupation of land becomes more strongly tied to agricultural use,[19] and although land speculation persists, it is increasingly linked to the marketable value of land. In the Brazilian Amazon, state subsidies explain much of the expansion of cattle ranching in the early stages of frontier development.[20] However, subsequent development has been driven by improved access to markets and profitable cattle ranching associated with the modernisation of cattle production systems.[21]

The modernisation of cattle ranching is a process of relative intensification of range fattening of cattle or calf production.[22] It is motivated by at least three factors: availability of new technologies and management practices, public regulations forcing producers to adopt new practices, and market demand for high quality beef. Research has yielded pastures better adapted to climatic conditions, rotational pasture management, and genetic improvement of animal stocks.[23] Public regulations that have triggered improved management practices are aimed at enforcing the traceability of animals, and the eradication of foot and mouth disease by sanitary regulations. Finally, slaughterhouses have demanded higher quality beef. To remain competitive, ranchers have had to improve their management practices.[24] Not all producers, however, are in a position to adopt the new technologies, particularly those with little access to investment capital.[25]

Being less efficient, extensive systems of cattle production require more land per animal. Suggestions that improvements in livestock technology will take pressure off forests assume that if ranchers can raise the same amount of cattle on less land they will not need to convert as much forest to pasture.[26] Thus, technologies that reduce pasture degradation will allow farmers to persist in an area rather than clearing additional forest for new pastures. However, in some cases, the reality could be the opposite.[27] The effect of capital- and labour-intensive technologies depends on the time scale involved. In the short run, new technologies will tend to reduce deforestation, as land managers concentrate more on a smaller area, but over time higher profits will attract additional labour and capital to the region, leading to a net increase in deforestation. The latter seems to be the case in the Amazon since deforestation takes place at the same time as the modernisation of beef production practices.

3. DYNAMICS OF LAND OCCUPATION IN THE REDENÇÃO AREA

Redenção is located in the southeastern portion of the south of the State of Pará.[28] It embraces the municipalities of Redenção, Santa Maria das Barreiras,

Agrarian Change, Cattle Ranching and Deforestation

and Santana de Araguaia to the south, and the municipality of Cumarú do Norte to the west, plus a portion of the southeast of São Félix de Xingu. The western border of the adopted zone is defined by the edge of the satellite imagery used for land-use classification (Figure 1). It is bordered by the State of Mato Grosso to the south, and the Araguaia River forms the border with the State of Tocantins to the east. The total area under consideration is 4.7 million ha. It was originally covered mostly by forests. Currently, about 2.8 million ha is still covered with forest, and most of the cleared land is under pasture.

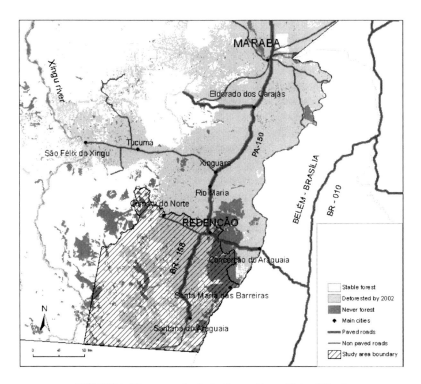

FIGURE 1. The research site in the southern State of Pará

The population of the Redenção region during the first decades of the last century was about 7,000 people. While the rural population grew in the 1970s to 89,000, it declined sharply during the 1980s by about half. The initial increase was a response to demand for manual workers to log and clear forest, as well as to an expansion of gold mines (*garimpos*). The subsequent decrease was driven by increasing use of machinery and the closure of *garimpos*. Much of the rural population moved to urban centres in the 1980s, mainly to Redenção which is the main city of the region. In parallel, commercial services demanded by the expansion of the livestock economy stimulated the growth of urban centres. The total urban population in 2000 was 108,000 people.

The occupation of land in the Redenção area occurred in four episodes. The first was the development of rubber (*caucho*) extraction in the late nineteenth century. The second, which began in the 1960s, was characterised by the expansion of corporate ranching by companies from southern Brazil, which in turn stimulated the arrival of medium- and large-scale traditional ranchers. The third episode is a depression cycle driven by the reduction of state incentives and the failure of corporate ranching during the mid-1980s and subsequently. The last began in the mid-1990s with the definitive collapse of corporate ranching accompanied by the expansion of medium-scale ranching, driven in part by the arrival of new investors, but also by the growth of smallholder settlements implanted by agrarian reform.

When the first episode of occupation of the Redenção area began in the late nineteenth century, the region was still populated by the indigenous Karajás and Kayapó groups. The village of Conceição do Araguaia was established during this period and was the most important urban centre, being the marshalling point for most of the flow of manufactured goods coming from other regions of the country, and rubber extracted in the region being transported to the external city of Belém.[29] The rubber industry was based on *caucho* (*Castiloca elastica*) rather than rubber trees (*Hevea brasiliensis*), and required the trees to be cut down in order to extract their resin. This fostered nomadic activity in the forest and did not support a large population.

Little remained during the first decades of the twentieth century after the extractive *caucho* economy collapsed. Scattered villages around the main rivers were associated with incipient agriculture based on small-scale farming that produced a diversified basket of crops (i.e. rice, corn, beans and manioc), and some small cattle herds. In contrast, cattle ranches were established and flourished in the eastern bank of the Araguaia River in *cerrado* (savanna) areas. These supplied beef to Conceição de Araguaia, by then a small city and still the most important urban centre. The forest cover constrained the expansion of cattle ranching to the west of the Araguaia River. From 1920 to 1950, agriculture was primarily limited to subsistence goals.

In the late of 1950s and early 1960s, the opening of the Belém-Brasília highway coincided with the first attempts to find gold in the Redenção area. A group of gold-seekers from the southern states of Brazil obtained information about the availability of public lands and claimed a portion of them.[30] This stimulated development of a livestock industry because it was followed by the arrival of large corporations from the south of the country supported by a programme of fiscal incentives put in place by SUDAM (Superintendence for Amazonian Development) in the early 1960s. This stimulated privatisation of land which until then had been the property of the State of Pará. Land was granted to corporations and individual investors for the purpose of agricultural development. Most of the land was sold through sealed tenders of land units (*glebas*) for a maximum of 900 *alqueires*[31] (Figure 2).

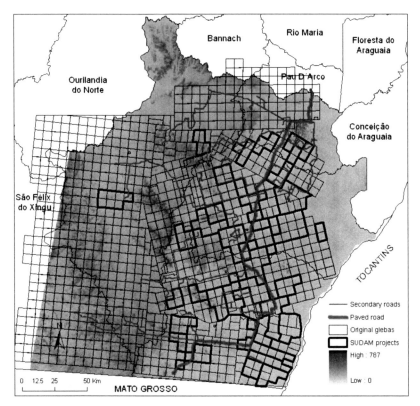

FIGURE 2. Redenção area: mosaic for land selling and SUDAM projects
Source: Elaboration by author based on data from SUDAM

By the mid-1970s, 74 projects were supported by SUDAM in the area south from Marabá, of which 52 were in the southeast portion of the state of Pará. Two main factors justified these corporate investments, neither of them related to revenue from cattle ranching: in the short run, the main source of profit were the institutional rents from state incentives (i.e., cheap credit and tax holidays); in the long run, they were motivated by the likelihood of profiting through land speculation.[32]

Much of the timber cut during forest clearing was wasted. It was not until the 1970s that timber companies arrived in the region to exploit mahogany (*Swietenia macrophylla*). This occurred mostly in the area between Marabá and Redenção, with less intensity to the south where the species is less common. Timber extraction was strongly associated with pasture expansion since the former facilitated access for the latter and allowed cattle ranchers to sell mahogany encountered during forest clearing. When the supply of mahogany was exhausted in the mid-1980s, several loggers became cattle ranchers to secure ownership of the areas they had cleared.

In the mid-1970s, some medium-scale entrepreneurs arrived to the Redenção area from southern states, motivated by the lower price of land.[33] While these ranchers did not benefit from SUDAM incentives as corporate ranches had, they were able to access PROTERRA loans for land acquisition, and this line of credit was instrumental in the expansion of non-SUDAM ranches.[34] About 60% of rural PROTERRA credit was applied to forest removal, pasture development, and acquisition of stock.[35]

The peak of public incentives and private investment was in the earlier 1970s, with a decline during the remainder of that decade. By 1978, about one billion dollars had been invested by SUDAM in corporate ranches, about US$ 2.7 million per ranch in direct investment.[36] Yet it has been argued that SUDAM funding did not achieve the anticipated pasture creation and expansion of cattle herds.[37] However, it did contribute to the introduction of high-quality livestock and an expansion of the road network.[38]

In the mid-1980s, most of the SUDAM ranches began to show signs of failure for several reasons. Firstly, the corporations were not involved directly in their administration, leaving much room for managers to take decisions that were not always in the corporate interest. Secondly, the productivity of pastures particularly of the *colonião* variety reached a crisis.[39] A new grass variety (*Brachiaria*) developed by the Brazilian Agricultural Research Agency (EMBRAPA) was introduced.[40] During the second half of the 1980s and in the 1990s, there were serious efforts by ranchers to convert their pastures to the productive *Brachiaria*. It was at this time that several (ex)SUDAM projects began to dissolve and the corporation sold their lands. In 1985, only 22 of the 52 SUDAM ranches were still supported with state incentives, these supporting only 40% of stock targets.[41]

In the early 1980s, there was an explosion of *garimpos* in the north of the municipality of Cumarú do Norte.[42] Most banking and associated financial services developed at this time. The livestock economy did not benefit much from gold mining, but the *garimpo* boom brought infrastructure and associated services to the city of Redenção. When gold mining collapsed in the late 1980s, most of the *garimpeiros* moved to Redenção, putting pressure on public services.

A portion of these immigrants invaded some large landholdings which became agrarian reform settlements in the mid-1990s.[43]

The 'Real Plan' implemented by the administration of Fernando Henrique Cardoso (FHC) (1995–2003) in the mid-1990s reduced the atmosphere of insecurity characteristic of the 1980s and promoted new investment in cattle ranching. Furthermore, the FHC administration promised agrarian reform and in so doing invigorated the movement of people into southern Pará.[44] Cattle ranchers had access to resources from the Constitutional Fund of the North (FNO), and from specific programmes targeted at smallholders such as the Special Credit Program for the Agrarian Reform (PROCERA) and the National Program for Strengthening Family Agriculture (PRONAF).[45] About 4.5% (US$ 35 million) of total FNO monies from 1989 to 2002 was allocated to the municipalities of the Redenção area. Over half of this was used to acquire 90,000 breeding stock. About 17% was used for pasture establishment and maintenance and equipment. Only 9% was used to finance dairy production, the rest going to support other activities.[46]

4. CATTLE RANCHING, LAND TENURE AND POWER RELATIONS

Much of the development of the Redenção area can be explained by the interactions taking place between demand for land, expansion of land markets, and the development of cattle production systems. However, a parallel political process has arisen with the arrival of smallholders.

Main cattle production systems

The rural economy in the Redenção area has traditionally been dominated by extensive cattle ranching, and agriculture has been important only since the expansion of smallholder settlements due to the fact that they combine subsistence agriculture with cattle raising. The dominance of extensive cattle ranching was linked to the availability of cheap land and the interest of investors and squatters to legitimise land ownership through conversion of forest to pasture and development of cattle production. Although SUDAM projects implemented modern cattle production techniques, traditional cattle ranchers arriving from southern states (Minas Gerais, Goiás and Paraná) reproduced their customary extensive systems of cattle production with a self-reproducing herd and little selection of animals. This was due to the low price of land, difficulties in finding qualified labour, and isolation from the main markets.

The cattle-raising system changed over time for two reasons. The first was the development of roads that allowed cattle producers to reach more distant markets in which they had to compete in terms of quality and price with suppliers from other regions. The quality of herds was improved by replacing mixed-breed cattle with introduced certified pure Nelore breeds, a type of zebu, and discarding cows with lower pregnancy rates. The second trigger to change was the gradual degradation of pasture productivity due to inappropriate grasses, over-grazing and compaction of soils.[47] Pasture management was improved by rotation.

According to official statistics, the number of cattle in the Redenção area has increased from 776,000 head in 1990 to 1.1 million in 2002, and 1.9 million in 2006.[48] This was about 12.6% of total bovine population in the state of Pará in 1990, and 11.2% in 2006. Although the reported cattle population declined in the first half of the 1990s, the pace of growth began to accelerate from 1996 to a rate of growth of 11.2% per year by 2002, rising to 15.1% per year between 2003 and 2006. Nothing suggests that this rate of expansion will decline in the future.

Milk production was almost nonexistent until the late 1990s but expanded with the proliferation of small-scale landholdings as a result of land invasions. By the middle of the first decade of the twenty first century, the Redenção area produced around 5 to 7% of all milk produced in the state of Pará in the middle of the first decade of the twenty-first century. Dairy production often occurs in small-sized dual-purpose cattle herds, with relatively low productivity. A few smallholders have adopted specialised cattle production for beef, and a significant number only grow annual crops due to their lack of capital to purchase cattle.

Land markets and trends in the fragmentation of tenures

As mentioned earlier, most of southern Pará was privatised through planned sales in the 1960s conducted by the Colonization and Land Department of the state of Pará. Some individuals served as brokers, taking advantage of connections with the public agency and linkages with companies interested in acquiring land. While some investors acquired their titles in this way, others merely squatted on the land, particularly some medium-scale ranchers who took advantage of the land rush to obtain land easily.[49] Most of the cattle ranchers were able to legitimise their ownership of land by acquiring titles from the state government, though some still persist in a semi-legal condition.[50]

The sale of land in southern Pará promoted the concentration of land in the hands of a few speculators who arrived first. They began to sell land to corporations, interested in obtaining SUDAM incentives in the 1960s and 1970s. As mentioned, there were about 52 of these projects operating in this area (Figure 2). Individual *glebas* were also sold to families or individual investors. It is reasonable to assume that land speculation decreased as subsidies shrank.

Figure 3 shows the major waves of land clearing, which can be equated with occupation of land over the last forty years. While there was an intense wave of settlement up until the 1980s, only 40% of the area allocated as *glebas* was cleared by the end of this period. In this same period, almost a half of all *glebas* occupied by SUDAM projects had in part been cleared for pasture es-

Agrarian Change, Cattle Ranching and Deforestation

tablishment. It is important to note that an area of 821,000 hectares was *glebas* settled without SUDAM support in the same period, almost double the area occupied by SUDAM projects in that period. Most of the land occupied was located around the main road that crosses the zone in a north-south direction, from which cattle ranching expanded further but gradually to the areas less connected to the main road network.

The dynamic of land occupation after the mid-1980s was as intense as that in the earlier years. In the period from the mid-1980s to 2002, about 60% of the total *glebas* were occupied (Table 1). Occupation moved westward, running perpendicular to the main road. Figure 3 shows large areas in the centre of the study area that were not effectively occupied until the late 1990s, a major part of which were SUDAM ranches that were partially abandoned in the 1980s

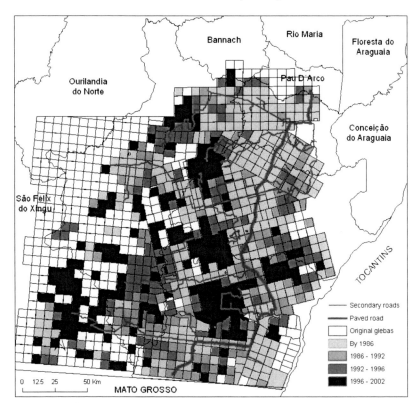

FIGURE 3. Timing of initial forest clearing, 1986–2002.

Note: Considers a *gleba* (a land unit of 4,356 ha) as initially deforested only when 10% or more of it was cleared.

Source: Elaboration by author based on data from SUDAM and estimates of deforestation based on LANDSAT imagery.

and invaded by squatters and landless people. Much of this central area became INCRA settlements. The ranch *Rio Cristalino* (formerly owned by Volkswagen) is a special case since it was invaded by a large number of small- and large-scale squatters, and it is still under dispute, facing expropriation by INCRA.[51]

TABLE 1. Cleared *glebas* acreage according to selected periods (thousand ha) (a)

	Until 1986	1986-92	1992-96	1996-02	Total
SUDAM projects	486.2	125.8	91.3	164.9	868.2
No-SUDAM projects	821.6	363.1	485.1	688.6	2,358.4
Total period	1,307.8	488.9	576.4	853.6	3,226.6
In percents	40.5	15.2	17.9	26.5	100.0

Source: Elaboration by author based on data from SUDAM and estimates of deforestation based on LANDSAT imagery.

TABLE 2. Evolution of medium and large landholdings.

	Planned (a)	1978 (b)	1998 (c)	2003 (d)
Number	1,068	148	210	338
Mean (ha)	4,309	23,557	18,077	12,462
St. dev. (ha)		± 30,960	± 26,379	± 18,728
Maximum (ha)		199,895	153,840	97,494
Minimum (ha)		2,936	1,123	855

Notes: a) Original grid obtained from ITERPA registers; b) 1978 information derived from a map elaborated by *Setentrional – Agrimensura e Topografia Ltda.* based on field information from P. Eleres taken in the period 1973/75; c) information for 1998 is based on a map made available by staff of *Agropecuária Versátil* based on their own field information; d) information for 2003 is based on a map elaborated by the author based on INCRA registers, interviews with landholders, and information provided by real estate brokers operating in the city of Redenção.

TABLE 3. INCRA settlements in the Redenção area according to creation date.

	1985–89	1990–94	1995-2000	Total
No. settlements	3	2	25	30
Thousand hectares	65.5	34.8	386.6	487.0
No. of families installed	212	230	5,644	6,086
No. of families capacity	716	1,040	8,676	10,432
% area under ex-SUDAM projects	100.0	100.0	55.6	64.8

Source: Author's calculations based on land registers from INCRA – SR 027

Anecdotal evidence suggests that a new wave of investors in cattle ranching arrived in the early 1990s from the northeastern states of Brazil. These new investors purchased some ranches that were for sale due to financial difficulties, and began to establish profit-sharing arrangements with local cattle ranchers to fund the acquisition of better quality animals. The increasing demand for land from this new wave of outside investors reinvigorated the cattle ranching economy and contributed to keeping land prices at relatively high levels. While a major concentration of land ownership occurred with the establishment of corporatist ranches during the mid-1970s, a counter-trend towards the fragmentation of land ownership developed as the frontier matured. In 1978, the average medium/large ranch was about 23,500 ha, and the largest around 200,000 ha. Twenty years later, the average area was 18,000 ha, and the maximum about 150,000 ha, a downward trend that continued into the early twenty-first century (Table 2 and Figure 4). There has been a trend towards incorporating new *glebas* into ranching, but through smaller-sized ranches, in parallel to the fragmentation of ranches.

The fragmentation of large estates parallels an increase in the price of land. This was a result of development of infrastructure and a greater demand for land. It is also likely that fragmentation of tenures is associated with economies of scale, and also that large-scale cattle ranches are under siege by movements of landless people.

Allocation of land by sealed tenders precluded smallholders from the frontier during the initial period of development. Land invasions began in the early 1990s, but in the late 1980s, three small colonies – *Arraiapora I and II*, and *Capitinga* – were created with the support of the land agency of the State of Pará (ITERPA). In turn, the federal land agency (INCRA) became more active in the study area in the early 1990s, but it was not until the FHC administration that the formal process of land expropriation and the formalisation of land invasions began in earnest (Table 3). Almost two thirds of the area affected by these processes was previously held by ranchers that had received SUDAM resources. These ranchers were bought out with debt securities, some receiving compensation far above market prices.[52] Expropriation was thus potentially profitable and some ranchers promoted invasion of their landholdings.

In the invaded areas, INCRA granted 50 ha lots to each household. Several families were able to claim more than one lot in the name of different family members. Some squatted or otherwise acquired land that exceeded the maximum allowed, triggering a partial concentration of land ownership within the INCRA settlements. Today, the average size of rural properties within these settlements was estimated to be 76 ha (± 53 s.d.).[53] The infrastructure available in INCRA settlements varies greatly, with some enjoying some access to services and others being relatively isolated without all-weather roads.

According to the Agricultural Census of 1995/96 carried out by the Brazilian Institute of Geography and Statistics (IBGE), there were about 4,000 establishments in this region embracing an area of 2.3 million ha. A major portion of them (65.9%) were of less than 100 hectares, whereas 2% of establishments covered 63% of the area. Table 4 and Figure 5 show my estimates of land distribution based on land registers obtained from both the federal (INCRA) and state (ITERPA) land agencies. According to ITERPA, about 60% of the Redenção area is occupied by cattle ranches, while 10% is INCRA settlements projects (*Projetos de Assentamento*, PA). A small proportion of land is occupied by small farmers settled to the east of the city of Redenção. Another 10% is part of the indigenous jurisdictions of Kayapó to the north and Badjônkôre to the south. These indigenous territories have been encroached by loggers in the past, especially for mahogany extraction, and some frontier expansion is currently taking place within the indigenous jurisdiction to the north.[54]

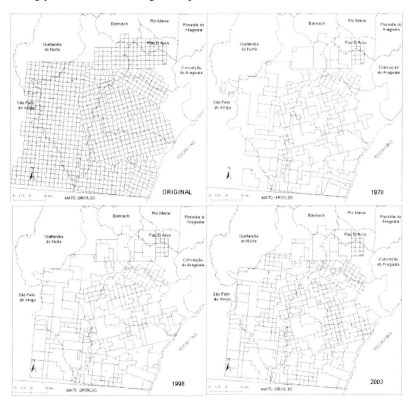

FIGURE 4. Evolution of land ownership in the Redenção area

Source: Original grid obtained from ITERPA registers. 1978 map elaborated by *Setentrional – Agrimensura e Topografia Ltda.* based on field information from P. Eleres for the period 1973/75. The 1998 map was made available by staff of *Agropecuária Versátil* and is based on their field information. The 2003 map was elaborated by the author based on INCRA registers, interviews with landholders, and information provided by real estate brokers operating in Redenção.

Agrarian Change, Cattle Ranching and Deforestation

TABLE 4. Types of landholdings in the Redenção area in 2002.

	INCRA settlements	Cattle ranches	Small farmers	No data (a)	Other areas (b)	Indigenous areas (c)	Total
Thousand ha	483	2,802	103	523	344	480	4,735
Per cent	10.2	59.2	2.2	11.0	7.3	10.1	100.0

Notes: a) Constitute areas within the property grid in which was not possible to obtain information about landholding limits (grey areas in Figure 5), yet much of this areas has already been occupied mostly by semi-legal or illegal means, b) constitute areas located outside of the original property grid (white areas in Figure 5), c) embraces the Kayapó indigenous territory to the north, and Badjônkôre to the south. Based on field interviews to people acting as real estate brokers, land registers from INCRA, and FUNAI

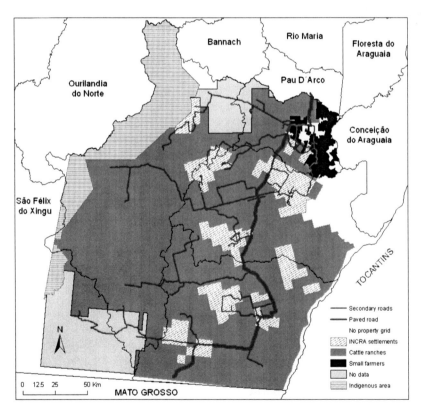

FIGURE 5. Land distribution in the Redenção area, 2002

Source: Based on information from INCRA, and information obtained in the process of compiling Figure 4.

Pablo Pacheco

Cattle industry and market restructuring

The market dynamics for cattle products were strongly linked to both road development and installation of processing industries, specifically slaughterhouses and dairies. During the first years of settlement until approximately the mid-1970s, most live cattle were sold in Belém,[55] competing with stock from Paragominas and Marajó. At that time, transportation to Belém was via Conceição de Araguaia to reach the Belém–Brasília highway, and was difficult. The opening of the Redenção–Marabá road in the second half of the 1970s improved access, with cattle being transported from Marabá along the Tocantins River to Belém.[56]

In the 1980s, before the widespread establishment of slaughterhouses, commerce in live cattle was monopolised by merchants, although a slaughterhouse was set up in 1980 near Santana de Araguaia that operated only to 1985 due to operational problems. The market situation in the study area changed when a slaughterhouse was set up in Redenção in 1996, and the slaughterhouse of Santana de Araguaia was purchased and re-opened by the same industrial group. In parallel, government measures that prohibited marketing of non-refrigerated meat led to a rapid expansion of the market for frozen meat, with slaughterhouses as key players.[57] As a result, these two slaughterhouses began to purchase most of the meat in the region, reducing transport costs and allowing access to distant markets mainly in the northeast and south of Brazil.[58]

To the extent that the cattle population expanded, so did the number of slaughterhouses that were set up in the region. In southern Pará, six more slaughterhouses were established from 1999 to 2004. This did not lead to improved prices for live cattle because it stimulated an increasing supply, and slaughterhouses started to demand higher quality animals. Most of these slaughterhouses are preparing to export meat once the barriers limiting it – linked to the Foot and Mouth Disease (FMD) free zone recognition – are removed, which will likely take place in the near future. Half of them already have a license to export beef. Export restrictions for beef do not apply to de-boned meat, but not all slaughterhouses have the technology to process and export this type of meat.

Milk production is increasing steadily, matching the expansion of smallholders in INCRA settlements. Rapid growth has taken place in the area surrounding the municipality of Xinguara where there has been an explosive increase in dairy processing plants, though most are of small capacity. About 13 new plants were set up in the area from Redenção to Marabá since the mid-1990s. Much of the production from these plants is marketed in the northeastern states.[59] There is only one dairy plant operating in Redenção. It failed previously, was reopened, and is currently administered by a co-operative. There are collection centres in Redenção for a large-scale dairy plant based in Conceição do Araguaia. Anecdotal evidence suggests that there is a large number of informal dairy plants that are not registered and evade tax obligations. It is likely that milk production will continue to expand in the future.

Politics, state regulations and local power relationships

The ascendance of cattle ranching in this and other regions in the Amazon was not accidental. Local and regional interests played an important role in shaping policy goals. During the military regime, aims of national sovereignty and regional economic growth were served by collusion between dominant elites and corporations in which the latter developed ranching operations in the Amazon in order to derive financial benefit from the sale of public land and the public resources flowing to the Amazon.[60] However, there were moments in the early 1970s in which social demands led the government to privilege populist instruments to populate and develop the Amazon through small-scale agriculture.[61]

In southern Pará in the early stages of the frontier development, there was little influence of the traditional elite *paraense*, as greater dominance was exerted by influential groups attached to corporations primarily based in the south of Brazil. For instance, the Association of Amazonian Entrepreneurs (AEA), based in São Paulo, was composed of corporate ranchers from southern Pará. The AEA lobbied government for policies favouring large-scale occupation based on the notion that this was the only rational means of occupying the region.[62] The incentives to ranching provided during the military regimes benefited such corporations. In exchange for fiscal incentives, the government transferred to these corporations the responsibility for infrastructure –mainly road building– that the government was not able to address.[63]

While corporations were important players in debating public policies with federal government agencies, they neglected local politics. Local ranchers organised rural associations (*sindicatos rurais*) to represent their interests against regularisation of land by state and federal governments, and to defend their lands against growing pressure from local peasant movements. In the 1980s, three ranchers' associations were created in the Redencão area, namely in Redenção, Xinguara and Rio Maria.[64] These associations became powerful actors in the local political arena and in negotiating policy instruments with government, though their main concerns were the land disputes that arose with the emergence of a strong movement of landless people. The rural associations remain as advocates of conservative paradigms of rural development based on cattle ranching and as strong opponents to government initiatives for agrarian reform.

The policy arena became more complex when the landless movement entered the regional scene. A tradition of violence around contested land claims in southern Pará persists[65] because land tenure remains highly skewed.[66] When the state or its institutions do not work for, or do not make decisions that satisfy the cattle ranchers, violence is used to impose their individual or group interests.

Violence has in some cases followed land invasions when cattle ranchers attempt to repel encroachers. Yet, a greater presence of public institutions in the region, mainly linked to agrarian reform programmes, and the eradication of slave labour, have stimulated rancher associations to negotiate with federal agencies.

Influential groups of ranchers and other local elites have also been able to erode mechanisms for enforcement of the land-use and environmental regulations, though that is changing slowly over time. Issued in 1995, the regulations aim to reduce deforestation by specifying that rural landowners should set aside 80% of the forest on their landholding as a Legal Reserve[67]. Since then, this has been the most contentious issue in the environmental policy for the Brazilian Amazon among medium- and large-scale landholders.[68]

The state environmental agency (IBAMA) has difficulty exerting effective control and cattle ranchers continue to clearcut forest without constraints.[69] In the State of Pará and elsewhere, government has tried to establish alternative mechanisms to enforce compliance with the Legal Reserve requirements, but with little success. Entrepreneurial landholders continuously attempt to change the law to allow more flexible land use. Although environmental law specifies that rural landowners can revegetate cleared land to meet their Legal Reserve obligations and avoid environmental penalties, this is rarely accomplished in practice.[70] Compliance with environmental regulations is against the interests of smallholders as well as large-scale landholders.

5. THE IMPRINT OF FRONTIER DEVELOPMENT ON THE LAND-SCAPE

Development of the frontier has had strong implications for the landscape as the expansion of cattle ranching in the region has been the primary cause of conversion of forest to other land uses. The rate of conversion has increased over time. The rate of conversion of forest to pasture was not great until the mid-1980s notwithstanding incentives for the development of corporate ranches. It began to grow in the 1990s and reached a peak around 2004, after which it started to slow down. Thus, it is reasonable to assume that market-driven development of cattle ranching has played a large role in deforestation in the Redenção area, and likely in the whole southern Pará region.

In 1986, about 84% of the study area was forest, almost 4 million ha, a small portion near Conceição do Araguaia was *cerrado*, and about 9% (416,000 ha) was pasture that had been converted from forest. The situation has changed dramatically since then. In 2002, about a third of the total area had been converted to pasture (1.4 million ha), with 61% remaining as forest (2.9 million ha), the area of pasture area having increased by a factor of 3.3 since 1986. Most deforested land was converted to pasture. Although forest regrowth was about 200,000 ha in the second half of the 1990s, it decreased to 79,000 ha in 2002.

These numbers counter the idea that cattle ranching leads to abandonment of pasture and reforestation since much of the converted pasture remains in use.

The annual rate of deforestation from the early 1970s to 1986 was about 26,000 to 34,000 ha/year, though some pasture expansion took place in the *cerrado*. Rates of deforestation increased to 53,000 hectares/year for the period 1986 to 1992, to 93,000 hectares/year for the period 1992 to 2000, and to 150,000 hectares/year from 2000 to 2002. In relative terms, the annual rates of deforestation grew from 1.37% in 1986-92 to 5.27% in 2000-02. Since then, deforestation has been in the order of 114,000 to 119,000 ha/year and the rate is decreasing, but these are a relatively high proportion of the remaining forest cover (Table 6).

Temporal trends in deforestation in the Redenção area are similar to those observed for the State of Pará and the Brazilian Legal Amazon as a whole.[71] In these two cases, the annual rates of deforestation were relatively low until the mid-1980s (0.4%), and they doubled until the early 2000 when reached about 0.8% in 2003-04. In 2005-06, these rates shrank to 0.6% in the State of Pará, and 0.5% for the whole Amazon region. Although the reasons for this reduction are not clear, the Brazilian government suggests that state actions such as credit constraints and greater monitoring have had a positive impact.

Much of the deforestation in the Redenção area has been driven by medium- and large-scale cattle ranching (see again Table 6) rather than smallholders. In this regard, there has been a lot of controversy about the contribution of SU-DAM-sponsored projects to deforestation. In the study area, SUDAM-sponsored projects were responsible for about 37% of all land cleared by 1986 (see again Table 1). However, SUDAM projects also prompted deforestation in indirect ways, mainly through creating roads that subsequently attracted significant numbers of medium-scale investors to neighbouring areas.

TABLE 5. Land use in the Redenção area, 1986–2002.

	In thousand hectares					Percentage				
	1986	1992	1996	2000	2002	1986	1992	1992	2000	2002
Water	29	11	11	16	13	0.6	0.2	0.2	0.3	0.3
Forest	3,994	3,538	3,324	3,005	2,895	84.4	74.7	70.2	63.5	61.1
Never forest	283	326	217	241	224	6.0	6.9	4.6	5.1	4.7
Pasture	416	661	981	1,269	1,424	8.8	14.0	20.7	26.8	30.1
Regrowth	-	199	201	202	79	-	4.2	4.3	4.3	1.7
Clouds	12	-	0	0	100	0.3	-	0.0	0.0	2.1
Total	4,735	4,735	4,735	4,735	4,735	100.0	100.0	100.0	100.0	100.0

Source: Author's estimates based on LANDSAT TM and ETM+ imagery analysis.

Note: 'Clouds' indicates areas that could not be assessed because cloud cover obscured the image.

Pablo Pacheco

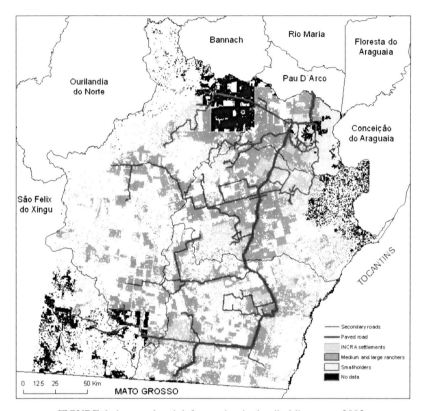

FIGURE 6. Accumulated deforestation by landholding type, 2002.

Source: Author's estimates based on LANDSAT TM and ETM+ imagery analysis, and land tenure information provided in Figure 5.

By the mid-1980s, 70% of accumulated deforestation was caused by medium- and large-scale cattle ranching. An additional 20% of deforestation occurred in areas for which land tenure is uncertain but which most likely were large-sized ranches, demonstrating that deforestation was driven overwhelmingly by medium- and large-scale ranchers in this region. Deforestation fostered by small-scale agriculture grew from 5% of the cumulative total in 1986 to 10% in 2002 and 17% in 2005. Many smallholders at the beginning of the settlements did not convert forest to pasture or crops because many INCRA smallholder settlements were in areas already converted to pasture on ranches that were previously large (mainly ex-SUDAM ranches). Nonetheless, increasing affluence led to an expanded contribution of smallholders to deforestation. Furthermore, the rapid growth of deforestation rates in INCRA settlements is due to there being little forest cover left in such lands.

TABLE 6. Deforestation in the Redenção area by agent, 1986-2006

| | Selected periods (a) | | | | | |
	1986–92	1992–96	1996–2000	2000–02	2003–04	2005–06
Annual deforestation (in thousand ha)						
Small farmers (b)	4.2	2.9	17.6	30.6	19.8	21.4
Cattle ranchers	39.2	72.4	56.9	77.1	72.1	70.8
Not identified (c)	10.0	18.2	20.0	42.1	22.8	27.1
Redenção area	53.4	93.6	94.5	149.7	114.8	119.4
State of Pará	455.2	608.8	543.8	641.1	775.9	561.8
Brazilian Amazon	1,407.9	2,070.5	1,652.4	1,921.0	2,633.1	1,639.9
Annual deforestation rate (in %) (d)						
Small farmers	0.9	0.7	4.8	10.4	12.8	20.7
Cattle ranchers	1.8	3.6	3.2	4.9	5.9	6.8
Not identified	0.8	1.6	1.9	4.3	3.2	4.2
Redenção area	2.0	3.9	4.7	7.6	5.5	6.7
State of Pará	0.4	0.6	0.5	0.7	0.8	0.6
Brazilian Amazon	0.4	0.6	0.5	0.5	0.8	0.5

Notes: a) the years corresponding to 1986 to 2002 are based on the author's estimates from LANDSAT TM and ETM+ imagery analysis, while the years from 2003 to 2006 are based on information provided by PRODES/INPE, b) includes smallholders outside the city of Redenção, and smallholders located in INCRA sponsored settlements, c) includes areas with no property grid information and the land corresponding to the indigenous territory located in the study area, d) deforestation rates are calculated for each class placing the annual deforestation for the period in the numerator, and the total remaining forest within the landholdings owned for each group in the denominator.

There is little likelihood that rates of forest removal will reduce in the future as markets for beef cattle are expanding and infrastructure and modernised beef production systems are in place. Indeed, the modernisation of ranching and fragmentation of tenures have not ameliorated deforestation thus far. For example, a large portion of the land in the southwest of the region where access is more difficult due to the poor quality of the roads, has been already converted to pasture. A few ranchers have converted some pasture to soybean production, mainly near the border with Mato Grosso, but it is uncertain how quickly this will develop in the future.

6. CONCLUSIONS

In this work, I have examined the conditions that shaped the development of typical corporatist frontiers as represented by the southeast portion of southern

Pará. This is a region dominated by medium- and large-scale cattle ranching, though there is an increasing presence of small farmers. Livestock production developed because private interests were able to obtain cheap land and incentives from the government. The incentives were intended to prompt the occupation of the Amazon for economic and geopolitical goals. Thus, military and corporatists interests colluded to motivate the change in question.

Most SUDAM-sponsored livestock projects failed to establish sustainable cattle ranching, although the primary objective of those that were sponsored was not necessarily to do so; some obtained land for speculative purposes. However, occupation of the frontier stimulated medium-scale entrepreneurs, mostly cattle ranchers from Minas Gerais, Paraná and São Paulo, to establish cattle ranches due to ease of access to land. Although this group did not benefit from SUDAM incentives, it did benefit from subsidised credit. In parallel, logging facilitated expansion of the frontier, but once the availability of mahogany was exhausted most loggers became cattle ranchers to secure land ownership. The expansion of cattle ranching, however, faced several constraints in the 1980s, including poor roads, distance to markets, and weak development of market chains.

Most of these constraints were overcome in the 1990s due to a more stable macroeconomic environment that depressed land speculation and created more incentives to invest. A new wave of investors arriving from the northeast injected capital into the region, invigorating land markets and promoting the fragmentation of large estates. Furthermore, growing investment in the cattle industry led to a restructuring of beef markets, making it easier for cattle ranchers to sell animals of improved quality. Demand led to an expansion of cattle herds which, in turn, stimulated expansion of processing capacity in the late 1990s. A similar phenomenon took place with dairy production because dairy plants set up in major urban centres favoured the expansion of milk production in smallholder settlements.

Land invasions became important in the early 1990s and have changed the land tenure situation and influenced local politics. Much of the invading population was attracted to the region as rural workers for forest clearing, logging, and by the *garimpos*, but they moved to the city as these endeavours collapsed. Invasions have been stimulated by landless people along with squatters and informal loggers. Land reform has legitimised this process, but it has been limited to regulating the dynamics of existing invasions rather than promoting a planned process of re-allocation of agricultural establishments with low productivity to landless people.

The landscape has undergone drastic change in southern Pará driven by conversion of forest to pasture. Medium and large ranchers, who hold a larger proportion of land in the study area, deforest more in absolute terms. Nevertheless, smallholders deforest a higher proportion of the land they occupy. This is explained by increasing affluence within INCRA settlements, and the relatively small-sized lots these settlers have access to.

Public policy has played a key role in defining the spatial and socio-economic configuration of this cattle ranching frontier. In this regard, economic incentives along with infrastructure development attracted new investments in the cattle industry, which in turn contributed to the consolidation of medium and large cattle ranches, and with it a drastic transformation of the landscape by conversion of forest to pasture. Nevertheless, policy shifts, infrastructure development, and increasing exposure to regional beef markets have altered the agrarian economy and forced cattle ranchers to compete in more demanding markets by improving pasture and herd management. However, the gains from market development in terms of arrival of slaughterhouses and the ability to sell in distant markets has provided new incentives for ranchers to continue deforestation to expand their pastures. Better infrastructure – e.g., electricity and roads – and increasing opportunities to make a living in the countryside, have also stimulated greater affluence and thus intensified land tenure struggles.

ACKNOWLEDGEMENTS

The author gratefully acknowledges the receipt of research funding from the Center for International Forestry Research (CIFOR), Bogor, Indonesia, and the support of the Institute of Environmental Research for Amazônia (IPAM), Belem, Brazil. I would like to thank Billie Turner, Dianne Rocheleau, Robert G. Pontius, and Sven Wunder for their comments on a previous version of this paper. I also want to thank to David Kaimowitz, Jean F. Tourrand, Marie G. Piketty, and Benoit Mertens for their collaboration when I was carrying out fieldwork. I am also grateful to those smallholders, cattle ranchers and ranch managers who patiently gave their time to answer my questions.

NOTES

1. Ronald R. Rindfuss et al., 'Developing a Science of Land Change: Challenges and Methodological Issues', *Proceedings of the National Academy of Sciences* 101, no. 39 (2004): 13976–13981, Garik Gutman et al., *Land Change Science: Observing, Monitoring and Understanding Trajectories of Change on the Earth's Surface*, Remote Sensing and Digital Image Processing Series, Vol. 6 (New York: Springer, 2004), Billie L. Turner et al., *Integrated Land-Change Science and Tropical Deforestation in the Southern Yucatan: Final Frontiers* (New York: Oxford University Press, 2004).

2. Jeffry A. Frieden et al., eds., *Modern Political Economy and Latin America: Theory and Policy* (Westview Press, 2000), Raymond Bryant and Sinéad Bailey, *Third World Political Ecology* (London: Routledge, 1997).

3. Deforestation data from INPE is available in the website: http://www.obt.inpe.br/prodes/

4. Benoit Mertens et al., 'Crossing Spatial Analyses and Livestock Economics to Understand Deforestation Processes in the Brazilian Amazon: The Case of Sao Felix Do Xingu in South Para', *Agricultural Economics* 27, no. 3 (2002): 269–294.

5. Hans P. Binswanger, 'Brazilian Policies that Encourage Deforestation in the Amazon', *World Development* 19, no. 7 (1991): 821–829.

6. Susanna Hecht, 'The Economics of Cattle Ranching in Eastern Amazonia', *Interciencia* 13, no. 5 (1988): 233–240.

7. David Kaimowitz, 'Frontier Theory and Practice' (unpublished draft, Bogor, Indonesia: 2002).

8. Robert Schneider, 'Government and the Economy on the Amazon Frontier. World Bank Environment Paper No. 11 .' (Washington D.C.: The World Bank Group, 1995).

9. Susanna Hecht, 'The Logic of Livestock and Deforestation in Amazonia', *Bioscience* 43, no. 10 (1993): 687–695.

10. Susanna Hecht and A. Cockburn, *The Fate of the Forest: Developers, Destroyers and Defenders of the Amazon* (London: Verso, 1989).

11. Kaimowitz, 'Frontier Theory and Practice'.

12. David Cleary, 'After the Frontier: Problems with Political Economy in the Modern Brazilian Amazon', *Latin American Studies* 25 (1993): 331-349, Anthony Hall, 'Environment and Development in Brazilian Amazonia: From Protectionism to Productive Conservation', in *Amazonia at the Crossroads: The Challenge of Sustainable Development*, ed. Anthony Hall (London: Institute of Latin American Studies, University of London, 2000), Daniel Nepstad et al., 'Frontier Governance in Amazonia', *Science* 295 (2002): 629–630, Robert Schneider, 'Government and the Economy on the Amazon Frontier'. World Bank Environment Paper No. 11.

13. Susanna Hecht, 'Soybeans, Development and Conservation on the Amazon Frontier', *Development and Change* 36, no. 2 (2005): 375–404, Pablo Pacheco, 'Populist and Capitalist Frontiers in the Amazon: Dynamics of Agrarian and Land-Use Change' (PhD dissertation, Clark University, 2005).

14. Kaimowitz, 'Frontier Theory and Practice'.

15. Schneider, 'Government and the Economy on the Amazon Frontier'.

16. Mohan Munasinghe, *Environmental Impacts of Macroeconomic and Sectoral Policies* (Washington DC.: ISEE-World Bank-UNEP, 1996).

17. Cynthia S. Simmons, 'Local Articulation of Policy Conflict: Land Use, Environment and Amerindian Rights in Eastern Amazonia', *The Professional Geographer* 54, no. 2 (2002): 241–258.

18. Susanna Hecht, 'Environment, Development and Politics: Capital Accumulation and the Livestock Sector in Eastern Amazonia', *World Development* 13, no. 6 (1985): 663–684.

19. Thomas Rudel, 'Changing Agents of Deforestation: From State-Initiated to Enterprise Driven Processes, 1970–2000', *Land Use Policy* 24, no. 1 (2007): 35-41.

20. Hecht, 'The Economics of Cattle Ranching in Eastern Amazonia'.

21. Sergio Margulis, 'Causes of Deforestation of the Brazilian Amazon'. Report No. 22, (Washington, D.C.: World Bank, 2004).

22. Eugenio Arima and Christopher Uhl, 'Ranching in the Brazilian Amazon in a National Context: Economics, Policy, and Practice', *Society and Natural Resources* 10, no. 5 (1997): 433–451.

23. Jonas Bastos da Veiga et al., 'A Amazônia Pode Virar uma Grande Região de Pecuária Bovina Sustentável?', (Belem, Brasil: Empresa Brasileira de Pesquisa Agropecuaria (unpublished draft), 2001).

24. David Kaimowitz and Arild Angelsen, 'Will Livestock Intensification Help Save Latin America's Tropical Forests?' *Journal of Sustainable Forestry* 27, no. 1–2 (2008):6–24.

25. Jonas Bastos da Veiga et al., 'Milk Production, Regional Development and Sustainability in the Eastern Amazon', (Belém, Brasil: CIRAD-Empresa Brasileira de Pesquisa Agropecuaria, 2001).

26. Douglas Southgate, *Tropical Forest Conservation: An Economic Assessment of the Alternatives in Latin America* (Oxford: Oxford University Press US, 1998).

27. Kaimowitz and Angelsen, 'Will Livestock Intensification Help Save Latin America's Tropical Forests?',

28. The region known as southern Pará constitutes a relatively large territory that embraces the southeastern portion of the State of Pará in the Brazilian Amazon. It ranges from the municipality of Maraba in the north to the border with the State of Mato Grosso in its extreme south.

29. Octavio Ianni, *A Luta Pela Terra: Historia Social da Terra e da Luta Pela Terra numa Área da Amazonia*, 2da ed. (Petropolis, RJ: Vozes, 1979).

30. Jonas Bastos da Veiga et al., *Expansão e Trajetórias da Pecuária na Amazônia* (Brasilia: Editora UNB, 2004).

31. The most common land unit in southern Pará is the *alqueire goiano* which equates to 4.84 ha. According to land legislation of the state of Pará, the maximum unit that a private investors could claim from the state to be granted private ownership by buying it was 900 *alqueires*, or 4,356 ha.

32. Hecht, 'The Logic of Livestock and Deforestation in Amazonia'.

33. da Veiga et al., *Expansão e Trajetórias da Pecuária na Amazônia*.

34. PROTERRA was available for both investment and land purchases. Investment credits were lent with 6 years grace period, interest rates of 10–14%, 12 years to amortise fixed investment, and 8 years for semi-fixed investment with 4 years grace. Land credits were 12% of interest rate, 20 years to amortise with 6 years grace-period.

35. Susanna Hecht, 'Cattle Ranching Development in the Eastern Amazon: Evaluation of Development Strategy' (Ph.D dissertation, University of California, 1982).

36. Ibid.

37. Clando Yokomizo, 'Incentivos Financeiros e Fiscais na Pecuarizacão da Amazônia. Texto Para Discussão No. 22', (Rio de Janeiro, Brazil: Instituto de Planejamento Econômico e Social (IPEA), 1989).

38. Cattle ranching, through subsidised SUDAM projects, helped to develop most of the primary road infrastructure. Due to the large occurrence of mahogany amidst Redenção and Xinguara, pressures emerged to open a road towards Marabá in the early 1970s. By then, most timber and cattle had to be transported from Redenção to Conceição do Araguaia to take the Belém-Brasília highway passing through the state of Tocantins (then part of Goias) and Maranhão to get to the Belém market in the northeast portion of Pará. The Redenção-Marabá road (PA 150) was constructed between 1973 and 1975, and extended to Santana de Araguaia between 1977 and 1982, offering a more direct connection with the Belém market. This road was paved in the early 1990s.

39. First-generation grasses on recently formed pastures were mainly *colonião* (*Panicum maximum*), and Brachiaria (*Brachiaria decumbens*), which are high productivity varieties, but are vulnerable to weeds and invader species. Much of the areas planted with this pasture variety degraded severely after ten or fifteen years of use in the mid-1980s. As a result, large areas became unproductive for cattle breeding, and had to be abandoned since it was not profitable to recover these areas for production.

40. The most popular grass variety introduced in the Amazonia, due to its resistance to climatic conditions, is Braquiarão (*Brachiaria brizantha*), though some other Brachiaria varieties have also been adopted to avoid pasture degradation. *Braquiarão* is vigorous, provides good ground cover to suppress weeds, and resists spittlebugs, which severely affected the previously used grasses varieties. Nigel Smith et. al., *Amazonia: Resiliency and Dynamism of the Land and its People* (New York: United Nations University Press, 1995)

41. Unfortunately, there is only fragmented information about implemented projects by ranch. This information comes from SUDAM registers included in DESMAT, a dataset combining socio-economic data to model deforestation at the municipal level developed by IPEA.

42. The *garimpos* that emerged in the first half of the 1980 decade in the south of Pará were diverse. The largest was Serra Pelada, closer to Marabá. The best known *garimpos* installed near Redenção were Cumarú, Forquilha, Maria Bonita and Malvinas e Garrapato.

43. Pacheco, 'Populist and Capitalist Frontiers in the Amazon: Dynamics of Agrarian and Land-Use Change'.

44. Philip M. Fearnside, 'Land-Tenure Issues as Factors in Environmental Destruction in Brazilian Amazonia: The Case of Southern Para', *World Development* 29, no. 8 (2001): 1361–1372.

45. FNO, PROCERA and PRONAF constitute subsidised credit lines specifically targeted to promote small-scale agriculture. A portion of these resources is also intended to consolidate and promote agricultural development in the projects of agrarian reform implemented by the state land agency, the National Institute of Colonization and Agrarian Reform, INCRA.

46. Pacheco, 'Populist and Capitalist Frontiers in the Amazon: Dynamics of Agrarian and Land-Use Change'.

47. Smith et al., *Amazonia: Resiliency and Dynamism of the Land and its People.*

48. IBGE. 'Pesquisa Pecuária Municipal (PPM)' (Rio de Janeiro, Brasil, 2007).

49. da Veiga et al., *Expansão e Trajetórias da Pecuária na Amazônia.*

50. It is almost impossible to identify the real situation of land titling in southern Pará as the land transaction registers held in the notaries are not public. The act of registering a transaction in the notary renders it legal even if it is based on fraudulent titles.

51. According to anecdotal evidence, the Volkswagen company created one of the largest cattle herds in the region on a single ranch (around 140,000 head). After selling most of the cattle and exhausting the ranch's forest reserve, they sold it to its current owner, the Matsubara group from Paraná, in the mid 1990s. This group promoted the invasion of the ranch by squatters as a way to facilitate its sale to INCRA as part of the program of a market-based agrarian reform. However, a contentious process impeded INCRA from acquiring the area, and now most of this ranch is illegally occupied by squatters. To complicate matters, uranium has been discovered in this area, which led the Ministry of Science and Technology to declare it as 'nuclear monopoly area', restricting occupation by outsiders.

52. Fearnside, 'Land-Tenure Issues as Factors in Environmental Destruction in Brazilian Amazonia: The Case of Southern Para'.

53. Pacheco, 'Populist and Capitalist Frontiers in the Amazon: Dynamics of Agrarian and Land-Use Change'.

54. Ibid.

55. Walmir Hugo dos Santos, 'Matadouros – Frigorificos: O Abate para Carne em Belém', (Belém: Ministerio da Agricultura, SUDAM, Governo do Estado do Pará, ACAR-Pará, 1976).

56. Veiga et al., *Expansão e Trajetórias da Pecuária na Amazônia.*

57. René Poccard-Chapuis et al., 'Cadeia Produtiva de Gado de Corte e Pecuarização da Agricultura Familiar Na Transamazônica. Documentos No. 106', (Belém: Empresa Brasileira de Pesquisa Agropecuaria, 2001).

58. Rene Poccard-Chapuis, 'Les Reseaux de la Conquete Filiere Bovine et Structuration de L´Espace sur les Fronts Pionniers D´Amazonie Oriental Bresilienne' (Université de Paris X – Nanterre, 2004).

59. Poccard-Chapuis et al., 'Cadeia Produtiva de Gado de Corte e Pecuarização da Agricultura Familiar na Transamazônica. Documentos No. 106.'

60. Hecht, 'Cattle Ranching Development in the Eastern Amazon: Evaluation of Development Strategy', John Browder and Brian Godfrey, *Rainforest Cities, Urbanization, Development and Globalization of the Brazilian Amazon* (New York: Columbia University Press., 1997).

61. Emilio Moran, 'Private and Public Colonization Schemes in Amazonia', in *The Future of Amazonia: Destruction or Sustainable Development?*, ed. D. Goodman and A. Hall (New York: St. Martin's Press, 1990), Emilio Moran, *Developing the Amazon* (Bloomington: University of Indiana Press., 1981).

62. Malori J. Pompermayer, 'Strategies of Private Capital in the Brazilian Amazon', in *Frontier Expansion in Amazonia*, ed. Marianne Schmink and Charles Wood (Gainesville, FL: University of Florida Press, 1984).

63. Hecht, 'Environment, Development and Politics: Capital Accumulation and the Livestock Sector in Eastern Amazonia.'

64. Marcionila Fernandes, *Donos de Terras: Trajetórias da União Democrática Ruralista, UDR* (Belem, Pará: UFPA, NAEA, 1999).

65. Marianne Schmink and Charles Wood, *Contested Frontiers in Amazonia* (New York: Columbia University Press, 1992).

66. Cynthia S. Simmons et al., 'The Changing Dynamics of Land Conflict in the Brazilian Amazon: The Rural-Urban Complex and Its Environmental ' *Urban Ecosystems* 6, no. 1–2 (2002): 99–121.

67. The 1965 Forest Code stipulated that at least 20% of any property area covered by forest should be kept as forest – thus allowing the owner to clear up to 80% of its area. In the Amazon, 50% of each property in areas covered by native forest had to be designated as Legal Reserve. In 1995, the Brazilian government used a Presidential Transitory Act to strengthen the Forest Code, raising the Legal Reserve area for the Amazon region from 50 to 80% of the property in an attempt to reduce deforestation.

68. Hall, 'Environment and Development in Brazilian Amazonia: From Protectionism to Productive Conservation.'

69. Ane Alencar et al., 'Desmatamento Na Amazônia: Indo Alem da Emergência Crônica', (Belem, Brasil: IPAM, WHRC, 2004).

70. Bernardo Mueller and Lee J. Alston, 'Legal Reserve Requirements in Brazilian Forests: Path Dependent Evolution of de Facto Legislation' (paper presented at the Anais do XXXV Encontro Nacional de Economia, Sao Paulo, 2007).

71. The Amazon Basin of Brazil has been defined by government decree in 1953, and it is referred as the Brazilian Legal Amazon (BLA). It covers an area of 5,000,000 km^2 embracing the six 'North' states (Acre, Amapá, Amazonas, Pará, Roraima and Rondônia), plus part of three others (Tocantins, north of the 130° parallel; Mato Grosso, north of the 160° parallel; and Maranhão, west of the 44° meridian). In 1977, the entire State of Mato Grosso was included in the BLA by the complementary law n°31 (art. 45). The main reason for the creation of the Legal Amazon was to define an area for the administration of economic development, rather than to delimit a region according to a uniform ecosystem. That is the reason why the BLA includes, besides forested land, extensive areas of natural savanna (*cerrado*), and open forest in the transition zone between closed forest and *cerrado*.

Farming at the Forest Frontier: Land Use and Landscape Change in Western Madagascar, 1896–2005

Ivan R. Scales

1. INTRODUCTION

Over the last twenty years Madagascar has become a poster child for global biodiversity conservation, with hundreds of millions of US dollars spent on finding solutions to habitat destruction and the loss of biodiversity.[1] It has been labeled one of the world's 'hottest biodiversity hotspots' and highest conservation priorities.[2] Whilst a range of threats to wildlife have been identified, the environmental discourse of Madagascar has tended to focus on the issue of deforestation.[3] The dominant narrative is one of severe forest loss driven by subsistence slash-and-burn agriculture. The blame for deforestation is usually placed on poor rural households, who are seen to practice forest clearance out of bleak necessity – caught in a Malthusian spiral of increased population and decreased land productivity:

> the poverty that afflicts Madagascar's people threatens to destroy what remains of this unique biology … widespread poverty, increasing population, and the absence of resources and techniques to improve the productivity of agricultural and pasture lands have led to massive deforestation.[4]

A World Bank report states that:

> Madagascar has already lost 80 percent of its original forest cover, and the rest is under severe pressure for reasons that relate principally to poverty…Traditional forms of itinerant agriculture, which are relied on by the poor because they have no incentives to intensify production, result in the burning of savanna and forests.[5]

According to this view, poverty results in environmental degradation, degradation leads to further poverty and a vicious circle ensues. This narrative is pervasive and has become deeply entrenched in the academic literature, so that papers on biodiversity conservation in Madagascar often frame their research in terms of dramatic deforestation and biodiversity loss due to poverty.[6] As the following quotation from a British national newspaper shows, this powerful narrative has also made its way into the media:

> Millennia-old trees and rare wildlife in Madagascar are vanishing as hungry families, their crops shrivelled by drought, sell bags of charcoal to survive … A forgotten famine is reducing one of the world's richest stores of biodiversity, the rainforests of Madagascar, to ash.[7]

Environment and History **17** (2011): 499–524.

Given the urgency expressed in the received wisdom on forest loss, it is not surprising that there is a growing body of research on deforestation in Madagascar. Unfortunately, it has tended to suffer from a number of limitations. Firstly, there has been a temporal bias to research, with little work addressing change before the 1970s (when satellite data first became widely available). Few researchers have drawn on historical sources to investigate how land use and landscape change have played out in different parts of Madagascar.[8]

Secondly, there has been a geographical bias to research, which has focused mostly on the eastern tropical humid rainforests, leaving biomes such as the western dry-deciduous forests relatively under-studied.[9] It has also tended to lump Malagasy farmers together into a single category, ignoring the biophysical, political, economic and cultural diversity of the world's fourth biggest island. In Madagascar's environmental discourse, the term *tavy* – a form of slash-and-burn agriculture – has become shorthand for the country's environmental problems. *Tavy* in fact refers specifically to forest clearance agriculture with fire to produce rain-fed rice (*Oryza sativa*), as carried out on hillsides in the eastern rainforests of the country.[10] Despite this specificity, it has been incorporated into the deforestation narrative and the complexity of rural land use is thus reduced to a single word that has become unhelpfully loaded with untested assumptions.

Finally, there have been few attempts to understand the underlying factors driving land use. There is now a small but growing body of work that has revealed more complex relationships between politics, economics and agriculture in Madagascar[11] but most research has tended to simplify the process of deforestation by focusing on the proximate factors leading to forest loss.

In this paper, I focus on the drivers of land use and landscape change in the Central Menabe region of western Madagascar between 1896 and 2005 and in particular on the relationship between agriculture and forest loss. To investigate how and why the landscape of western Madagascar has changed during this time period, I draw on a diverse range of methods and sources including: i) research in the French colonial archives; ii) the analysis of aerial photographs and maps from the late colonial period; iii) the analysis of imagery from Landsat satellites; iv) interviews with rural households; and v) a review of the project reports, as well as interviews with current and former employees, of government and non-government organisations (NGOs) working on agricultural development and forest use since the 1960s. I show the importance of booms in the cultivation of cash crops in shaping the region's landscape. I also show that economic and political factors operating at a range of spatial levels have been critical in stimulating the expansion of these crops. I argue that research and policy in Madagascar needs to pay more attention to the factors that drive land use, rather than assuming *a priori* the importance of poverty and population growth as drivers of deforestation.

Ivan R. Scales

2. STUDY AREA

Central Menabe (Map 1) is sandwiched between the Mozambique channel to the west and Madagascar's highland plateau to the east. Its northern border is marked by the Tsiribihina river and its southern border by the Morondava river, which gives its name to the region's administrative and economic centre. Its flat sandy soils are dissected by fertile river valleys. The landscape of Central Menabe is made up of a complex heterogeneous mosaic of tropical dry-deciduous forest, shrubland and various types of herbaceous formations, as well as a variety of agricultural land cover types.

The region has a semi-arid tropical climate with two clearly distinct seasons – a cooler dry season from April to November and a hot rainy season from December to March. The majority of rain falls in January and February with a mean

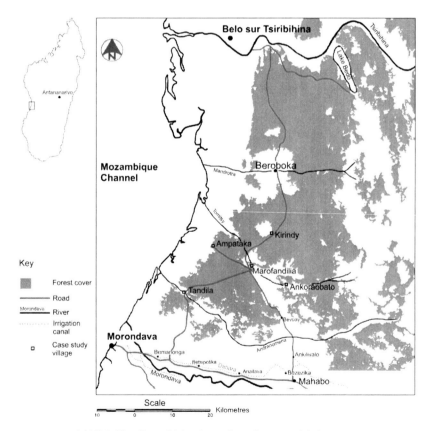

MAP 1. The Central Menabe region of western Madagascar

of 762.6 mm. However, this average masks considerable variability. Between 1902 and 2004, total annual rainfall ranged from a minimum of 204.8 mm to a maximum of 1688.1 mm.[12] To put this in context, maize (*Zea mays*) – the most common rainfed crop grown in the region – requires a minimum of 500 mm of annual rainfall, with best yields produced with 1,200 to 1,500 mm of rain per year.[13] During the period between 1902 and 2004, thirteen rainy seasons failed to reach the 500 mm threshold and the threshold for optimum yields was surpassed in only four of the years.

The region's economy is dominated by agriculture, with over eighty per cent of the population involved in the cultivation of crops.[14] Agriculture takes a variety of forms, from rice paddies to large-scale sugar cane plantations. Away from the river valleys, the majority of households rely on rain-fed agriculture.[15] This takes two forms in Central Menabe. The first is cultivation using *hatsake*, an agricultural system involving forest clearance by fire for the purposes of growing maize. Apart from drought years, maize is generally well suited to the region's climate. However, the region's soils are nutrient-poor and require additional inputs. The burning of forest provides nutrient rich ash and light for crops. Whilst forest clearance for agriculture has been illegal since 1987,[16] the practice of *hatsake* continues – a household survey I carried out showed that 44 per cent of cultivating households had cleared forest for the cultivation of maize during the most recent growing season.[17]

The other option for households without access to irrigated land is the cultivation of *monka* – land that has been cleared through *hatsake* and left fallow for at least three years. The relatively nutrient poor soils of *monka* mean that maize cannot be grown on this land. Instead, households grow groundnut (*Arachis hypogaea*) and manioc (*Manihot esculenta*).

The Central Menabe landscape, with unpredictable rainfall and nutrient poor soils, clearly imposes certain constraints on the land use decisions of rural households, especially those without access to irrigated land. The availability of forested land for clearance means that the issue of deforestation is a key part of the regional environmental discourse, mirroring national and international concerns. As in many other parts of Madagascar, previous studies of deforestation in the region have been limited to measuring relatively recent forest loss.[18] Given the environmental constraints, it might be tempting to assume that the modern landscape is simply the product of the limits placed by an unforgiving environment on a poor rural population. However, as I show in this paper, to think of forest loss as resulting simply from the actions of rural households responding to environmental factors and a lack of alternatives is to ignore the importance of large-scale commercial agriculture, as well as a diverse range of political and economic factors.

3. METHODS AND SOURCES

Defining forest cover and deforestation

In this paper, I define forest as an area of land covering at least 0.5 hectares and dominated by trees, with a canopy cover of at least ten per cent. I define trees as woody vegetation with a diameter at breast height (dbh) greater than 10 cm and a height greater than five metres. This fits the Food and Agriculture Organization of the United Nations definition of forest cover.[19] The definition of deforestation poses more of a problem, since the literature contains a wide range of definitions, from the total removal of forest cover to small changes in forest composition and structure, by selective logging for example.[20] The forests of Central Menabe have been selectively logged for over a century, which is likely to have had an impact on the structure and species composition of the forest. It is therefore probable that there have been two parallel processes during the twentieth century – the loss of forest to agriculture and a reduction in canopy cover and change in forest structure and species composition due to logging. For the purposes of this paper I focus of the process of slash-and-burn agriculture and I define deforestation as the total removal of canopy cover.[21]

Understanding land use and land cover change during the colonial period (1896–1960)

Reconstructing landscape change during the colonial period poses some challenges. The first aerial photographs of the region were taken in 1949 by the colonial mapping service and are now available from the Foiben-Taosarintanin' Madagasikara (FTM) – the National Geographical and Hydrological Institute of Madagascar in Antananarivo – but many have deteriorated to the extent that they are unusable. In order to reconstruct landscape change before 1954, I rely on a combination of sketch maps and reports from the French colonial archives at Le Centre des Archives d'Outre-mer in Aix-en-Provence which contain records of imports, exports and concessions awarded. As well as figures from the port of Morondava and measurements of the areas covered by different crops, the archives also contain accounts written by governors of their administrative tours of the region. Whilst these data do not allow me to measure forest cover precisely, they do offer an indication of where areas of forest were located, where agriculture was taking place and the type of crops that were being grown. The governors' reports also give an indication of the government policies driving these changes.

Reconstructing 1954 forest cover

In 1954, the French colonial government flew aerial missions over the whole of Madagascar in order to produce detailed maps of the country. As well as providing useful information on the features of the 1954 landscape such as settlements, irrigation canals and plantations, these maps can also be used to measure forest cover.[22] In order to do this for Central Menabe, I acquired four 1:100,000 scale maps from FTM based on 1:20,000 aerial photographs taken in 1954. I digitised them and imported the Tagged Image File Format (TIFF) files into ERDAS Imagine image analysis software.[23] I then georeferenced the images using latitude/longitude coordinates collected in the field with a Global Positioning System (GPS) at twenty easily identifiable points (mostly road junctions). Following the georeferencing of the maps, I stitched them together using ERDAS Imagine software to make one scene for 1954. I then classified this scene using a supervised maximum likelihood classification to reduce the original map into a simple forest/non-forest map. The end product was a digital image for 1954, in the same format as subsequent classifications of satellite imagery from 1973 onwards, that allowed me to measure forest cover

Reconstructing forest cover change from 1973 to 2005

From 1973 onwards, it is possible to measure forest cover using satellite imagery. To measure changes in forest cover between 1973 and 2005, I carried out a supervised maximum likelihood classification of Landsat satellite images from 1973, 1978, 1992, 1999 and 2005.[24] To classify the images, I selected 207 training sites using stratified random sampling. For each image, I obtained spectral signatures from the training sites and merged them into land cover signatures. I carried out a separability assessment to identify which bands gave the best separability between forest and non-forest.[25] Following land cover classification I carried out accuracy assessments of the forest cover classifications (Table 1. Accuracy assessment results for the land cover classifications of satellite imagery, 1973 to 2005), using 100 ground reference points.[26]

Table 1. Accuracy assessment results for the land cover classifications of satellite imagery, 1973 to 2005

Year	Producer's accuracy (%)		User's accuracy (%)	
	Forest	Non-forest	Forest	Non-forest
1973	86.0	92.0	91.5	86.8
1978	88.0	90.0	89.8	88.2
1992	92.0	92.0	92.0	92.0
1999	94.0	90.0	90.4	93.8
2005	90.0	94.0	93.4	90.4

Ivan R. Scales

Understanding the drivers of land use change after 1954

To understand the socio-economic and political context of the land cover change measured after 1954 I gathered data using i) Rapid Rural Appraisal (RRA) methods in five villages (Map 1);[27] ii) 45 semi-structured interviews with key informants; and iii) a review of the project reports of government agencies and non-governmental organisations (NGOs) working on agricultural development and forest use in the region since Madagascar's independence. I carried out field-work between July 2005 and August 2006, with research in villages carried out primarily between July and September 2005 and between May and August 2006.

4. RESULTS

4.1. LAND COVER AND LAND USE CHANGE BETWEEN 1896 AND 1954

The arrival of French colonialism

The arrival of French colonialism was a period of tremendous political and economic upheaval in western Madagascar. During the eighteenth and nineteenth centuries Central Menabe was controlled by the Sakalava kingdom, one of the most powerful political entities in an as yet un-unified Madagascar.[28] It had risen to power through links with Indian and Swahili traders, exchanging slaves and zebu for muskets and gunpowder.[29] The region's economy was based largely on extensive zebu pastoralism, with shifting cultivation carried out mostly by slaves.[30] The landscape was thus one shaped largely by pastoralism rather than settled agriculture.

Madagascar became a French colony in 1896, although it was not until 1902 that control was established over Central Menabe. Once control had been established, the government rapidly went about trying to make the region profitable. Early reports show that the colonial government saw the Central Menabe landscape as potentially highly productive but under-exploited. An economic report from 1904 states that 'Everything leads us to believe that the littoral plains can become vast and productive cotton fields'.[31] The regional governor in 1918 was convinced that Menabe would 'within a quarter of a century become the richest region in Madagascar',[32] whilst a report of 1916 states that Menabe has 'excellent agricultural land; all that is needed is labour'.[33]

The government's aim was to increase the economic productivity of the region and boost tax incomes by focusing on agriculture. However, it was frustrated by the lack of willingness of rural Sakalava households, whose livelihoods were still primarily based on extensive cattle pastoralism, to undertake waged labour and participate in the cash economy:

In groups of three, four or five families, the Sakalava travel through the forest, following their whims ...They find an area in the middle of nowhere and build themselves vile little huts, where they will quite happily live, and when we need them to pay their taxes or require their labour, we can't find anyone.[34]

In order to achieve its plans the French colonial government's policies focused on five key areas: i) the introduction of taxes to be paid in cash, in order to encourage rural households to participate in the economy and to switch from extensive subsistence pastoralism to sedentarised livelihoods based on cash cropping;[35] ii) a system of forced labour (*corvée*), usually without pay, in order to develop infrastructure (especially canals and dams to allow the irrigation of land);[36] iii) a policy of minimum cultivation per village which required each household to cultivate at least three hectares of land;[37] iv) the introduction of a system of private land tenure and land registration to replace local customary land tenure;[38] and v) the awarding of large concessions to foreign individuals and companies.[39] Initially, the colonial government saw these concessions as the ideal way of boosting productivity and by 1905 more than 50,000 hectares, mostly in the fertile and seasonally flooded river valleys, had been given to expatriates.

For the system of concessions and plantations to work, large amounts of labour were required. The government saw the low population density of the region and resulting lack of labour as particularly problematic, with one governor reporting that 'it is a shame that they [the Sakalava] are so few in numbers for such vast areas'.[40] The government believed that improved irrigation and transport, as well as the opportunity to work in concessions, would encourage migrants to settle in the region.

The rice boom and Central Menabe's first land rush

Within the first decade of colonialism, government policy was already having a significant impact on the region's land use and landscape through the conversion of land to rice paddies. Irrigated rice cultivation was a relatively new technique in Central Menabe, having been introduced to the region by Merina and Betsileo migrants from the highlands of Madagascar during the course of the nineteenth century.[41] However, given the semi-arid climate, it was limited to areas in close proximity to the Morondava and Tsiribihina rivers where households had built small canals. It was not until large-scale irrigation projects were undertaken by the colonial government, using forced labour, that rice cultivation could expand. As a result of improvements in infrastructure, such as the Hellot canal which was finished in 1901, the area covered by rice cultivation increased from 2,000 hectares in 1901 to 14,130 hectares in 1904 – a sevenfold increase in three years. By 1921 over 40,000 hectares of land had been converted to paddies.[42]

The registration of land titles became possible following Décret du 16 juillet 1897, whilst the state ownership of non-titled land was formalised by Décret du

28 septembre 1897. Given the constraints placed on agriculture by the limited water availability in semi-arid Central Menabe, it is not surprising that the newly created rice paddies were in great demand. However, the distribution of this land tended to favour some groups over others. Sakalava households largely missed out on the rush for irrigated land, whilst French settlers and Malagasy elites with close ties to the colonial government were able to secure access to this newly created irrigated land. Another group to benefit from land titling were *Karana* traders, a group of Indian origin who settled in Madagascar during the eighteenth century and have dominated the region's import and export business ever since.

The new laws, ignoring customary land tenure, thereby created a legal framework for the appropriation of land. The physical and political changes to the landscape brought about by irrigation, the privatisation of land and the awarding of large concessions to expatriates set the scene for more dramatic changes during the 1920s and 1930s.

The butter bean boom of the 1920s

The boom in rice cultivation was followed by a boom in the cultivation of butter beans (*Phaseolus lunatus*), known as *Pois du Cap* in French. The land under butter bean cultivation went from 4,835 hectares in 1914 to 16,120 hectares in 1920 – more than a threefold increase in six years.[43] Whilst the rice boom had been limited by the availability of permanently irrigated land, the boom in butter bean cultivation saw the spread of agriculture onto the seasonally flooded alluvial soils around the region's main rivers – the Morondava and Tsiribihina.

The boom was largely stimulated by an increase in demand from Europe, in particular London, with French and British ships sailing to Morondava specifically to purchase butter beans.[44] Whilst the increase in demand was the trigger, other factors facilitated the boom. Firstly, the trading houses of Morondava and Belo played a key part in connecting rural households to international markets, by acting as middlemen between them and foreign merchants sailing to Morondava. Secondly, the large concessions awarded to expatriates underpinned a system of sharecropping, where concession owners offered the use of part of their land in return for a fifty per cent share of the crops produced. This meant they could convert large areas of the region's fertile river valleys to the cultivation of butter beans, with minimal effort, by relying on migrant labour. Furthermore, both traders and concession owners facilitated the incorporation of new migrants into the agricultural economy of the region by offering loans to allow them to purchase tools and seeds.

Although the government was satisfied by the increase in the cultivation of cash crops, it began to express concerns over the concession system. Firstly, it noticed that the Sakalava had been largely marginalised by the privatisation of land, commenting that 'the expansion of European colonisation has tended to appropriate all the best land of the river'.[45] It also criticised the system of sharecropping, accusing concession owners of relying on local labour and traditional agricultural methods, without introducing new techniques or intensifying production.[46] It is important to note that Sakalava households were not entirely passive in this process. Some managed to use their status as *tompontany* [masters of the land] to allow migrants to farm on their land in return for a percentage of their harvest.[47]

In contrast to rice production, this trend did not continue and the cultivation of butter beans almost totally collapsed in 1927. The collapse was the result of a substantial drop in prices in the London markets, which resulted in rural farmers abandoning the crop and forced trading houses in Morondava to close. This was tied to the global agrarian crisis that occurred in 1925, which was followed by the stock market crash of 1929 and the inter-war Great Depression.[48]

The maize boom of the 1930s

Whilst the rice and butter bean booms were significant in the conversion of Central Menabe from a landscape based on extensive pastoralism to one based on the cultivation of crops, their impacts were limited by the availability of water. During the 1930s a third agricultural boom occurred, this time in the cultivation of maize. This boom was particularly important in terms of landscape change due to the fact that unlike rice or butter beans, maize does not require irrigation or flooded alluvial soils. Using *hatsake*, high yields can be obtained on cleared forest and maize can be grown in areas unsuitable for other crops. The maize boom therefore impacted the forest in a way that the previous agricultural booms had not.

Whilst the practice of slash-and-burn agriculture had existed in the region prior to the arrival of French colonialism, it had been primarily for subsistence and carried out largely by slaves.[49] The 1930s saw a change in the underlying drivers of *hatsake* from subsistence needs to the cultivation of maize as a cash crop grown for export. The records in the colonial archives show that maize exports passing through the port of Morondava went from 7,000 tonnes in 1935 to a peak of 29,500 tonnes in 1939 – more than a fourfold increase in four years.[50] Like the butter bean boom, the initial stimulus was an increase in maize prices. Once again, the system of concessions, sharecropping and trading houses played a key role in facilitating the expansion of maize cultivation. Also critical to the boom was the construction of a railway line (completed in 1925) that allowed the cheap transportation of maize from more remote forest areas.[51]

The possibility of sharecropping cash crops on forested land drew more migrants to the region, mainly from the south of Madagascar. This not only contributed to deforestation, as migrants were allowed to clear forest by landowners in return for half of their crop, but had a considerable impact on the region's demography and resulted in the establishment of new villages. Whilst in 1910 the population had been 40,537 it had risen to 165,835 by 1941, with Sakalava households making up less than 35 per cent.[52]

The maize trade collapsed, however, as rapidly as it had expanded. By 1940, maize was no longer being exported. The collapse in maize production was largely the result of the removal of the railway line, due to the government's belief that it could be put to better use elsewhere in the country. The removal of the railway resulted in a ten-fold increase in transportation costs, making the cultivation of maize unprofitable.[53]

Unfortunately, it is not possible to know the exact extent of forest cover before or during the majority of the colonial period since accurate mapping was not undertaken until 1954. This makes it impossible to calculate precise figures for forest loss between 1896 and 1954. A report from 1941, based on visits to affected areas, stated that 55,000 hectares had been lost to slash-and-burn agriculture.[54] To put this figure in context, Central Menabe covers approximately 900,000 hectares. The estimate of forest loss must of course be treated with caution. However, it is possible to use another proxy, in the form of maize export figures from the port of Morondava. Between 1936 and 1940, a total of 90,500 tonnes of maize passed through the port of Morondava.[55] Since maize was (and still is) grown only through the clearance of forest, it is possible to estimate how much forest clearance was involved in the maize boom. Modern agricultural experiments have shown that one hectare of cleared forest produces 2.2 tonnes of maize over a period of three years, after which the land is exhausted.[56] The 90,500 tonnes of maize exported during the boom therefore corresponds to approximately 45,000 hectares of cleared forest.

As Map 2 shows, the French colonial policy to convert the landscape to the cultivation of cash crops for export was successful. By 1943, the region was producing rice, butter beans, maize and tobacco. This transition occurred through a series of booms, as well as some spectacular busts, and was underpinned by important changes in land tenure, demography and infrastructure. As I show in the next section, the changes of the colonial period were to have a significant impact on the region's landscape, and in particular its forests, after independence.

4.2. LAND COVER AND LAND USE CHANGE, 1954 TO 2005

Rates and pattern of forest loss

From 1954 onwards, it is possible to measure changes in forest cover more accurately. Between 1954 and 2005, total forest cover declined from 184,347 hectares to 151,269 hectares, a total loss of 33,078 ha over 51 years, which equals 17.9 per cent of the total forest area and a mean annual loss of 0.4 per cent.

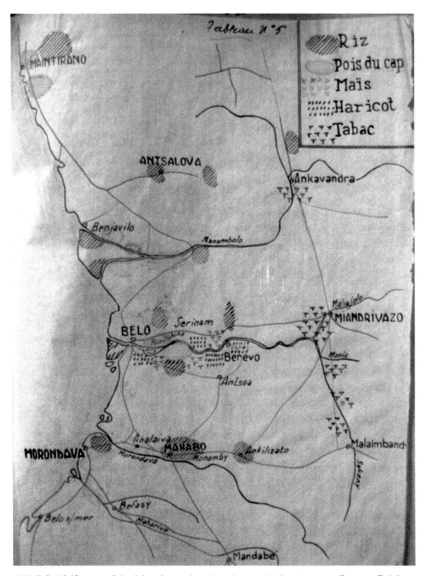

MAP 2. 1943 map of the Menabe region showing agricultural zones. Source: Région de Morondava, Circonscription Agricole, Rapport 1943. Aix 2D 177

Table 2 shows that rates of forest loss have been uneven – low during the 1960s and 1970s, increasing during the 1980s and increasing dramatically after 1999.

Ivan R. Scales

TABLE 2. Forest loss in study area, 1954 to 2005

Time period	Forest loss (ha)	Annual rate (ha/year)	% loss	Annual % loss
1954 to 1973	6,515	342.9	3.5	0.2
1973 to 1978	796	159.2	0.5	0.1
1978 to 1992	11,737	838.4	6.6	0.5
1992 to 1999	4,528	646.9	2.7	0.4
1999 to 2005	9,502	1,583.7	5.9	1.0
1954 to 2005	33,078	648.6	17.9	0.4

As well as being uneven in time, forest loss has been uneven in space. Map 3 shows that rather than steadily gnawing away at the edges of the forest, deforestation has been concentrated around certain villages. Beroboka and Marofandilia together account for 32.4 per cent of all the forest lost in the study area between 1954 and 2005.

The drivers of land use and forest loss between 1954 and 2005

Many of the villages that have been foci of forest loss since 1954 (such as Ankoraobato, Beroboka, Marofandilia and Tandila) owe their origins to colonial concessions. The underlying drivers of forest loss post-independence must therefore be partly understood in relation to the political, economic and demographic changes of the colonial period that resulted in the establishment of large agricultural concessions and the influx of migrant labourers, who often decided to stay and establish household farms.

Beroboka was originally a 40,000 hectare forest concession awarded to the French company Société Foncière et Minière in 1910. It was left undeveloped until logging operations began in 1937. By 1942 the concession was supplying two thirds of the wood sold in Morondava, relying on Antandroy and Betsileo workers, who established a village and rice paddies around the Mandrotra river.[57]

The village of Marofandilia also owes its origins to migrant workers employed on a logging concession established in 1910 by a French national, Monsieur Homsi, who was awarded a 3,500 hectare area of forest by the colonial government. Migrant workers established themselves close to the concession and created new settlements. They logged the forest to the east of Marofandilia, close to the Tomitsy river, so logs could be floated down the river to supply the saw mill. As well as supplying timber, they established rice paddies, using water from the Tomitsy to supply small irrigation canals. The production of rice was boosted by the construction of a rice decorticating machine, which used power from the saw mill.[58]

Once independence from France came in 1960, most concessions were abandoned. Many of the households who had come to work as waged labourers stayed in the villages and focused on subsistence agriculture.[59] Following independence, the newly formed Malagasy government inherited the French legal system and land tenure rules. Much like the colonial government before them, the post-colonial government saw traditional agricultural methods as unproductive and focused their efforts on large projects designed to boost rice and cotton production. In Menabe, the 1960s and early 1970s were a period of substantial investment in large-scale agricultural projects. The Société de Developpement de Morondava (SODEMO) was created in 1972 to focus on the irrigation of 10,000 hectares of rice and cotton. Other significant projects included an orange grove at Bezezika, which was started in 1968 with Israeli technical assistance, as well as experimental tobacco and cotton fields at Ankilivalo. However, these large-scale projects were blighted with problems, mostly due to the failure of irrigation systems resulting from a lack of maintenance and damage from cyclones.[60]

The government's projects were focused mainly in the valley of the Morondava river and had little impact on more remote forest villages such as Marofandilia and Beroboka. Correspondingly, rates of forest loss were relatively low between 1954 and 1973. A total of 6,515 hectares of forest were cleared, at a rate of 0.2 per cent per year. Although the overall rate of forest loss was low, there was one major exception to the general pattern. In 1960, the concession at Beroboka was taken over by the De Heaulme family, who originated in Réunion and emigrated to Madagascar in 1928. They cleared a total of 3,034 hectares to establish a sisal (*Agave sisalana*) plantation.[61] This was a significant deforestation event, contributing 46.6 per cent of the forest lost between 1954 and 1973 and 9.2 per cent of all the forest lost between 1954 and 2005.

Whilst the 1960s and early 1970s were periods of considerable government activity, the rest of the 1970s saw extreme economic stagnation and political turbulence. In 1975, Didier Ratsiraka was elected President and established a socialist party – l'Avant-Garde de la Revolution pour Madagascar (AREMA). The government's agricultural policies focused on increased intervention in markets, and government bodies were established with monopoly powers over the collection, transportation, processing and distribution of agricultural commodities.[62] However, government policy focused on the major rice growing areas of the highlands and on massive investment in developing the industrial sector. This was a particularly dark period in the region's history, with massive unemployment and the failure, due to neglect, of the large-scale agricultural projects of the early post-colonial period.[63] The period between 1973 and 1978 saw the lowest rates of forest loss of the post-colonial period, at 0.1 per cent per year, with the majority of households relying on subsistence agriculture, unable to find a market for cash crops.

Ivan R. Scales

MAP 3. The pattern of forest loss between 1954 and 2005. Light grey areas show forest cover at the end of the time period. Dark grey areas show forest loss during the time period.

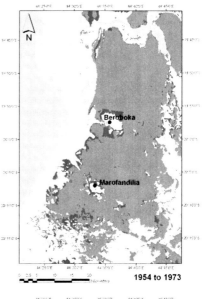

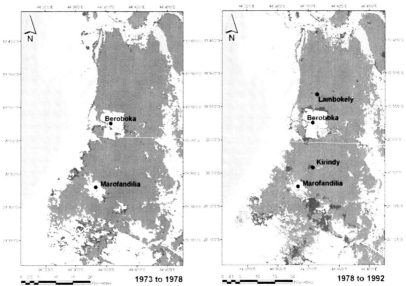

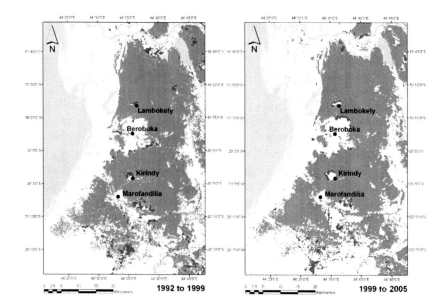

1992 to 1999   1999 to 2005

The period between 1978 and 1992 saw an increase in forest loss, with a mean rate of 0.5 per cent per year. This was the result of an increase in the cultivation of maize as a cash crop, triggered by an increase in prices, which quadrupled between 1978 and 1983:[64]

> In the 1960s, the forest was very close and you could hunt tenrecs close to the village ... there was still forest between Marofandilia and Bevoay and women could not walk alone because of the forest spirits. There were more Sakalava back then. But other groups started to arrive in the late 1970s, setting up new villages and clearing land to grow maize.[65]

Map 3 shows that whilst Beroboka and Marofandilia were again key hotspots of deforestation, the boom in maize cultivation in the 1980s saw the establishment of new settlements, most notably Kirindy and Lambokely. Families from Beroboka, who had migrated to the region from the south of Madagascar in the 1940s and 1950s, were given permission to clear land to the south and north by the Circonscription de l'Environnement, des Eaux et Forêts (CIREEF) in Morondava. The families eventually started new villages next to these new clearings, and were soon joined by migrant households from the south of Madagascar. The villages have grown steadily in size during the last twenty years.

The period between 1999 and 2005 has seen the highest rates of forest loss since 1954. This acceleration occurred despite the fact that forest clearance has been illegal since 1987. Interestingly, an analysis of socio-economic data collected using household surveys shows that there was no significant difference between rich and poor households in the proportion involved in forest clearance agriculture.[66]

Whilst the underlying driver of the increase in forest loss since the early 1980s has been the demand for maize as an export cash crop, there have been important facilitators.[67] During the 1970s trade in crops was controlled by the state through the Société Malgache de Collecte et Distribution (SOMACODIS), which purchased agricultural products. This changed dramatically in 1980 when Madagascar faced economic collapse, having borrowed heavily to fund a range of large infrastructural and large-scale agricultural projects. Madagascar was forced to borrow from the International Monetary Fund and accept conditions of a Structural Adjustment Programme, including the removal of trade barriers and the devaluation of the Malagasy franc.[68] Karana traders based in Morondava filled the trade gap and benefited from the new possibilities for free export trade. Much like during the colonial period, they have played a key role by connecting rural households to international markets, and providing loans of money and seeds to rural households (including migrants).[69] Traders have representatives in many of the forest villages of Central Menabe, including the five villages I surveyed, who buy crops from farmers and store them until a lorry comes from Morondava to collect the produce.

5. DISCUSSION: THE DRIVERS OF LAND USE AND LANDSCAPE CHANGE IN CENTRAL MENABE

Looking at over a century of landscape change in Central Menabe has highlighted a complex set of factors that have influenced land use. Whilst the specific drivers have changed over time, there are four key themes I wish to focus on:

i) Hatsake – *making the soil productive*

The most important crop in terms of forest loss has been maize. It has played a key part in the agricultural and landscape history of Madagascar because of the fact that, using *hatsake*, it can be grown away from the fertile and irrigated soils of the region's river valleys. This has made it especially useful to migrant households, who have been able to gain access to forested land and produce a crop using minimal labour and no additional inputs. Looking at deforestation over the entire period between 1896 and 2005, the importance of cash crops as well as wealthy landowning and trading elites in the process of forest loss challenges the dominant narrative, which attributes deforestation to poverty driven subsistence agriculture. Whilst *hatsake* is clearly an attractive option to poor households, it has also enabled some groups to accumulate wealth. A system of sharecropping has allowed elites – which over the last century has included expatriate concession owners and traders, as well as established rural households who control access to forested land – to profit from the exportation of maize as a cash crop.

ii) Connecting rural households to global markets

The cultivation of cash crops does not occur in a socio-economic vacuum, and there have been key factors that have allowed changes in demand for agricultural commodities to affect land use and land cover in Central Menabe and have created specific patterns of forest loss. Infrastructure in particular has played a key part. The establishment of a port in Morondava during the colonial period helped to connect Central Menabe to global commodity markets and helped to stimulate booms in butter bean and maize cultivation, whilst the construction of a railway in the 1920s helped to stimulate the first maize boom. As well as infrastructure, the establishment of trading houses in Morondava has played a significant part in connecting rural households to global markets. *Karana* businesses have controlled imports and exports since the late nineteenth century, acting as middle-men between national and international buyers and rural farmers, and have helped to transform Morondava into an important centre for trade.

iii) Migration

As well as infrastructural factors, the transformation of Central Menabe from a pastoralist to an agricultural landscape has depended on the influx of migrants. This migration has been driven primarily by economic opportunities, including waged labour in the region's logging concessions and plantations and the possibility of clearing forest to establish household farms in the region's villages. The *Karana* trading houses have again played a key part in the establishment of new farms by immigrants. In the absence of financial institutions, they have offered credit and seeds and have thereby facilitated agricultural expansion, especially by migrant households who do not have access to the family and clan support mechanisms available to established households.

iv) Government policy

The factors discussed so far have themselves often been underpinned by decisions made by the government, whether colonial or post-colonial. It was the colonial government that offered large concessions to foreigners, monetised the economy, imposed taxes, introduced a system of compulsory labour and improved certain parts of the regional infrastructure to encourage the expansion of exportable cash crops and increase international trade. The introduction of private land tenure and subsequent distribution of alluvial soils and newly irrigated land, tended to favour European and Karana entrepreneurs, as well as Malagasy elites with ties to the colonial government. It was also the colonial government that built the railway line making the cheap transportation of maize from more remote forest areas possible and subsequently removed it, leading to the collapse of the maize trade.

The post-colonial period has been turbulent, and this is reflected in Central Menabe's landscape. Independence saw a renewed vigour in agricultural policy, with numerous large-scale projects. However, the orange groves and cotton fields of the 1960s and 1970s have been abandoned and are now just scrubland. The economic doldrums of the 1970s, when Central Menabe was largely ignored by the state, saw a return to subsistence agriculture, whilst the economic reforms of the 1980s and 1990s saw the re-linking of the region to the global economy and a return to cash cropping, with a corresponding increase in the rate of forest loss.

6. CONCLUSIONS

I began this paper by looking at the dominant narrative of forest loss in Madagascar, which attributes deforestation to poverty driven subsistence agriculture and is based on temporally and regionally biased research. By drawing on a wide range of methods and sources, I have shown that Central Menabe has a long history of forest loss. To put rates of forest loss in Central Menabe in context, the mean figure of 0.4 per cent per year between 1954 and 2005 is relatively low when compared other parts of Madagascar, for example 0.7 per cent per year in the Masoala peninsula (in the north-east of the country) between 1957 and 1991 and 1.5 per cent per year between 1950 and 1985 for the whole of the eastern rainforests.[70] The most recent rates of annual forest loss are however higher than the national mean of 0.3 per cent, the African mean of 0.6 per cent, and the global mean 0.2 per cent for the period of 2000 to 2005.[71]

Looking at the drivers of land use and land cover change, I have shown that the Central Menabe landscape has, for over a century, been shaped by social, political and economic factors operating at a range of spatial levels. Whilst rural households have played a significant role in deforestation, it is important to remember the contribution of large concessions and plantations, not only to deforestation (the impact of the sisal plantation being particularly noteworthy) but also to the demography and political economy of the region. The ripples of these radical changes can still be seen and felt today.

Even when rural households have been the primary agents of landscape change, for example over the last twenty years, it is vital to understand the underlying drivers of their land use decisions. My research has provided an insight into some of the drivers, showing that households have responded to increases in crop prices, as well as various government policies. Furthermore, it has shown that wealthy as well as poor households have cleared forest and that some rural households have been able to benefit from the influx of migrants through sharecropping. It is important to note that government economic policies are not the only political dimensions of the land use decisions of rural households. Research elsewhere in Madagascar has shown that local institutions have also often played important roles in shaping access to and use of natural resources.[72] The evolution of Madagascar's forest laws, and their impact on resource use, have been well documented.[73] More research is now required in Central Menabe to better understand how both formal and informal rules and norms influence present day land use. This is made all the more urgent by the recent and dramatic increase in forest loss revealed in this paper.

Finally, this paper has highlighted the importance of distinguishing between proximate land uses (in this case forest clearance for agriculture) and their underlying social, political and economic drivers. It is important to understand not only what different land users are doing, but the factors that enable and encourage them to make particular land use decisions. Rather than assuming *a priori* the importance of rural poverty as a driver of land cover change, it is time that research and policy in Madagascar shifted away from the dominant narrative to more nuanced and historically informed understandings of land use and landscape change.

ACKNOWLEDGEMENTS

I would like to acknowledge the financial support of the Economic and Social Research Council and Natural Environment Research Councils of the United Kingdom (Research Award PTA-036-2004-00023), the Philip Lake Fund (Department of Geography, University of Cambridge) and King's College (University of Cambridge). I would like to thank Bill Adams, Bhaskar Vira and Tim Bayliss-Smith for their guidance whilst carrying out this research. I would also like to thank the two anonymous reviewers for improving this paper. For permission to carry out research in Madagascar, I would like to thank the Malagasy Ministry for Foreign Affairs, the Ministry for Higher Education and the University of Antananarivo. Finally, thanks to Dr Bruno Ramamonjisoa and Dr Gabrielle Rajoelison at the Ecole Supérieure des Sciences Agronomiques, University of Antananarivo.

NOTES

1.   Kull 1996.
2.   Ganzhorn *et al.* 2001; Goodman and Benstead 2005.
3.   Kull 2004; Klein 2002.
4.   Sussman *et al.* 1994: 334.
5.   World Bank 1996, p. 10.
6.   For example Hannah *et al.* 1998; Smith 1997; Whitehurst 2009.
7.   Carroll, R. 'Starving farmers destroy rainforest to buy food'. *The Observer*, London, June 29 2003.

8. Notable exceptions are Jarosz 1993; Kaufmann 2001; Middleton 1999; Kull 2004
9. Bollen and Donati 2006.
10. Styger *et al.* 2007; William-Hume 2006.
11. For example Blanc-Pamard *et al.* 2005; Casse *et al.* 2004; Minten and Meral 2006; Réau 2002.
12. Rainfall data from Morondava, Service Météorologique, Antananarivo.
13. FAO 2000a.
14. UPDR 2001.
15. Of the 228 households I surveyed, 71% grew only rain-fed crops.
16. Décret numéro 87-143, 20 April 1987.
17. Based on a survey of 228 households, of which 193 had cultivated crops in the twelve months prior to the survey
18. For example Smith 1997; Tidd *et al.* 2001; Whitehurst *et al.* 2009.
19. FAO 2000b.
20. Downton 1995; Geist and Lambin 2002.
21. FAO 2000b.
22. The FTM maps do not specify how forest cover was defined. However, by looking at the original aerial photographs on which the FTM maps are based, it is clear that the category corresponds to dense dry-deciduous forest. The photographs are sufficiently detailed to be able to distinguish dense forest from other forms of land cover. A small number of other studies (e.g. Elmqvist *et al.* 2007) have used FTM maps to measure forest cover in the 1950s. To verify the accuracy of the FTM maps, I used a sample of 10 randomly selected copies of the original aerial photographs obtained from FTM.
23. I used an Umax Mirage IIse flatbed scanner and MagicScan (Umax Systems GmbH) scanning software.
24. The details of the satellite images are: 1973 – Multi-Spectral Scanner (MSS) image (path 172, row 74) acquired on 15 June. 1978 – MSS image (path 172, row 74) acquired on 23 September. 1992 – Thematic Mapper (TM) image (path 160, row 74) acquired on 21 May. 1999 – Enhanced Thematic Mapper (ETM) image (path 160, row 74) acquired on 6 September. 2005 – TM image (path 160, row 74) acquired on 22 March. All images were orthorectified to a Universal Transverse Mercator coordinate system / World Geodetic System 84 datum.
25. For the classification of the 1973 and 1978 Landsat MSS images, these were bands 1 (green), 2 (red) and 3 (near infrared). For Landsat TM and ETM these were bands 1 (blue), 2 (green), 5 (near-infrared) and 6 (thermal).
26. For the purposes of accuracy assessment, it is important that at least 50 ground reference samples are collected per land cover class and that the data are independent of any training data (Congalton & Green 1999).
27. McCracken *et al.* 1988; Chambers 1992. The main RRA techniques I relied on were based on focus group activities. I selected group participants to represent a cross-section of each village in terms of age, gender, ethnicity and migration history. We discussed the history of each village and key issues such as agriculture and economic activities of households.
28. Feeley-Harnik 1978.
29. Campbell 2005.
30. Chazan-Gillig 1991; Fauroux 1980; Goedefroit 1998.
31. Colonie de Madagascar, Cercle de Morondava. Rapport Economique annuel résumant la période de 1898 à 1904. Centre des Archives D'Outre-Mer, Aix-en-Provence, Aix 2 D 172.
32. Gouvernement Général de Madagascar, Direction des Affaires Indigènes, Province de Morondava, Rapport Politique et Administratif 1919. Centre des Archives D'Outre-Mer, Aix-en-Provence, Aix 2 D 174.

33. Province de Morondava, Rapport Général de 1916. Centre des Archives D'Outre-Mer, Aix-en-Provence, Aix 2 D 173.

34. Gouvernement de Madagascar et Dépendances, Cercle de Morondava. Rapport politique, administrative et Economique, Année 1909. Centre des Archives D'Outre-Mer, Aix-en-Provence, Aix 2 D 172 B.

35. Corps D'Occupation de Madagascar, Territoire Sakalave, Cercle de Morondava. Rapport Politique, 2eme trimestre 1903. Centre des Archives D'Outre-Mer, Aix-en-Provence, Aix 2 D 171 B.

36. Province de Morondava, Rapport Géneral de 1915. Archives D'Outre-Mer, Aix-en-Provence. Centre des Archives D'Outre-Mer, Aix-en-Provence, Aix 2 D 173.

37. Région de Morondava, Circonscription Agricole, Rapport 1943. Centre des Archives D'Outre-Mer, Aix-en-Provence, Aix 2 D 177.

38. Gouvernement de Madagascar et Dépendances, Cercle de Morondava, Rapport annuel Politique et Administratif 1907. Centre des Archives D'Outre-Mer, Aix-en-Provence, Aix 2 D 171 B.

39. Madagascar et Dépendances, Cercle de Morondava. Rapport économique 1905. Centre des Archives D'Outre-Mer, Aix-en-Provence, Aix 2 D 172.

40. Rapport sur la situation de la province de Morondava au 31 décembre 1914. Centre des Archives D'Outre-Mer, Aix-en-Provence, Aix 2 D 175.

41. Le Bourdiec 1980.

42. Rapport Economique Annuel, 1896 to 1929. Centre des Archives D'Outre-Mer, Aix-en-Provence, Aix 2 D172 and 2 D175.

43. Rapport Economique Annuel 1913 to 1929. Centre des Archives D'Outre-Mer, Aix-en-Provence, 2 D175.

44. Province de Morondava, Rapport Economique 1921. Centre des Archives D'Outre-Mer, Aix-en-Provence, Aix 2 D 175.

45. Region de Morondava, Rapport Economique 1935. Centre des Archives D'Outre-Mer, Aix-en-Provence, Aix 2 D 176.

46. Province de Morondava, Rapport Economique 1929. Centre des Archives D'Outre-Mer, Aix-en-Provence, Aix 2 D 175.

47. Gouvernement Général de Madagascar, Direction des Affaires Indigènes, Province de Morondava, Rapport Politique et Administratif 1926. Centre des Archives D'Outre-Mer, Aix-en-Provence, Aix 2 D 174.

48. Patnaik 2003.

49. Fauroux 1980.

50. Inspection des Affaires Administratifs, Rapport No 6-cf au sujet des tavy dans la region de Morondava, 1938. Centre des Archives D'Outre-Mer, Aix-en-Provence, Aix 3 D 266.

51. Henry-Chartier 1994.

52. Région de Morondava, Rapport Annuel, 1941. Centre des Archives D'Outre-Mer, Aix-en-Provence, Aix 2 D176.

53. Région de Morondava, Rapport Annuel 1942. Centre des Archives D'Outre-Mer, Aix-en-Provence, Aix 2 D 176.

54. Région de Morondava, Rapport Annuel, 1941. Centre des Archives D'Outre-Mer, Aix-en-Provence, Aix 2 D176.

55. Inspection des Affaires Administratifs, Rapport No 6-cf au sujet des tavy dans la region de Morondava, 1938. Centre des Archives D'Outre-Mer, Aix-en-Provence, Aix 3 D 266.

56. Ramboarison 1998

57. Rapport Annuel de la Circonscription Forestière du Sud-Ouest à Morondava 1942. Centre des Archives D'Outre-Mer, Aix-en-Provence, Aix 2 D177.

58. Province de Morondava, Rapport Economique 1918. Centre des Archives D'Outre-Mer, Aix-en-Provence, Aix 2 D175. Province de Morondava, Rapport Economique 1920. Centre des Archives D'Outre-Mer, Aix-en-Provence, Aix 2 D175.
59. Interviews with Prosper Fanany (Marofandilia village, 4 June 2006); Relimby (Tandila village, 27 June 2006); Frause (Marofandilia village, 28 June 2006); Monja(Ankoraobato village, 10 July 2006); Bernard Noely (Ankoraobato village, 12 July 2006).
60. Henry-Chartier 1994.
61. Sisal is a succulent plant and member of the Agavaceae family. Fibres from the plant are turned into rope. The sisal grown in Central Menabe during the 1960s and 1970s was exported to France.
62. Barrett 1994.
63. Fauroux 1995.
64. Interview with Prosper Fanany, Marofandilia village, 30 June 2006.
65. Interview with Frause, Marofandilia, 28 June 2006.
66. Chi-squared = 7.59 (p = 0.108, n = 193, df = 4). The result means there is no significant difference between wealth categories at the 95 per cent confidence interval in the proportion of households cultivating maize. This was based on a classification of households into five categories (from very rich to very poor) using a participatory wealth ranking exercise (Grandin 1988).
67. For an in-depth account of maize export in Madagascar since the 1980s, and some of the factors that have contributed to an increase in exports, see Minten and Meral 2006.
68. This led to a new focus on liberalisation and agricultural markets, as well as the devaluation of Malagasy franc and an increase in foreign trade (Shuttleworth 1989).
69. Interviews with Jean Noely (Ankoraobato village, 10 July 2006); Prosper Fanany (Marofandilia village, 30 June 2006)
70. Kremen *et al.* 1999; Green and Sussman 1990.
71. FAO 2007.
72. Elmqvist *et al.* 2007; McConnell and Sweeney 2005.
73. For example Bertrand 1999; Bertrand and Sourdat 1998; Bertrand *et al.* 2007; Kull 2004.

BIBLIOGRAPHY

Barrett, C. B. 1994. 'Understanding uneven agricultural liberalisation in Madagascar'. *The Journal of Modern African Studies* **32**: 449–476.

Bertrand, A. 1999. 'La gestion contractuelle, pluraliste et subsidaire des ressources renouvelables à Madagascar (1994–1998)'. *African Studies Quarterly* **3**.

Bertrand, A. and Sourdat, M. 1998. *Feux et déforestation à Madagascar, revues bibliographiques.* Antananarivo: CIRAD/ORSTOM/CITE.

Bertrand, A., N. Horning, A.S. Rakotovao, R. Ratsimbarison and V. Andriatahiana 2007. 'Les nouvelles idées de gestion locale des ressources renouvelables et le processus de promulgation de la loi 96-025: histoire du cheminement d'une évolution majeure de la politique environnementale à Madagascar', in *Le transfert de gestion à Madagascar, dix ans d'efforts: Tanteza (tantanana mba hateza: gestion durable)*, eds. P. Montagne, Z. Razanamaharo and A. Cooke. Montpellier: CIRAD.

Blanc-Pamard, C., P. Milleville, M. Grouzis, F. Lasry and S. Razanaka 2005. 'Une alliance de disciplines sur une question environnementale: la déforestation en forêt des Mikea (Sud-Ouest de Madagascar)'. *Natures Sciences Sociétés* **13**: 7–20.

Bollen, A. and G. Donati 2006. 'Conservation status of the littoral forest of south-eastern Madagascar: a review'. *Oryx* **40**: 57–66.

Campbell, G. 2005. *An Economic History of Imperial Madagascar, 1750–1895: The rise and fall of an island empire*. Cambridge: Cambridge University Press.

Casse, T., A. Milhøj, S. Ranaivoson and J.R. Randriamanarivo. 2004. 'Causes of deforestation in southwestern Madagascar: What do we know?' *Forest Policy and Economics* **6**: 33–48.

Chambers, R. 1992. *Rural appraisal: Rapid, relaxed and participatory – IDS Discussion Paper 311*. Brighton: Institute of Development Studies.

Chazan-Gillig, S. 1991. *La Société Sakalave*. Paris: Karthala.

Congalton, R. G. and K. Green. 1999. *Assessing the Accuracy of Remotely Sensed Data*. New York: Lewis Publishers.

Downton, M. A. 1995. 'Measuring tropical deforestation: Development of the methods'. *Environmental Conservation* **22**: 229–240.

Elmqvist, T., M. Pyykönen, M. Tengö, F. Rakotondrasoa, E. Rabakonandrianina and C. Radimilahy. 2007. 'Patterns of loss and regeneration of tropical dry forest in Madagascar: The social institutional context'. *PloS ONE* **5**: e402.

FAO. 2000a. *Tropical Maize: Improvement and production*. Rome: Food and Agriculture Organization of the United Nations.

FAO. 2000b. *Forest Resources Assessment 2000*. Rome: Food and Agriculture Organization of the United Nations.

FAO. 2007. *State of the World's Forests 2007*. Rome: Food and Agriculture Organization of the United Nations.

Fauroux, E. 1980. 'Les rapports de production Sakalava et leur évolution sous influence coloniale (Région de Morondava)', in *Changements Sociaux dans l'Ouest Malgache*, eds. R. Waast, E. Fauroux, B. Schlemmer, F. Le Bourdiec, J. P. Raison and G. Ganday). Paris: ORSTOM.

Fauroux, E. 1994. 'Les échanges marchands dans les sociétés pastorales de l'ensemble méridional de Madagascar'. *Cahiers des Sciences Humaines* **30**: 197–210.

Fauroux, E. 1995. *Les Maladresses de L'Etat, Acteur de Dévéloppment dans une Région Isolée de Madagascar*. Paris: Centre National de la Recherche Scientifique.

Feeley-Harnik, G. 1978. 'Divine kingship and the meaning of history among the Sakalava of Madagagascar'. *Man* **13**: 402–417.

Ganzhorn, J. U., P. P. Lowry II, G. E. Shatz and S. Sommer. 2001. 'The biodiversity of Madagascar: one of the world's hottest hotspots on its way out'. *Oryx* **35**: 346–348.

Geist, H. J. and E. F. Lambin. 2002. 'Proximate causes and underlying driving forces of tropical deforesation'. *BioScience* **52**: 143–150.

Goedefroit, S. 1998. *A L'Ouest de Madagascar: Les Sakalava du Menabe*. Karthala, Paris.

Goodman, S. M. and J. P. Benstead. 2005. 'Updated estimates of biotic diversity and endemism for Madagascar'. *Oryx* **39**: 73–77.

Grandin, B. 1988. *Wealth Ranking in Smallholder Communities: A Field Manual*. London: Intermediate Technology Publications Ltd.

Green, G. M. and R.W. Sussman. 1990. 'Deforestation history of the eastern rain forest of Madagascar from satellite images'. *Science* **248**: 212–215.

Hannah, L., B. Rakotosamimanana, J. U. Ganzhorn, R. A. Mittermeier, S. Olivieri, L. Iyer, S. Rajaobelina, J. Hough, F. Andriamialisoa, I. Bowles and G. Tilkin. 1998. 'Participatory planning, scientific priorities, and landscape conservation in Madagascar'. *Environmental Conservation* **25**: 30–36.

Henry-Chartier, C. 1994. *L'Histoire mouvementée d'un périmetre irrigué: Le réseau hydroagricole de Dabara*. Unpublished.

Jarosz, L. 1993. 'Defining and explaining tropical deforestation: Shifting cultivation and population growth in colonial Madagascar (1896–1940)'. *Economic Geography* **69**: 366-79.

Kaufmann, J. C. 2001. 'La question des raketa: Colonial struggles with prickly pear cactus in southern Madagascar, 1900–1923'. *Ethnohistory* **48**: 87–121.

Klein, J. 2002. 'Deforestation in the Madagascar highland – Established "truth" and scientific uncertainty'. *GeoJournal* **56**: 191–199.

Kremen, C., V. Razafimahatratra, R.P. Guillery, J. Rakotomalala, A. Weiss. and J.S. Ratsisompatrarivo. 1999. 'Designing the Masoala National Park in Madagascar based on biological and socioeconomic data'. *Conservation Biology* **13**: 1055–1068.

Kull, C. A. 1996. 'The evolution of conservation efforts in Madagascar'. *International Environmental Affairs* **8**: 50–86

Kull, C.A. 2000. 'Deforestation, erosion, and fire: Degradation myths in the environmental history of Madagascar'. *Environment and History* **6**: 423–50.

Kull, C.A. 2004. *Isle of Fire: The political ecology of landscape burning in Madagascar.* Chicago: The University of Chicago Press.

Le Bourdiec, F. 1980. 'Le développement de la riziculture dans l'ouest Malgache', in *Changements Sociaux dans l'Ouest Malgache*, eds. R. Waast, E. Fauroux, B. Schlemmer, F. Le Bourdiec, J. P. Raison and G. Ganday. Paris: ORSTOM.

McConnell, W.J. and S.P. Sweeney. 2005. 'Challenges of forest governance in Madagascar'. *The Geographical Journal* **171**: 223–38.

McCracken, A., W. Pretty and G. R. Conway. 1988. *An Introduction to Rapid Rural Appraisal For Agricultural Development.* London: International Institute For Environment And Development.

Middleton, K. 1999. 'Who killed "Malagasy Cactus"? Science, environment and colonialism in Southern Madagascar (1924–1930)'. *Journal of Southern African Studies* **25**: 215–248.

Minten, B. and P. Méral. 2006. *International trade and environmental degradation: A case study on the loss of spiny forest in Madagascar.* Antananarivo: World Wild Fund For Nature.

Patnaik, U. 2003. 'Global capitalism, deflation and agrarian crisis in developing countries'. *Journal of Agrarian Change* **3**: 33–66.

Ramboarison, R. 1998. 'La deforestation en pays Sakalava (Ouest Malgache)'. Geographie. (PhD diss.,Université Louis Pasteur, Strasbourg).

Réau, B. 2002. 'Burning for zebu: The complexity of deforestation issues in western Madagascar'. *Norwegian Journal of Geography* **56**: 219–229.

Shuttleworth, G. 1989. 'Policies in Transition: Lessons from Madagascar'. *World Development* **17**: 397–408.

Smith, A. 1997. 'Deforestation, fragmentation, and reserve design in western Madagascar', in *Tropical Forest Remnants: Ecology, Management, and Conservation of Fragmented Communities*, eds. W. Laurance and R. Bierregaard. Chicago: University of Chicago Press.

Styger, E., H. M. Rakotondramasy, M. J. Pfeffer, E. C. M. Fernandes and D. M. Bates. 2007. 'Influence of slash-and-burn farming practices on fallow succession and land degradation in the rainforest region of Madagascar'. *Agriculture, Ecosystems and Environment* **119**: 257–269.

Sussman, R. W., G. M. Green and L. K. Sussman. 1994. 'Satellite imagery, human ecology, anthropology, and deforestation in Madagascar'. *Human Ecology* **22**: 333–354.

Tidd, S. T., J. E. Pinder and G. W. Ferguson. 2001. 'Deforestation and habitat loss for the Malagasy Flat-Tailed tortoise from 1963 through 1993'. *Chelonian Conservation and Biology* **4**: 59–65.

UPDR. 2001. *Monographie de la région du Menabe.* Morondava: Unité de Politique pour de Dévéloppement Rural.

Whitehurst, A. S., J. O. Sexton and L. Dollar. 2009. 'Land cover change in western Madagascar's dry deciduous forests: a comparison of forest changes in and around Kirindy Mite National Park'. *Oryx* **43**: 275–283.

William-Hume, D. 2006. 'Swidden agriculture and conservation in eastern Madagascar: Stakeholder perspectives and cultural belief systems'. *Conservation and Society* 4: 287–303.

World Bank. 1996. *Madagascar Poverty Assessment*. Washington: World Bank.

Soil Erosion, Scientists and the Development of Conservation Tillage Techniques in the Queensland Sugar Industry, 1935–1995

Peter Griggs

INTRODUCTION

Soil erosion has long been associated with the pursuit of cane growing. By the 1660s, twenty years after the commencement of sugar cane cultivation on Barbados, soil erosion was contributing to declining sugar cane yields. By 1710, much of the island had temporarily been abandoned for cane cultivation because of soil loss. Loss of soil in hilly cane growing areas throughout Jamaica and St Kitts is mentioned in contemporary accounts during the eighteenth century. Nearby on Cuba, declining yields during the mid-nineteenth century were caused by the 'disappearance into the sea' of thousands of tons of soil through erosion. In South Africa's sugar growing regions, the 'menace of soil erosion' was reported in the late 1940s as becoming 'a matter of importance to all farmers'.[1]

In Queensland's sugar cane growing lands, soil erosion is not mentioned in the official documents until the 1930s, almost seventy years after the establishment of the industry in the 1860s. Its presence, however, probably occurred long before the State's agricultural officers became concerned enough to mention the problem. This paper sets out to provide the context surrounding the recognition of this problem in the 1930s and to explain the location and extent of this soil loss throughout Queensland's sugar cane growing districts after 1945. Attempts to slow this soil loss are also considered, although the analysis will highlight that the techniques recommended to reduce soil erosion (i.e. contour tillage, terraces and grassed waterways) were only slowly adopted, with many canegrowers continuing to suffer soil loss either knowingly or unwittingly.

The relationship between soil erosion, conservation efforts and small European canegrowers is particularly interesting for three main reasons. First, the topic has not been considered previously, with the historical accounts of the Queensland sugar industry failing to mention soil erosion, probably because most focus upon the period prior to 1930 and deal mainly with the issue of the industry's labour supply.[2] Historical studies dealing with more recent developments in the Queensland sugar industry have concentrated on documenting the introduction of mechanical harvesting, not the land management practices adopted by Queensland's canegrowers.[3] Therefore, this paper aims to redress this omission in the historiography of the Australian sugar industry. Second, the paper will highlight that many Queensland canegrowers between 1930 and 1980 employed agricultural practices that were environmentally unsound (e.g. leaving fields fallow during the summer rainy season; planting in furrows aligned downslope; cultivating land with slopes of greater than ten per cent). Third, the analysis will illustrate how the transfer of models from other crops and physical environments is not always successful or possible. The agricultural scientists and extension officers were restricted in the types of soil conservation techniques that they could recommend to Queensland canegrowers. Sugar cane in Queensland is cultivated as an intensive monoculture in high rainfall localities and the farmers lacked alternative suitable crops that were as profitable as sugar cane. Soil conservation methods appropriate for lighter rainfall areas which were cropped with mixtures of cereals, legumes and grasses could not be easily adopted in Queensland's sugar producing lands.

I begin this account by briefly providing an overview of the Queensland sugar industry. The second section contains information on how farming practices contributed to soil erosion in Queensland's sugar cane producing lands. Details about the extent of soil loss throughout Queensland's sugar cane producing lands are presented in the third section of this article, although the discussion will show that the figures on the magnitude of the problem need to be interpreted with caution. In the fourth section, the conservation tillage techniques developed by the agricultural scientists are outlined. Information about the adoption of these techniques by Queensland canegrowers is provided in the final section, although it will be highlighted that the State's canegrowers were slow to adopt the practices that saved their soil until the 1980s.

THE QUEENSLAND SUGAR INDUSTRY

The Queensland sugar industry was established in the 1860s, almost two hundred years after other colonial sugar industries were founded in the Caribbean, Louisiana and Brazil. Initially, the Queensland sugar industry mirrored the established production model, being based on plantations, although the field workers were indentured Melanesians or Asians, not slaves. The practice of recruiting Melanesians for the Queensland sugar industry became known as 'blackbirding' and was increasingly opposed by sections of the Australian community after 1880. This opposition to the employment of non-European workers in the Queensland sugar industry and the implementation of the White Australia Policy after 1900 led to a transformation in its production structure during the 1890s and 1900s. Large numbers of small, European-owned family farms supplying sugar cane to cooperative or proprietary central sugar mills took the place of plantations.[4] This arrangement still existed in the early 1990s, with the basis of the Queensland

Peter Griggs

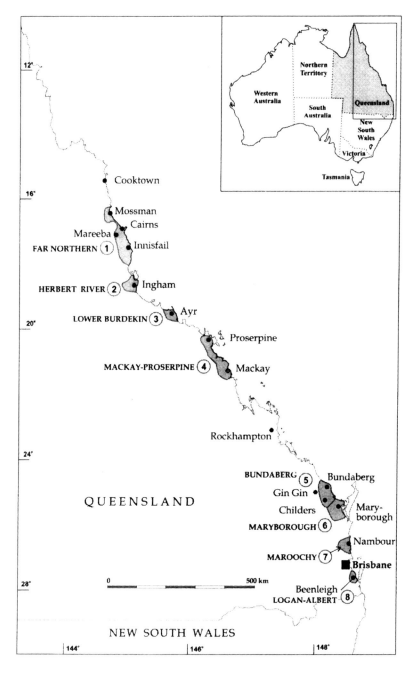

FIGURE 1. Sugar cane growing regions of Queensland in 1995.

sugar industry being approximately 6,300 farms, most of which were cultivated with between 30 and 90 hectares of cane, and 25 mills.

Growth in the area under sugar cane cultivation was steady between 1860 and 1920, followed by a particularly large expansion in the 1920s. During this period, sugar cane cultivation spread across generally well-watered fertile alluvial coastal plains, containing some sloping terrain, and the eight main non-contiguous sugar cane growing districts were established (Figure 1). Expansion in the area cropped with sugar cane slowed during the 1930s, due to fears about overproduction and some attempts at regulating growth (see below). Growth resumed after World War Two as new farms in existing sugar cane growing localities were settled by eligible ex-servicemen. A further expansion was sanctioned by the Queensland Government in the early 1950s, leading to the area under sugar cane increasing substantially during the 1960s (see Figure 2), although its cultivation did not spread to any new districts in Queensland. Between 1970 and 1995, the Queensland sugar industry continued to expand.[5] Essentially, however, this expansion was a 'filling in' process, making use of previously uncultivated land in most mill areas. Unfortunately, the most suitable land for cane growing was already in use, so the land occupied by this expansion, particularly in northern cane growing districts, was either sloping or poorly drained.[6]

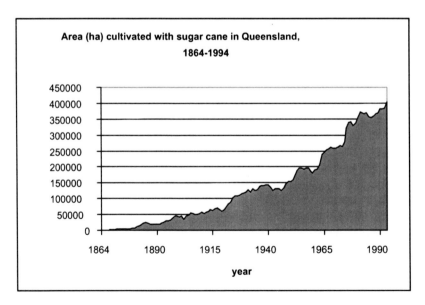

FIGURE 2. The area cultivated with sugar cane in Queensland, 1864–1994.

Source: Assembled from figures published in the *Statistics of the Colony of Queensland* and the *Australian Yearbook*

Peter Griggs

Another distinguishing feature of the Queensland sugar industry, until quite recently, was the high degree of regulation over all aspects of production. During the early 1910s, tensions developed between canegrowers and millers over the prices paid for cane and the inclination of some canegrowers to send their cane to different mills each year. Legislation was introduced in 1915 (and subsequently amended several times over the next fifty years) to create the Central Cane Prices Board. This organisation controlled the areas of land on which individual farmers grew sugar cane or the location of land 'assigned' to cane cultivation. By 1930, Queensland canegrowers were granted an entitlement which allowed them to deliver to a mill for payment, cane grown on a number of hectares of a specific amount of land assigned to an individual mill (i.e. an assignment). After 1929, millers operated under what became known as the Peak Year Scheme, and were allocated a set tonnage of raw sugar they could produce each year. The price canegrowers received for their cane and the price millers received for their sugar was determined by legislation. The domestic market was protected by a ban on imports of sugar and all sugar produced by the mills was compulsorily acquired by the Queensland Sugar Board, which handled the marketing of sugar domestically and overseas. Removal of many of these regulatory controls has occurred during the 1990s and early 2000s.[7]

CAUSES OF SOIL EROSION IN QUEENSLAND'S CANE GROWING LANDS

Soil erosion is the loosening and removal of soil from its previous resting place, through the agency of wind and water. Wind erosion is usually experienced in drier regions. The topsoil becomes loose and powdery and the wind carries it away. Sugar cane cultivation, however, is mostly concentrated in high-rainfall localities, so water erosion is more a problem. Rainfall causes two types of erosion: sheet and gully. The latter is the most easily discernable form of erosion and is commonly found on sloping land: generally the steeper the slope the more frequent the gullies. These gullies occur in two forms. The commoner is a complex of small gullies (sometimes known as rills or grooves) that can be crossed and reclaimed with the usual farm implements. The other type of gully has deep and steep sides, often one to three metres in depth (see Figure 3). Such gullies are difficult to reclaim with ordinary farm machinery, but are usually rare.[8] In contrast, sheet erosion is not easily detected. When heavy rain falls on unprotected soil, absorption takes place at a rate governed by the soil type, its structure and texture and existing level of soil moisture. As soil becomes saturated or when the rate of precipitation exceeds that of absorption, free water develops on the surface. When free surface water develops on even gently sloping land, it moves towards lower levels at a speed dependent upon the slope

Soil Erosion in the Queensland Sugar Industry

FIGURE 3. Serious gully erosion on a cane farm at Childers, 1950.
Source: Vallance 1950: 28.

gradient. Soil particles dislodged by the raindrops are carried away suspended in the moving water. The result is the loss of finer soil particles, often over the entire field, and its effects often pass unnoticed until gullying may commence.[9]

Soil erosion is caused when farmers cultivate inappropriate locations. The steepness or gradient of the land has a very direct influence on the degree of erosion present. Land with a slope greater than ten per cent is very difficult to cultivate. In the Isis district (near Childers), J.L. Tardent, a local forester, calculated in 1938 that 80 per cent of the area planted in the district was located on slopes between 6 and 20 per cent, and that damage from erosion reached an 'alarming total'.[10] In 1945, an investigation into soil erosion in Queensland cane growing regions found that slopes with angles of 25 degrees in the Mackay district were being cultivated and the resulting erosion was 'very far advanced'. The Queensland Bureau of Investigation which surveyed soil erosion in Queensland during 1946, concluded that the 'cultivation of slopes of from 15 to 20 per cent is not uncommon, while in some instances slopes exceeding 25 per cent have been used for the production of sugar cane'.[11]

Soil erosion can also be caused when farmers employ incorrect farming techniques. Agricultural practices used by Queensland canegrowers had improved between 1890 and 1920, with some canegrowers now resting parts of their farms under soil enriching leguminous crops, improving on-farm drainage and arresting soil deterioration by applying both natural and artificial manures.[12] However, contour tillage appears to have been unknown in Queensland sugar cane growing districts before 1945 (or present but not recorded in any official documents), and the long straight drills went up and down hills with no thought given to how excess water might form channels seeking lower levels. Land was often left fallow during summer months and fields with young crops suffered from heavy falls of rain; scouring and small rills can occur on land with even a slight slope. In 1939, Charles Young, the General Manager of the Fairymead Sugar Company Ltd., for example, told a Royal Commission on Sugar Peaks and other Cognate Matters, that the organisation he represented had been forced to throw out of cultivation 260 acres (105 hectares) 'largely on account of soil erosion whilst it was fallow'.[13] In addition, historically fields with ratoon sugar cane crops (i.e. cane regrowth from the stalks left in the ground after a crop has been harvested) in Queensland were often exposed to summer rains. The conservation of trash (i.e. cane tops and leaves or the residue left after harvesting) was not commonly employed in the Queensland sugar industry and even less so after 1930 due to the introduction of widespread pre-harvest burning of crops. This practice was adopted as a health precaution to reduce the incidence of Weil's disease amongst cane cutters, thereby avoiding strike action by cane cutters, and post-harvest burning of crop residues because of the concerns that this material harboured pests and disease.[14]

Queensland's regulated system of sugar cane production compounded the above poor practices. As mentioned above, Queensland's canegrowers after 1930 had an assignment or an entitlement that allowed them to deliver to a mill for payment cane grown on a number of hectares situated within the boundaries of a designated block of land assigned to a particular mill. Farmers were expected to produce a crop of cane every year, although until 1965 they were restricted to harvesting no more than 75 per cent of their assigned area, thereby ensuring annually that a quarter of the assignment was rested and cropped with soil-regenerating cover crops.[15] Moreover, canegrowers could not easily get permission to change assignments, even within existing farm boundaries. Such arrangements caused the continued use of eroded and eroding lands.[16] The Queensland government in the 1970s recognised that this undesirable feature of Queensland's regulated system of sugar production was contributing to soil erosion in at least the Isis district. Working in conjunction with all sectors of the sugar industry in the Isis district, the Queensland government assisted seventy canegrowers farming sloping land to move their assignments to more level farms created from vacant Crown land or areas surrendered from local State Forests.[17]

OFFICIAL CONCERN ABOUT SOIL EROSION THROUGHOUT QUEENSLAND'S CANE GROWING LANDS

Soil erosion was noted in other parts of Australia long before its presence was recorded in the sugar cane growing lands of Queensland. Wind erosion and drifting sand dunes emerged as a concern in the semi-arid, lighter soils districts of South Australia and Victoria during the late nineteenth century. By the late 1920s and early 1930s, residents of Sydney were increasingly complaining about the heavy dust storms hitting their city. Widespread use of bare fallow practices in the drier parts of Australia's wheat zone was leading to the creation of Australia's own dustbowl. In wetter areas, agricultural scientists reported that heavy summer rainstorms contributed to severe sheet and gully erosion throughout the wheat growing areas.[18] Thus, reports of soil erosion in Queensland's cane growing regions in the late 1930s were part of an Australian-wide concern for the loss of soil and form part of an intensified worldwide concern for soil erosion that arose following the creation of the 'Dust Bowl' in the southern Great Plains of the United States in the early 1930s.

The loss of Australia's soil became the topic of several books during the 1930s and 1940s. Amongst these publications was the now classic work *Soil Erosion in Australia and New Zealand* by James Macdonald Holmes, which included a map that categorised the seriousness of soil erosion in Australia. The sugar cane growing lands of Queensland were identified as suffering the greatest risk of sheet erosion and gullying, although Holmes did not specifically mention soil loss throughout these districts in his text.[19] Moreover, as soil loss grew widespread, governments in Australia began to take the problem more seriously. New South Wales led the way, forming a government committee in 1933 to monitor the problem and in 1938 a soil conservation service was created to survey the extent of the problem and begin ameliorative action. In South Australia, a Soil Conservation Committee had prepared maps on the extent of soil erosion in the State by 1937 and had commenced promoting better farming techniques. Victoria was slower to respond, despite the evidence of widespread wind erosion in the Mallee region during the early 1930s. A government committee to investigate the matter was established in 1936, but a State-wide survey of the problem did not commence until a Soil Conservation Board was formed in 1940.[20]

Loss of soil throughout Queensland's cropping lands, especially on the Darling Downs, was recognised as a problem in the early 1930s. In 1935, A.F. Skinner, a cadet in the Agriculture Branch of the Queensland Department of Agriculture and Stock, surveyed and constructed a contour bank over 25 hectares of land on the northern outskirts of Toowoomba. This first government soil conservation work done in Queensland encouraged several farmers on the Darling Downs to install contour banks on their properties. This interest in soil conservation was halted by World War Two, when all work was directed towards the production of food and fibre. Soil conservation work resumed in 1947 when Jasper Ladewig was appointed a Soil Conservation Officer in the Queensland Department of Agriculture and Stock, thus creating the nucleus of a soil conservation service in Queensland. Eventually legislation was introduced in 1951, requiring the Queensland Department of Agriculture and Stock to assess the extent of soil erosion throughout the State and to assist landholders with the reduction of the extent of soil loss. Under this legislation districts could be declared areas of soil erosion hazard.[21]

The first official report about soil erosion in Queensland's cane growing lands that could be found in the extant records was made in 1938 by J. L. Tardent, who estimated that 800 ha in the Isis district once grew sugar cane but 'are absolutely derelict land now or carry very poor crops'. He attributed this situation to the extensive soil loss occurring in the district. Interestingly, Edward Knox Junior, the General Manager of the Colonial Sugar Refining Company – Australia's largest sugar miller and refiner – had hinted at this problem much earlier. In 1899, he visited one of the firm's Queensland sugar mills at Isis, near Childers. As part of this visit, he toured the district, inspecting the farms of those European settlers who supplied the Isis Sugar Mill with sugar cane. In his report to the Company's Board of Directors, Knox expressed concern at what he had witnessed, writing, 'already some drawbacks are showing themselves. Many of the hills are very steep and in course of time trouble will be caused through the soil being washed into the gullies in heavy rains'.[22]

On the eve of World War Two, Dr Henry William (Bill) Kerr and Arthur Bell from the Queensland Bureau of Sugar Experiment Stations (hereafter BSES) wrote, 'many fields of valuable land are being rapidly deprived of their fertile surface soils'. They singled out the Isis and South Johnstone (near Innisfail) districts as the two Queensland sugar cane producing regions suffering the most soil erosion.[23] World War Two interrupted any official action to deal with the issue of soil loss. However, in March 1945, the Queensland Cane Growers' Council asked the Prime Minister of Australia to survey the extent of the problem in Queensland. The Commonwealth Government acceded to this request. During the latter months of 1945, officers from the Division of Soils, Council for Scientific and Industrial Research, and the BSES, toured half the sugar producing regions of Queensland, with the focus of their investigations being northern localities. The report provided a qualitative assessment of the extent of the problem. Serious field erosion was reported from all districts visited, except Ingham, with the deep red soils found in the Isis and South Johnstone districts being the worst affected.

The first quantitative estimate of the extent of the cultivated areas needing soil erosion measures in the sugar cane producing lands in Queensland are attached as an appendix to a letter from the Hon. Frank Nicklin, Premier of Queensland, to Hon. Robert Menzies, Prime Minister of Australia in 1959. The author of these figures is not stated and the archival document provides no clue as to why they were assembled. Clearly, these figures need to be interpreted cautiously, as there is no explanation as to what criteria were used to determine which cane growing lands needed soil conservation measures. No definition was provided as to what constituted a soil conservation measure (i.e. contour banks or reducing the amount of bare fallow land). With these caveats in mind, the figures suggested that overall, approximately 31 000 ha in sugar producing shires in Queensland or nearly a third of the cultivated land in these shires required some soil conservation measures. Local authorities in the Maryborough district suffered the greatest amount of soil erosion, with between 60 and 80 per cent of the cultivated area experiencing some form of soil loss. Between 20 and 30 per cent of the cultivated area in a group of mostly southern Queensland shires around Bundaberg and Nambour needed soil conservation measures. In these districts, sugar cane growing was often conducted on hilly land. Around 10 per cent of the cultivated areas in the remaining shires needed soil conservation measures. In the Mackay and Innisfail districts, soil erosion occurred because of the extensive cultivation of sugar cane on sloping land.[24]

As the above expansions progressed, the presence of soil erosion in Queensland's sugar cane producing lands continued to be mentioned in official publications produced by the Queensland Department of Primary Industries.[25] By 1975, the Isis and Gin Gin districts, two of southern Queensland's sugar cane growing districts, had been declared areas of soil erosion hazard. In addition, local Shire and River Trust Engineers began expressing concern that soil eroding from sugar cane growing lands was contributing to siltation in the lower reaches of the major rivers along the Queensland coast and exacerbating flooding hazards and negating drainage improvements undertaken in flood-prone areas.[26] This perceived worsening soil erosion problem in Queensland's sugar cane growing lands led to the Queensland Department of Primary Industries commencing investigations into quantifying how much soil was being lost from sloping paddocks cultivated with sugar cane. The first of these studies conducted in the Mackay district during the wet seasons of 1976/77 and 1977/78 concluded that losses approximately equivalent to 42 to 227 tonnes/ha per year were occurring on paddocks with slopes of between six and eight per cent. These amounts were extremely high, as maximum losses of 12.5 tonnes/ha per year were considered acceptable.[27] Further studies during the early 1980s investigated the magnitude of soil loss on cane paddocks under conventional cultivation in Far North Queensland. Average annual losses were calculated to be 150 tonnes/ha, with a range of annual measurement being 70–500 tonnes/ha.[28] With the loss of these amounts of soil, it was not surprising that local Shire and River Trust engineers were expressing concern about river siltation.

As the above studies were being undertaken, the Division of Land Utilisation of the Queensland Department of Primary Industries also completed a series of

Peter Griggs

studies on land suitability for sugar cane cultivation throughout Queensland.[29] As part of these investigations, details on the extent of eroded cane land throughout Queensland were gathered from soil conservation staff. Again these figures need to be interpreted cautiously, as no explanation was presented outlining what criteria were used to determine eroded land. With this caveat in mind, the figures suggested that in 1983 approximately 107 900 ha or a third of Queensland's assigned sugar cane producing lands were eroded or eroding. This overall figure is much the same as that calculated in 1958, with the most widespread erosion continuing to be concentrated in the Isis/Maryborough and Bundaberg districts. However, the 1983 figures suggest higher levels of erosion in the Northern and Mackay districts compared to the 1958 figures. Such an assessment would seem plausible, given that sugar cane cultivation had spread onto more marginal lands during the 1970s and early 1980s expansions.

Since the above figures were published, major changes have occurred in the on-farm land management practices employed by many Queensland cane-growers. Green cane harvesting, trash blanketing and minimum tillage practices have become widely used throughout some sugar cane producing districts (see below). These new practices have clearly reduced soil erosion, but concerns remain over the vulnerability of soil to erosion during the planting phase or in districts with a low level of trash blanketing (e.g. Burdekin).[30] Recently claims have been made that soil erosion remains a major problem throughout the sugar cane growing lands of Queensland, although none of these authors provide any evidence to support their claims other than references to the studies done at Mackay and Innisfail in the late 1970s and early 1980s.[31] Moreover, inquiries associated with the research for this paper have led to the conclusion that after 1983 the collection of figures on the extent of cane growing lands requiring soil conservation measures was sporadic and had ceased altogether by the early 1990s. Any figures collected during the 1980s probably remain with the government agencies and are unpublished. Thus, the safest conclusion that can be drawn from these inquires is that the extent of cane growing lands requiring soil conservation measures in Queensland's cane producing lands in the mid 1990s was unknown.

PREVENTING EROSION

Measures to reduce soil loss fall into three broad categories (Table 1). The first category incorporates mechanical field methods such as contour tillage, terracing and waterways, which are all used to either reduce the velocity of the flow of water across the soil surface or convey water at a non-erosive velocity to a suitable disposal point. Practices to manage the soil more effectively make up the second category. These measures maintain soil structure, allow infiltration

TABLE 1. Practices used for soil conservation

Category	Practice	Description
Mechanical means	Terracing	Large earth embankments constructed across slopes to intercept surface runoff and to convey it to an outlet at a non-erosive velocity.
	Contour bunds	Small earth banks, mostly hand-constructed, thrown across slopes to act as a barrier to surface runoff.
	Contour tillage	Completing ploughing, planting and cultivation on the contour, instead of furrows aligned downslope.
	Waterways	Structures designed to convey runoff at a non-erosive velocity to a suitable disposal point. They include grass waterways, diversion channels and terrace channels.
	Stabilisation structures	Small dams are built across gullies to trap sediment and slow surface runoff. They can be used in gully reclamation and gully erosion control.
Soil management	Improved drainage	Increasing the rate of subsurface water movement by installing mole drains or the breakup of compacted sub-surface layers by subsoiling. Such practices reduce the amount of surface runoff.
	Minimum tillage	Concentrating the number of operations (e.g. fertilising; weed control) as much as possible in one pass and/or by restricting them to the row where the plant grows.
Agronomic measures	Crop rotation	Alternating row crops with legumes and/or grasses. Erosion under row crops is counteracted by low rates under other crops.
	Cover crops	Growing cover crops (often legumes) during fallow periods or as ground protection under trees.
	Strip cropping	Combining row crops and protection effective crops in alternating strips aligned on the contour. Erosion is limited to the row crops and the soil removed from these is trapped in the next strip downslope.
	Mulching	Covering the soil with crop residues (e.g. straw; maize stalks; palm fronds; standing stubble) to protect if from raindrop impact and to reduce the velocity of runoff and wind.

Source: Morgan 1979: 57–66.

Peter Griggs

and reduce surface runoff. Conventional tillage techniques often pulverise the soil near the surface, creating a compacted layer at plough depth and thereby reducing infiltration and increasing runoff. To lessen these effects, tillage operations are restricted by reducing their number and carrying out as many operations (e.g. fertilising, weed control) as possible in one pass. The aim is to concentrate these activities on the rows where the plant grows and leaving the inter-rows untilled. In addition, a reduction in runoff can be achieved by increasing the rate of subsurface water movement by installing mole drains and the breakup of compacted sub-surface layers by subsoiling. The third category includes those agronomic measures using plant cover to reduce erosion. Generally row crops (under which erosion is higher) and cover crops are cultivated either simultaneously or consecutively in rotation. Bare fields during the fallow period are also avoided by the use of cover crops or mulching.

Some of the above techniques such as crop rotation, contour tillage and terracing have a long history, being employed by the ancient Peruvians or used by the farmers in Ancient Italy.[32] Subsequently they have been modified, refined and used in many environments around the world. In contrast, a form of terracing known as 'cane-hole agriculture', was only developed in the early eighteenth century and specifically designed to arrest soil erosion in areas cultivated with sugar cane in the Caribbean. Its development occurred in response to the emergence of soil deterioration and soil loss on the sugar cane growing lands throughout the Caribbean during the late seventeenth century. The former could be ameliorated by the commencement of general manuring using appropriate materials produced on individual estates by means of the farm dung, trash, megass (i.e. the final crushed sugar cane fibre remaining after milling) when not consumed in the mill furnaces and unused molasses. The adoption of manuring techniques, however, did not stop soil loss, especially in localities where cane was cultivated on sloping land. The sugar cane planters realised that planting cane in trenches or furrows aligned downslope contributed to soil erosion, and as such sought an alternative system of planting.[33]

The essential feature of cane-hole agriculture involved the slaves using hand hoes to systematically subdivide the fields into squares approximately 1.5 metres in size. Within each square, a 'hole' or depression measuring approximately 0.5 metre to 1 metre long and 12.5 cm to 15 cm deep was dug, with the slaves raising the soil along the tops and sides of the depression into ridges known respectively as saddles and banks. Once excavated, the holes remained unused until planted with cane, but the presence in the landscape of a system of two-directional ridges prevented or contained any downslope loss of soil. The cane was eventually planted in the depressions packed with manure. As the cane plants grew taller they provided protection for the soil, thereby easing the threat of soil erosion. Cane holes were also advantageous, preserving more soil moisture than trenches. Cane-hole agriculture, however, was very labour intensive, with such work being regarded as the 'most taxing of all on sugar estates'.[34]

Galloway claims that cane-hole agriculture never spread very widely throughout the West Indies, and it was abandoned increasingly after 1850 in favour of plough agriculture, except on Barbados where the practice still existed in the 1970s. Soil loss throughout the West Indies was minimised increasingly by mulching using trash or the erection of contour bunds.[35] In Australia during the late nineteenth century, a variation of cane-hole agriculture was mentioned as being used by a few sugar cane planters, who were observed planting cane into holes about eighteen inches square and eight inches deep. However, this practice was never widespread, despite a plentiful supply of indentured labour being available to allow the sculpturing of the fields into squares, saddles and banks. Instead, the majority of sugar cane planters and the small canegrowers that took their place mostly practised plough agriculture, with apparent little regard for the slope of the land or the orientation of the furrows. Moreover, as mentioned above, mulching using trash was not an established practice.[36]

Thus, when Dr H. W. (Bill) Kerr, Director of the BSES, began considering ways to reduce soil erosion in Queensland's cane growing lands in the mid-1930s, he was faced with several challenges. First, he could not easily recommend cane-hole agriculture, one of the traditional methods used to reduce soil loss in other sugar cane growing countries. There was very limited experience of its adoption in Australia. In addition, soil loss in Queensland's cane growing lands had to be reduced on small family farms where a minimum amount of labour was employed to complete fieldwork. The canegrowers would have been loathe to return to a time-consuming, labour-demanding arrangement involving the creation of squares with hoes, when they were now using gasoline powered tractors and ploughs. Second, he could not recommend three of the agronomic practices listed in Table 1. Queensland canegrowers, because of legislative requirements, could not alternate sugar cane with other row or tree crops every few years, nor easily engage in strip cropping. Moreover, even if no legislative restriction upon the cultivation of other crops existed, there was the difficulty of finding a profitable alternative crop to grow on these farms. Rubber, rice, cotton, tea, coffee and bananas had all been tried in Queensland's coastal districts during the late nineteenth century, but all had proved unprofitable, except in the Innisfail region where a small banana industry co-existed alongside sugar cane cultivation. Mulching using the trash to protect the soil was also not an option as it was increasingly being burnt because of associated health and industrial concerns. Moreover, it was highly unlikely Kerr would be able to convince the authorities to reverse this practice on environmental grounds, even though officers from the BSES recognised that destroying trash was short-sighted and that its use was the best means available for restoring soil humus and fertility and protecting the soil from erosion. Third, cane growing in Queensland was being conducted on slopes over 10 per cent. According to world authorities on soil conservation, such land should have been used only for grazing or forestry. However, sugar cane cultivation could not be easily stopped on this land, so the agricultural scientists were required to develop erosion control methods suited to such steep gradients.[37]

Kerr's initial advice to Queensland canegrowers on soil erosion occurred in 1936, when he suggested a combination of some techniques mentioned in Table 2. First, he urged canegrowers to practise better soil management on their properties. The absorptive capacity of the soil could be improved by sub-drainage. Providing channels through which the water could pass was achievable by deep ploughing, subsoiling or growing deep-rooted crops (e.g. lucerne) on the land when it was not cultivated with cane. Second, mechanical measures should be adopted. The rows of cane needed to follow the contour or run parallel to the slope, avoiding the long straight drills that went up and down hills. When this method proved not adequate or suitable, the farmers should establish contour banks or terraces and grassed waterways on their land. Fields could be crossed by broad, shallow waterways following the contour, which were flanked on the down-hill by mounds or ridges of earth. The contour banks diverted run-off water to larger outlet channels before it had time to obtain sufficient velocity to erode the soil from the cultivated areas between the terraces. Finally, during the fallowing of land, the field should be protected by a green manure crop or a trash blanket.[38]

Kerr's advice was not supported by the results from any field experiments under Queensland conditions and the outbreak of World War Two curtailed any further investigations into this matter. However in 1945, C. Stephens from the Division of Soils, Council for Scientific and Industrial Research, alerted the BSES to potential difficulties in using terraces. He suggested that the contour banks used in the American soil conservation programs would not cope with the huge volume of water that accompanied such high intensity rainfall in Queensland and they would be seriously damaged.[39] Thus, immediately following the conclusion of World War Two, Norman King and L.G. Vallance of the BSES and the cotton expert W.G. Wells met with Isis canegrowers in September 1945 and agreed to show farmers how they could reduce soil loss on their properties. During the wet season of 1946–47, L.G. Vallance conducted a soil erosion control experiment in the Isis district. The trial was placed on a slope having a maximum gradient of 16 per cent. Part of the paddock was terraced with contour banks, and all the run-off was diverted by waterways into one main outlet. A considerable amount of soil movement occurred and the diversion waterways became almost completely silted up, but the contour banks had stopped the formation of deep gullies that were noticeable in nearby fields that were unprotected by contour banks.[40]

As the above experiments progressed, P.A. Yeomans, an Australian mining engineer who owned a small property near Sydney, was developing his Keyline planting method. This approach aimed to increase both the depth and fertility of the soil by remoulding the landscape, firstly by a proper assessment of the natural resources on a particular property, and secondly by special methods of planning design based on water control and land management. At the heart of this method was the Keyline, a very specific contour line that occurs in all valley and ridge topography. This contour line delineates the transition areas above which all contour tillage must proceed up the slope and below which all contour tillage must proceed down the slope. Terraces and waterways are established to channel run off water into a series of farm dams, thereby drought-proofing properties.[41] To improve soil infiltration, the ground is contour ripped parallel to the terraces.

Armed with the details gained from the earliest experiments in the Isis district and the approach adopted by Yeomans, the BSES scientists designed further trials during the early 1950s in order to refine the advice they provided to canegrowers. Contour banks were built on additional cane growing paddocks in the Isis district, using the Keyline method, although elements of the approach such as contour ripping and on-farm dams were omitted. The main aim of these trials was to determine the spacing of the terraces on sloping land. Canegrowers had expressed dissatisfaction with placing contour banks every 100 feet apart, as they did not deal adequately with the large amount of run-off. By the late 1950s, the BSES scientists had concluded that contour banks placed at distances between 70 and 90 feet apart achieved a greater reduction in the velocity of the flowing water on cane paddocks. In addition, they had evidence that the contour banks had restricted water flow across fields and increased water penetration, as there had been improved growth in the cane rows adjacent to the banks.[42]

The knowledge gained from these layouts contributed to a greater understanding of the problem as it applied to sugar cane growing conditions, especially on very steep land. Equipped with this information, a succession of BSES officers over the next two decades promoted soil conservation measures throughout Queensland cane growing regions. To slow the flow of water across their properties, canegrowers were urged to adopt one or more of the following measures: contour tillage (later known as contour row direction); erecting contour banks; encircling their fields on sloping land with grassed waterways or ditches to divert water away from paddocks; and the building of stabilisation structures across gullies to prevent their extension. Canegrowers were also urged to minimise the duration of the bare fallow during the summer season of high rainfall by planting a summer cover crop.[43]

The above efforts were supported by other organisations. From the early 1950s onwards, the Colonial Sugar Refining Company Ltd. (later changed its name to CSR Ltd.), Australia's largest sugar miller and refiner, conducted its own trials in the Innisfail area to determine the effectiveness of contour banks in slowing soil loss. Results from these trials convinced CSR that contour banks were an effective way to reduce soil loss, so the firm urged its canegrowers who cultivated sloping land in the Innisfail district to build contour banks and cover fallow paddocks with green manure crops.[44] The Queensland Department of Agriculture and Stock (later the Department of Primary Industries) also turned its attention to soil loss in the sugar cane growing lands. Soil conservation officers were appointed at Bundaberg and Mackay in 1955 and 1967 respectively and a soil conservation unit was established in 1980 at the South Johnstone Research Station, near Innisfail, in order to tackle the loss of soil in northern cane growing regions. By 1983, eleven field officers and five technical support staff were employed to service the State's cane producing areas.[45] A map of part of a canegrower's property modified in 1976 to reduce soil erosion following advice from Mackay officers of the Queensland Soil Conservation Branch is shown in Figure 4.

During the late 1970s and early 1980s, Queensland canegrowers were being provided with increasingly more refined advice on ways to reduce soil erosion. The suitability of land for cane growing was being determined via land resource surveys. In some instances, such as the expansion of cane growing to the Julatten district (near Mareeba) in the early 1980s, official confirmation of assigned areas was made conditional on satisfactory soil conservation measures being established from the outset. In addition, canegrowers were urged to complete plough out-replant operations or the replanting of fallow land early to ensure an adequate crop cover by the rainy season. Contour banks, diversion waterways and contour row direction continued to be promoted. However, the old layouts with sharp corners and many short rows were being superceded by layouts that contained slighter curves, parallel banks and fewer short rows.[46] A comparison between the layout of a property in the Bundaberg region showing contour banks following the exact grade line, and many short rows (top map) and the parallel layout approach, with fewer short rows (bottom map) is shown in Figure 5. The parallel layout approach is also illustrated in Figure 6.

Despite the above changes, N. Dawson, R. Berndt and B. Venz, officers from the Queensland Department of Primary Industries, expressed concern in 1983 that the efforts by the State's canegrowers to reduce soil erosion were not quick enough. They claimed that the adoption of soil conservation measures did not keep up with the rate of development of erosion prone land in Queensland's sugar cane producing districts, let alone make inroads into the 100,000 ha of existing land cropped with cane that had untreated erosion problems. In the absence of any further expansion in cane growing, the canegrowers would take 40–50 years to treat all the eroding land. The Queensland sugar industry could not rely on additional Department of Primary Industry staff to hasten the implementation of these measures.[47] The solution: some major change to the current soil conservation programs in sugar cane producing lands was needed if these lands were to be stabilised within a reasonable time. That change was

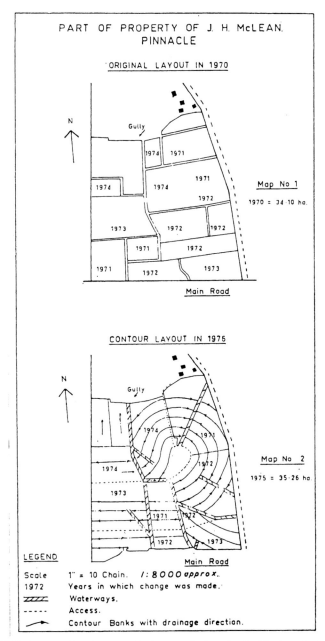

FIGURE 4. A comparison between the layout of a farm in the Mackay district before
and after contour banks had been established in 1976.
Source: Veurman 1977: 584.

Peter Griggs

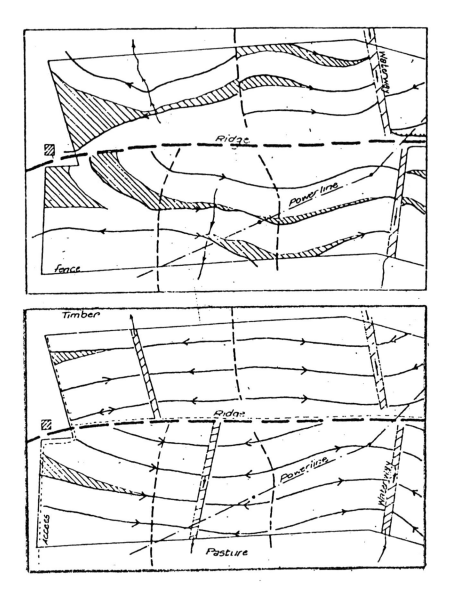

FIGURE 5. A comparison between the layout of a property in the Bundaberg region
showing contour banks following the exact grade line (top map) and the parallel
layout approach (bottom map). The shaded areas indicate short rows. The parallel
layout approach eliminates the many short rows making the design more acceptable to
mechanical harvesting.

Source: Pink 1975: 19.

Soil Erosion in the Queensland Sugar Industry

FIGURE 6. Parallel contour banks protecting the land from erosion and providing long runs for mechanical harvesters on a cane farm in the Nambour district, 1976. Source: *The Cane Growers' Quarterly Bulletin*, Vol. 39, No. 3 (January 1976), p. 77.

to be green cane harvesting (i.e. harvesting cane without prior burning), trash blanketing and minimum tillage techniques.

Before 1930, cane crops in Australia were mostly unburnt before manual harvesting. However, as mentioned above, pre-harvest burning of cane became standard practice in Queensland during the 1930s. After the introduction of mechanical harvesting in the 1960s, the health reason for the pre-harvest burning of cane crops vanished as workers were no longer exposed to Weil's disease. Yet the practice continued as pre-harvest burning reduced the amount of trash, as the earliest machines struggled to handle large amounts of trash. In addition, any trash generated during harvest was often raked into heaps by the canegrowers and burnt, thereby depriving canegrowers of vital mulch that could protect their paddocks from erosion.[48]

Interest in green cane harvesting re-surfaced in the mid-1970s because of the disruptions to the 1973 and 1975 harvest seasons by prolonged periods of wet weather. By 1979 green cane harvesting entirely had been adopted by at least six harvesting groups in North Queensland.[49] Green cane harvesting produced larger amounts of trash than burnt cane harvesting and not all canegrowers were inclined to burn this trash, allowing it to cover their fields. Thus, trials in North Queensland were initiated by the BSES in 1977 to identify any unexpected problems associated with trash retention. These and subsequent trials during the 1980s and early 1990s, including those conducted by CSR Ltd., confirmed that trash blanketing caused no agronomic problems such as reduced ratoon growth, reduced yields or an increased incidence of diseases.[50] In 1982, the Queensland Department of Primary Industries and BSES also commenced trials to determine the effectiveness of trash retention as a method of reducing soil erosion. These studies concluded that average annual soil loss over the period 1982–1987 was 5 t/ha under a green cane harvest with 100 per cent of the trash retained as a blanket cover. If no trash blanketing was practised, the soil losses grew to 15 t/ha. Thus, trash blanketing or trash blanketing and no tillage provided a substantial soil conservation measure regardless of soil type. These results were confirmed when additional studies into the magnitude of soil loss in sloping sugar cane land under conventional cultivation and various no-tillage operations were completed by the Queensland Department of Primary Industries in the early 1990s.[51]

As a result of these trials, canegrowers now had a soil conservation measure that did not solely involve contour tillage and/or building contour banks and ditches. Moreover, trash blanketing reduced weed growth. Therefore, the amount of chemicals applied to control weeds could be reduced and canegrowers did not have to make frequent passes over the paddocks for weed control, thereby reducing fuel bills and the wear and tear on cultivation implements and tractors. Hence, substantial savings on cultivation costs could be achieved if trash blanketing and minimum tillage were adopted.[52] In addition, canegrowers adopting such practices would be perceived as displaying a concern for the environment, as they demonstrated a reduction in both soil erosion and chemical use.

ADOPTING SOIL CONSERVATION MEASURES

Information on soil conservation measures practised in Queensland's cane growing lands before 1945 is not extensive. During the 1930s, some Italian farmers in the Finch Hatton district, near Mackay, were observed to have stone pitched the heads of gullies on their farms to stop them from getting any larger. Dams built across gullies to check the loss of soil from cultivated fields by causing silting of channels were also observed in the Childers and Innisfail district in the early 1940s. Dr H. Kerr, however, noted in 1936 that contour banks had not been constructed in any Queensland cane growing areas.[53] This fragmentary evidence suggests that the use of soil conservation measures was probably not widespread in Queensland sugar cane producing districts before 1945.

Attitudes to soil conservation methods amongst Queensland canegrowers changed only slowly after World War Two. In 1947, W. Sloan of the BSES noted, 'too many farmers still plant and cultivate up and down the slopes'.[54] In the Isis district where erosion was particularly severe, some canegrowers established contour banks on their farms. Yet R. Moller of the BSES observed that in 1958 only a 'small minority of farmers' had adopted soil erosion measures through-

TABLE 2. Extent of soil conservation measures in selected Queensland cane-growing shires, 1970-1976

Shire	Year of survey	Details about soil erosion and conservation measures	Source
Ayr	1973	'soil erosion is not a major hazard; very little soil conservation measures adopted'	Finlay & Cribb 1973: 4-11
Burrum	1971	'of the area of 20,800 acres under cultivation in 1969 approximately 14,500 acres are requiring intensive soil conservation measures; from January 1967 until May 1970 approximately 600 acres of agricultural land has been protected against erosion'	Middleton 1970: 4-3
Douglas	1974	'conservation measures are not practised widely'	Middleton & Barnes 1974: 4-3
Gooburrum	1971	'approx. 750 acres has been protected'	Hawley 1971: 4-8
Hinchin-brook	1975	'Shire not subjected to significant erosion'	Seton 1975: 4-3
Kolan	1972	'sheet, rill and gully erosion have taken a heavy toll of the limited land resources'	Hawley 1972: 4-10
Maroochy	1976	'about 20 % of cultivated area suffers from erosion; no control measures have been adopted as yet'	Butcher *et. al* 1976: 66
Mirani	1971	'23,000 acres require intensive soil conservation measures, but only 194 acres protected by soil conservation structures'	Everett 1971: 4-2
Pioneer	1971	'72 farmers employ soil conservation practices; area of land protected totals 1180 acres'	Filet, 1971(a): 4-2
Proserpine	1973	'about 800 ha need soil conservation measures; very little work has been carried out to date'	Cribb 1973: 4-1
Sarina	1971	'only 1.6 % of the area requiring protection (26,000 acres) had been treated'	Filet 1971(b): 4-2
Tiaro	1973	'soil conservation measures have been taken on a few farms; more needs to be done'	Beal 1973: 4-3
Woongarra	1975	'199 ha of land have been protected with contour banks, and 351 ha have been treated with simple conservation measures'	Adams 1975: 4-3

Sources: Based upon the authors mentioned in the table. See reference list for full details.

out the Isis district. He expressed concern that erosion control practices were not implemented until gully erosion had become quite serious.[55] In 1957, A.R. Taylor of the South Johnstone Cane Pest and Disease Control Board observed the sporadic use of contour ploughing and stone walls in the South Johnstone district. However, he also noted that tragically no attempt had been made to halt erosion on numerous farms and fields throughout the district. Taylor was particularly scathing of the State's canegrowers, claiming that despite repeated warnings about the need for soil conservation measures the 'subject seldom progresses beyond the discussion stage with the average Queensland cane farmer'.[56]

Soil conservation measures such as contour tillage and/or the building of contour banks were still not being widely used in many cane growing districts during the early 1970s (see Table 2). The Queensland Department of Primary Industries during its compilation of the Sarina Shire Handbook summarised the general attitude amongst local canegrowers: 'acceptance of soil conservation practices in sugar cane is very recent'.[57] Moreover in 1970, some of the severest soil erosion in the State's cane growing districts still occurred in the Isis district, but only 23 canegrowers or nine per cent of the district's canegrowers used soil conservation measures. The Isis Land Use Committee noted forlornly that if the number of canegrowers adopting soil conservation measures doubled from six per year 'it would be about twenty years before an acceptable level of soil conservation would be applied over the whole district'.[58]

By the early 1980s, BSES officers reported that some canegrowers farming sloping land had constructed head ditches to prevent excess water flowing onto their fields, but except in a few instances, further control measures had not been implemented. Similar observations were made by northern soil conservation officers who noted that in North Queensland 'little adoption of the more traditional soil conservation methods had been achieved'.[59] In 1986, for example, only 1,300 ha had been treated with soil conservation measures throughout the Innisfail district. This amount was a tiny fraction of out of a possible 23,000 ha with a slope above 3 per cent requiring contouring. In contrast, adoption of soil conservation methods amongst Maryborough, Childers and Bundaberg canegrowers had improved by the mid-1980s, with the local soil conservation officer estimating that a third of the existing cultivated land in 1987 was protected by soil conservation measures.[60]

Assessing the reasons for why there was such slow acceptance by some Queensland canegrowers of the facts about soil erosion and solutions to the problems is difficult. The agricultural scientists and extension officers have left many opinions in the literature about why the State's canegrowers continued to engage in such exploitive mismanagement of the soil. These opinions are not confirmed or refuted by the views from canegrowers due to the absence of detailed oral histories. Nevertheless, sixty canegrowers were surveyed at Innisfail in 1982 about soil erosion on their properties, so some information from the canegrowers' perspective can be presented in this narrative.

Acceptance that soil loss needed to be reduced required canegrowers to make two confessions: that a problem existed; and that this problem existed because of poor agricultural practices adopted currently or in the past. W. Sloan of the BSES claimed in 1947 that many canegrowers were still not fully 'cognisant of the dangers of soil erosion' or that they did not understand that a problem existed.[61] However, there is a contrasting view. In 1946, C.K. Simpson, the Technical Field Officer at Goondi Mill, noted that the local canegrowers knew about the dangers of soil erosion, but were 'apt to consider soil erosion was something that nothing can be done about.' The survey of Innisfail canegrowers in 1982 confirmed that they recognised erosion as a problem on their farms, but it was rated as a lesser problem than rising production costs and declining prices of sugar.[62] Thus, this fragmentary evidence suggests that some Queensland canegrowers knew that soil loss was a problem, but were slow to respond for other reasons.

Three reasons have been put forward in the literature to explain the slow uptake of soil conservation measures by Queensland's canegrowers. The first factor was that the loss of topsoil by soil erosion, except in the most extreme instances, had not transferred into lost productivity. The State's canegrowers reaped the benefits of scientific advances in other areas. Improvements in plant breeding provided them with sucrose-rich cane varieties suited to each district and the threat of diseases and pests were reduced by the BSES control programs. Greater use of fertilisers, including the addition of approximately 200 kg of nitrogen per hectare by the 1970s, reduced soil deterioration.[63] Moreover, in some cane growing districts, the eroding soils had sufficient depth to permit repeated turning up of a fresh layer for cultivation. Yet in the Mackay district by the early 1970s, so much topsoil had been removed that farmers were reduced to using the subsoil. [64]

The second reason advanced for the slow uptake of soil conservation measures was the perceived disadvantages of contour tillage and terraces. Canegrowers, including those surveyed at Innisfail in 1982, claimed that a great number of short rows almost always occurred between the contour banks at frequent intervals along their length. Such short rows were difficult to cultivate and became less acceptable following the widespread adoption of mechanical harvesting in the 1960s. Field layouts had to be designed to allow the use of heavy mechanical harvesters requiring well-defined turning points. In addition, contour banks absorbed an appreciable amount of scarce tillable land on their assignments, they became places where weeds and grasses accumulated, they silted up every few years and difficulties arose using traveling irrigators (i.e. large pieces of machinery that move across fields spraying water) on non-straight layouts.[65]

A third factor was the lack of equipment and staff with expertise. In 1959, R. Moller claimed that the construction of soil conservation measures in the

Peter Griggs

Isis district was hindered because earth-moving equipment was not readily available for hire. Even when earth moving equipment became more readily available, the Queensland Department of Primary Industries admitted in 1983 that canegrowers wishing to implement soil conservation measures were hindered by the following factors: lack of suitable topographic information about their properties; the ratoon system which inhibited implementation across an entire property as part of the farm is always under a crop; and a restricted implementation period (i.e. April to September). Under these constraints, soil conservation officers in cane producing districts had only been able to plan and survey soil conservation measures on approximately 250 ha per man-year. In contrast, rates of 7000 ha per man had been achieved in extensive grain cropping lands of central Queensland. [66]

As a few innovative canegrowers began using trash blanketing in the late 1970s, M. Sallaway of the Queensland Department of Primary Industries noted that the largest problem with the general acceptance of trash retention may be 'with the outlook of individuals'.[67] Sallway hinted at the innate conservatism of Queensland's canegrowers when it came to changing their land management practices. Yet the successful implementation of trash retention and minimum tillage techniques by these canegrowers and improvements in the green cane capability of harvesters accelerated the adoption of green cane harvesting. Furthermore, the slump in world sugar prices in the early 1980s encouraged more canegrowers to reduce their cultivation costs, achievable by adopting trash blanketing and reduced tillage.[68] Hence, by 1996, half the entire Queensland crop was produced using green harvesting, although its adoption was initially greater in cane growing districts north of Townsville. South of Townsville, the adoption of green cane harvesting has been much slower, although had increased substantially by 2001 (See Table 3). In the Burdekin region, the thick trash blanket posed difficulties for irrigation, while the rationing of cane under the colder conditions of southern Queensland presents significant problems. In

TABLE 3. Percentage of green cane harvesting in Australian cane producing regions, 1985 to 2001

Region	1985	1987	1989	1990	1992	1994	2001
Northern*	21	45	68	73	79	88	94
Burdekin	0	0	0	5	4	3	5
Central	0	0	11	12	12	27	87
South	0	0	18	15	17	24	58

* Sugar cane growing districts north of Townsville include Mossman, Cairns, Innisfail, Tully and Ingham.

Source: Prove and Hicks 1991: 69; Woods *et al.* 1997: 481; & personal communication, BSES Ltd.

addition, by 1991, an estimated 80 per cent of the sloping land (> 2 per cent) in Far North Queensland was being farmed with trash blanketing and zero tillage methods.[69] Thus, despite the different rates of adoption based upon latitude, the acceptance of trash retention and minimum tillage with its soil erosion reduction characteristics has occurred much faster than the implementation of traditional soil conservation measures.

Unexpected environmental consequences, other than reduced erosion, have also emerged following the adoption of green cane harvesting and trash blanketing. Some Queensland canegrowers have commenced reporting their general pleasure with the positive effect on soil conditions due to trash retention. The most apparent effects on the soil condition noted by canegrowers included improved soil structure, more earthworms and less damage by machinery wheel tracks following cultivation and harvesting. More wildlife on farms has also been observed following the cessation of cane burning. In addition, canegrowers in the Herbert River, Bundaberg and Mackay districts using trash blanketing have reported improved yields (i.e. tonnes of cane per hectare and tonnes of sugar per hectare).[70]

CONCLUSION

Soil erosion throughout parts of Queensland's sugar cane producing lands was first acknowledged as a significant problem by officials from the industry in the 1930s. By the early 1980s, at least a third of the assigned area cultivated with sugar cane in Queensland was identified as suffering some form of erosion, with the Isis (Childers), Bundaberg and Mackay districts being the worst affected. The discussion in this paper has highlighted that soil loss existed so long because many Queensland canegrowers between 1930 and 1980 employed exploitive land management techniques. They persisted in cultivating sloping land without adopting soil conservation techniques such as contour tillage, terraces and diversion waterways. Declining productivity which may have prompted more Queensland canegrowers to adopt soil conservation techniques did not emerge. The State's canegrowers reaped the benefits from improvements in plant breeding which provided them with sucrose-rich cane varieties suited to each district and the threat of diseases and pests was reduced by the BSES control programs. Soil deterioration was reduced by greater use of fertilisers. Together these measures maintained yields, masking the deleterious effects of any soil loss.

The analysis in this paper has also illustrated that the transfer of models from other crops and physical environments is not always successful. Sugar cane in Queensland was cultivated as an intensive monoculture in high rainfall localities under a highly regulated system of production. Queensland canegrowers, because of legislative requirements, could not alternate sugar cane with other row or tree crops every few years, nor easily engage in strip cropping to slow the velocity of water across their fields. Moreover, even if no legislative restriction upon the cultivation of other crops existed, repeated efforts had failed to find a profitable alternative crop to grow on these farms. Mulching using the trash to protect the soil was also not an option as it was initially burnt because of associated health and industrial concerns, and later to facilitate mechanical harvesting. The officials promoting soil conservation could only mostly recommend mechanical methods such as contour tillage, contour banks and waterways. However, even when modified to take into account the higher amounts of rainfall in tropical and semi-tropical districts, Queensland's canegrowers were reluctant to adopt these measures claiming that they hindered mechanical harvesting and irrigation and that the banks absorbed too much scarce, tillable land on their assignments.

Finally, the BSES and Queensland Government's soil conservation policies did not readily exacerbate soil erosion in the sugar growing lands of Queensland. However, due to the reluctance of the canegrowers to adopt the soil conservation measures recommended, they did little to ameliorate its effects between 1930 and 1980. The few techniques recommended can be categorised as top-down approaches, promoted by BSES scientists and agricultural extension officers from the Queensland Department of Primary Industries. Success in reducing soil loss only followed the introduction of green cane harvesting and trash blanketing in the 1980s, a bottom-up solution aimed not at reducing soil erosion, but eliminating delays in harvesting due to prolonged periods of wet weather which precluded pre-harvest burning of the crops to be harvested. Nevertheless, many canegrowers willingly adopted the practices promoted by the harvest operators (and eventually the BSES officials), partially because of the ease of implementation, but also due to the associated reduction in tillage costs and improved yields that followed their use. In doing so, they ensured that the State's sugar cane industry in the 1990s finally commenced using more sustainable land management practices, instead of persisting with the exploitive land management practices of the past decades.

NOTES

I am grateful to Ms Adella Edwards, cartographer, for drawing the map that appears in this paper. The assistance of the staff at the Noel Butlin Archives Centre, Australian National University, was greatly appreciated. CSR Ltd. kindly granted me permission to use the firm's archival records. My thanks also to the two reviewers who provided comments that contributed to the revision of this paper.

1.   Watts 1987: 222, 397 and 438; Richardson 1992: 30–31; Fraginals 1976: 92; Mechanisation Committee of the South African Sugar Association 1949: 9.

2. The main studies include Easterby 1932; Lowndes 1956; Saunders, 1982; Graves 1988; and Graves 1993.

3. See and Crouch 1963; Department of Labour and National Service 1970; Willis 1972; Burrows and Shlomowitz 1992; Kerr and Blyth 1993.

4. These changes are discussed fully in Griggs 1997 and Griggs 2000.

5. For overviews of the pre-1945 expansions in the Queensland sugar industry see Graves 1988: 144–54; Graves 1993: 11–19; and Shlomowitz 1979: 100–110; for details on more recent expansions in the Queensland sugar industry see Courtenay 1982: 129–31 and Robinson 1995: 217–18.

6. Wilson, Wissemann and Dwyer 1982: 1; Prove 1991: 29.

7. Graves 1988: 144–9; deregulation is discussed by Robinson 1995.

8. Kerr 1936: 26.

9. King, Mungomery and Hughes 1956: 221.

10. J. Tardent, 1938. 'Soil erosion in Queensland', p. 12. Queensland Forestry Files, SRS 5213/1, Box 12, Item 65: Soil Erosion, 1935–1966, Queensland State Archives, Brisbane, Queensland (hereafter QSA).

11. Stephens 1945: 3; Kemp 1947:651.

12. For a full account see Griggs 2004.

13. Charles Young, 1939. 'Evidence' in Minutes of Evidence before the Royal Commission on Sugar Peaks Scheme and Cognate Matters, p. 97. In Colonial Sugar Refining Company Records (hereafter CSRR), Z/109, Box 48, Noel Butlin Archives Centre, Canberra (hereafter NBAC).

14. Kerr 1940: 222; King, Mungomery and Hughes 1956: 48, 50 and 219; Penrose 1998: 134–5.

15. This limit was relaxed for the 1964 season and extended to 85 % and removed entirely in 1975.

16. Stephens 1945: 6; Vallance 1947: 120; Isis Land Use Committee 1971: 4-2; Industry Commission 1992: 37–8.

17. Isis Land Use Committee 1971: 7-3 and 7-4; Kerr 1996: 136–7.

18. For more detailed discussions see Bolton 1981: 138–40; McTainish and Leys 1993: 193–200 and Williams 1974: 303–5.

19. See for example Alldis 1937; Ratcliffe 1938; Bank of New South Wales 1939; Pick 1942; and Holmes 1946: 28.

20. Williams 1974: 305–6; Soil Conservation Authority, Victoria 1953: 6–9.

21. Skerman, Fisher and Lloyd 1988: 207–9.

22. J. Tardent, 1938. 'Soil erosion in Queensland', p. 12. In Queensland Forestry Files, SRS 5213/1, Box 12, Item 65: Soil Erosion, 1935-196612, QSA ; Edward W. Knox, 'Notes on a visit to the Isis district, 21 July 1899', pp. 1–2, in 'Memoranda and Reports for the Board of Directors 1887–1899', CSRR, 142/2753, NBAC.

23. Kerr and Bell 1939: 109.

24. Figures calculated from statistics provided in an Appendix attached to a letter from Hon. Frank Nicklin, Premier of Queensland to the Hon. Robert Menzies, Prime Minister of Australia, 30 July 1959. Queensland Department of Co-ordinator General Files, SRS1043/1, Box 718, Item 2158: Soil erosion and conservation flood mitigation; QSA.

25. For example see Everett 1971: 4-2; Filet 1971a: 4-2; Adams 1975: 4-2.

26. Capelin and Prove 1983: 88.

27. Sallaway 1979b: 130; Sallaway 1981.

28. Capelin and Prove 1983; Prove and Hicks 1991: 68.

29. See for example Anon 1974; Capelin 1979; Holz, 1979; and Holz and Shields, 1985.

30. Commonwealth Scientific and Industrial Research Organisation 2002: 27.

31. Johnson *et al.* 1998: 98; Johnson and Bellamy 2000: 165.

32. Bennett 1939: 32–3 and 48.

33. Watts 1987: 399–401 and 435; Ormrod 1979: 160–2.

34. Watts 1987: 403–4.

35. Galloway 1989: 102; Watts 1987: 512 and 547; Davy 1851: 116.

36. For planting methods in the nineteenth century Australian sugar industry see Griggs 2004: 6–8.

37. For views on the use of trash see King 1934: 127–8, Kerr and Bell 1939: 114 and Kerr 1940: 222; opinions on the cultivation of steeply sloping land are found in Moller 1958: 93 and Moller 1959: 89.

38. Kerr 1936: 28–33.

39. Stephens 1945: 6.

40. Kerr 1996: 133; Vallance 1947: 121–6.

41. For an abbreviated account of this method see Holmes 1946: 138–9; for more details see Yeomans 1965.

42. Kerr 1996: 134; Smith 1955: 103; Moller 1958: 94–5.

43. Taylor 1957; King 1958: 28; Moller 1958; Rosser 1961; Linedale 1970; Wright 1970.

44. For examples of this advice see CSR Ltd., 'Agricultural Circulars for Goondi Mill, 1951 to 1970', in CSRR, N 126/131, NBAC.

45. Smith 1955: 103; Amiet and Jones 1970:153; Dawson *et al.* 1983: 50; Prove 1991: 31.

46. See for example Veurman 1975: 94; Capelin 1979; Capelin and Prove 1983: 89–90; Holz 1979; and Holz and Shields 1985.

47. Dawson *et al.* 1983: 50.

48. King, Mungomery and Hughes 1965: 128–9.

49. Baxter 1983: 33; Ridge *et al.* 1979: 89.

50. Matthews and Makepeace 1981; Ridge *et al.* 1979; Hardman *et al.* 1985; Wood 1991; Dick 1993.

51. Bureau of Sugar Experiment Stations 1984; Prove *et al.* 1986: 79; Prove and Hicks 1991: 68–9; Prove, Doogan and Truong 1995.

52. Mackson 1983: 22–3; Prove *et al.* 1986: 81.

53. Stephens 1945: 3; Kerr 1991: 66; C. Simpson, Technical Field Officer, Goondi Mill, to General Manager, CSR Ltd., Sydney, 9 October 1946, Letter No. 255N, re: visit of Mr Skinner, Queensland Lands Department, CSRR, N/126/131, NBAC; Kerr 1936: 33.

54. Sloan 1947: 160.

55. Moller 1958: 93; Kerr 1996: 134–5.

56. Taylor 1957: 149.

57. Filet 1971b: 4-2.

58. Isis Land Use Committee 1970: 7-3.

59. Matthews and Makepeace 1981: 43; Wilson, Wissemann and Dwyer 1982: 5; Prove and Hicks 1991: 69.

60. Prove, Truong and Evans 1986: 78; *Australian Canegrower*, vol. 9, 10 (October 1987): 26

61. C. Simpson, Technical Field Officer, Goondi Mill, to General Manager, CSR Ltd., Sydney, 27 November 1946, Letter No. 258, re: soil erosion, CSRR, N126/131, NBAC; Sloan 1947: 157.

62. Wilson, Wissemann and Dwyer 1982: 11.

63. Sloan 1947: 157; Prove 1991: 31; Hogarth and Allsopp 2000: 162–5.

64. Veurman 1977: 582

Soil Erosion in the Queensland Sugar Industry

65. Smith 1955: 104; Moller 1958: 94; Wright 1970: 138; Sypkens 1970: 137; Amiet and Jones 1970: 152; Veurman 1977: 582; Wilson, Wissemann and Dwyer 1982: 9 and 13; Klein 1984: 4–5.

66. Moller 1959: 90; Dawson *et al.* 1983: 50.

67. Sallaway 1979a: 133.

68. Wood 1991: 71.

69. Quabba 2000: 160; Prove and Hicks 1991: 70.

70. Gutteridge, Haskins and Davey 1996: 112; Kalpana 1996:96; Quabba 2000: 160; Chapman, Larsen and Jackson, 2001

ARCHIVAL SOURCES

Noel Butlin Archives Centre, Australian National University, Canberra

Colonial Sugar Refining Company Records (CSRR):

Agricultural Circulars for Goondi Mill, 1951 – 1970, CSRR, N 126/ 131.

Simpson, C., Technical Field Officer, Goondi Mill, to General Manager, CSR Ltd., Sydney, 9 October 1946, Letter No. 255N, re: visit of Mr Skinner, Queensland Lands Department, CSRR, N 126/131.

Simpson, C., Technical Field Officer, Goondi Mill, to General Manager, CSR Ltd., Sydney, 27 November 1946, Letter No. 258, re: soil erosion, CSRR, N126/131.

Knox, Edward W., 'Notes on a visit to the Isis district, 21 July 1899', in 'Memoranda and Reports for the Board of Directors 1887–1899', CSRR, 142/2753.

Young, Charles 1939. 'Evidence' in 'Minutes of Evidence before the Royal Commission on Sugar Peaks Scheme and Cognate Matters', pp. 96–8, in CSRR, Z/109, Box 48.

Queensland State Archives, Brisbane, Queensland (QSA):

Tardent, J. 1938. 'Soil erosion in Queensland'. *Queensland* Forestry Files, SRS 5213/1, Box 12, Item 65: Soil Erosion, 1935–1966.

Nicklin, Frank. Letter to the Hon. Robert Menzies, Prime Minister of Australia, 30 July 1959. Department of Co-ordinator General Files, SRS1043/1, Box 718, Item 2158: Soil erosion and conservation flood mitigation.

REFERENCES

Adams, N. 1975. *Woongarra Shire Handbook*. Brisbane: Queensland Department of Primary Industries.

Alldis, V. 1937. *Soil Erosion*. Young, New South Wales: Witness Press.

Amiet, P. and Jones, H. 1970. 'Soil conservation in relation to canegrowing'. *Proceedings of the Queensland Society of Sugar Cane Technologists*, 47–52.

Anon. 1974. *Moreton Region Non-urban Land Suitability Study*. Division Land Utilisation Technical Bulletin No. 11. Brisbane: Queensland Department of Primary Industries.

Bank of New South Wales. 1939. *Conserve your Soil: A Simple Guide to Soil Erosion*. Sydney: Bank of New South Wales.

Baxter, B. 1983. 'Green cane harvest review'. *Australian Canegrower*, February, 32–5.

Beachey, R.W. 1957. The *British West Indies Sugar Industry in the Late Nineteenth Century*. Oxford: Blackwell.

Beal, D. 1973. *Tiaro Shire Handbook*. Brisbane: Queensland Department of Primary Industries.

Bennett, Hugh. 1939. *Soil Conservation*. New York and London: McGraw-Hill Book Company Inc.

Bolton, G. 1981. *Soil and Spoilers. Australians Make Their Environment 1788–1980*. Sydney: George Allen & Unwin.

Bureau of Sugar Experiment Stations. 1984. 'A review of results of trials with trash management for soil conservation'. *Proceedings of the Australian Society of Sugar Cane Technologists*, 6: 101–6.

Burrows, Geoff and Shlomowitz, Ralph. 1992. 'The lag in mechanisation of the sugar cane harvest: some comparative perspectives'. *Agricultural History*, 66: 61–75.

Butcher, F., Parsons, R. and Beal, D. 1976. *Maroochy Shire Handbook*. Brisbane: Queensland Department of Primary Industries.

Capelin, M. 1979. *Moreton Mill Area – A Sugar Cane Land Suitability Study*. Division Land Utilisation Technical Bulletin No. 37. Brisbane: Queensland Department of Primary Industries.

Capelin, M. and Prove, B. 1983. 'Soil conservation problems of the humid tropics of North Queensland'. *Proceedings of the Australian Society of Sugar Cane Technologists*, 5: 87–93.

Chapman, L., Larsen, P. and Jackson, J. 2001. 'Trash conservation increases cane yield in the Mackay district'. *Proceedings of the Australian Society of Sugar Cane Technologists*, 23: 176–84.

Commonwealth Scientific and Industrial Research Organisation. 2002. *Unlocking Success Through Change and Innovation: Options to Improve the Profitability and Environmental Performance of the Australian Sugar Industry*. Submission to the Independent Assessment of the Australian Sugar Industry, Townsville, 5 April. Website, http://www.affa.gov.au/corporate_docs/publications/pdf/sugar_assessment/csiro.pdf. Accessed 15 October 2004.

Courtenay, Philip. 1982. *Northern Australia. Patterns and Problems of Tropical Development in an Advanced Country*. Melbourne: Longman Cheshire.

Cribb, I. 1973. *Proserpine Shire Handbook*. Brisbane: Queensland Department of Primary Industries.

Davy, John. 1851. *The West Indies Before and Since Slave Emancipation*. London: W. & F. Cash.

Dawson, N., Berndt, R. and Venz, B. 1983. 'Land use planning – Queensland canelands'. *Proceedings of the Australian Society of Sugar Cane Technologists*, 5: 43–52.

Department of Labour and National Service, Australia. 1970. *Men and Machines in Sugar Cane Harvesting*, Employment and Technology Series No. 7. Melbourne: Department of Labour and National Service, Australia.

Dick, R. 1993. 'Minimum tillage agriculture in southern Queensland'. *Proceedings of the Australian Society of Sugar Cane Technologists*, 15: 312–15.

Easterby, H. 1932. *The Queensland Sugar Industry: An Historical Review*. Brisbane: Government Printer.

Everett, M. 1971. *Mirani Shire Handbook*. Brisbane: Queensland Department of Primary Industries.

Filet, C. 1971a. *Pioneer Shire Handbook*. Brisbane: Queensland Department of Primary Industries.

Filet, C. 1971b. *Sarina Shire Handbook*. Brisbane: Queensland Department of Primary Industries.

Finlay, M. and Cribb, I. 1973. *Ayr Shire Handbook*. Brisbane: Queensland Department of Primary Industries.

Fraginals, M. 1976. *The Sugar Mill: The Socioeconomic Complex of Sugar in Cuba, 1760–1860*. New York and London: Monthly Review Press.

Galloway, J.H. 1989. *The Sugar Cane Industry. An Historical Geography from its Origins to 1914*. Cambridge: Cambridge University Press.

Graves, Adrian. *Cane and Labour. The Political Economy of the Queensland Sugar Industry 1862–1906*. Edinburgh: Edinburgh University Press, 1993.

Graves, Adrian. 'Crisis and change in the Australian sugar industry, 1914–1939' in Bill Albert and Adrian Graves (eds), *The World Sugar Economy in War and Depression 1914–1940*, pp. 142–56. London: Routledge, 1988.

Griggs, Peter. 1997. 'The origins and early development of the small cane farming system in Queensland, 1870–1915'. *Journal of Historical Geography*, 23 (1): 46–61.

Griggs, Peter. 2000. 'Sugar plantations in Queensland, 1864–1912: origins, characteristics, distribution and decline'. *Agricultural History*, 74 (3): 609–47.

Griggs, Peter. 2004. 'Improving agricultural practices. Science and the Australian sugarcane grower, 1864–1915'. *Agricultural History*, 78 (1): 1–33.

Gutteridge Haskins and Davey Pty Ltd. 1996. *Environmental Audit of the Queensland Canegrowing Industry. Report to CANEGROWERS*. Brisbane: Gutteridge Haskins and Davey Pty Ltd.

Hardman, J., Tilley, L. and Glanville, T. 1985. 'Agronomic and economic aspects of various farming systems for sugarcane in the Bundaberg district'. *Proceedings of the Australian Society of Sugar Cane Technologists*, 7: 147–53.

Hawley, G. 1971. *Gooburrum Shire Handbook*, Queensland Department of Primary Industries.

Hawley, G. 1972. *Kolan Shire Handbook*. Brisbane: Queensland Department of Primary Industries, Brisbane.

Hogarth, M. and Allsopp, P. 2000. *Manual of Cane Growing*. Brisbane: Queensland Bureau of Sugar Experiment Stations.

Holmes, J.M. 1946. *Soil Erosion in Australia and New Zealand*. Sydney: Angus and Robertson.

Holz, G. 1979. *Rocky Point – A Sugar Cane Land Suitability Study*. Division Land Utilisation Technical Bulletin No. 38. Brisbane: Queensland Department of Primary Industries.

Holz, G. and Shields, P. 1985. *Mackay Sugar Cane Land Suitability Study*. 2 vols. Brisbane: Queensland Department of Primary Industries.

Industry Commission. 1992. *The Australian Sugar Industry*. Canberra: Australian Government Publishing Service.

Isis Land Use Committee. 1971. *Report on a Land Use Study of the Isis District*. Brisbane: Queensland Department of Primary Industries.

Johnson, Andrew, McDonald, Geoffrey, Shrubsole, Dan and Walker, Dan 1998. 'Natural Resource Use and Management in the Australian Sugar Industry: Current Practice and Opportunities for Improved Policy, Planning and Management'. *Australian Journal of Environmental Management*, 5: 97–108.

Johnson, Andrew and Bellamy, Jennifer 2000. 'Managing for ecological sustainability: Moving from rhetoric to practice in the Australian sugar industry'. In Peter Hale, Anita Petrie, Damian Moloney and Paul Sattler (eds), *Management for Sustainable Ecosystems*, pp. 163–74. Brisbane: Centre for Conservation Biology, The University of Queensland.

Kalpana, Parthasarathy, 1996. 'Economics of sustainable sugarcane production: the case of Bundaberg'. Master of Agricultural Economics Studies, University of Queensland, Brisbane.

Kemp, J. 1947. 'Annual Report of the Bureau of Investigation for 1946'. *Queensland Parliamentary Papers*, 2 (1947–48): 639–60.

Kerr, H. 1936. 'Soil erosion'. *The Cane Growers' Quarterly Bulletin*, 4 (1): 26–33.

Kerr, H. 1940. 'Some agricultural problems of the Mackay district'. *Proceedings of the Queensland Society of Sugar Cane Technologists*, 221–3.

Kerr, H. and Bell, A. 1939. *The Queensland Agricultural and Pastoral Handbook. Vol. 4: Sugar Cane and its Culture*. Brisbane: Queensland Government.

Kerr, John. 1991. *Top Mill in the Valley. Cattle Creek Sugar Mill, Finch Hatton, 1906–1990*. Mackay: Mackay Sugar Cooperative Association Limited.

Kerr, John 1996. *Only Room for One. A History of Sugar in the Isis District*. Childers: Isis Central Sugar Mill Company Limited.

Kerr, W. and Blyth, K. 1993. *They're All Half Crazy. 100 Years of Mechanical Cane Harvesting*. Brisbane: Canegrowers.

Klein, J. 1984. *Erosion Awareness Survey. Hummock Area, Bundaberg*. Project Report Q084010. Brisbane: Queensland Department of Primary Industries.

King, Norman 1934. 'Trash conservation'. *The Cane Growers' Quarterly Bulletin*, 1 (1): 126–32.

King, Norman 1958. *Annual Report of the Queensland Bureau of Sugar Experiment Stations*. Brisbane: Queensland Department of Primary Industries.

King, Norman, Mungomery, Reginald and Hughes, Cecil 1956. *Manual of Cane-Growing*. Sydney: Angus and Robertson.

King, Norman, Mungomery, Reginald and Hughes, Cecil 1965. *Manual of Cane-Growing*. Revised edition , 1956; rpt. Sydney: Angus and Robertson

Linedale, T. 1970. 'Combating erosion in the Moreton area'. *The Cane Growers' Quarterly Bulletin*, 39 (3): 77–9.

Lowndes, A. (ed). 1956. *South Pacific Enterprise: The Colonial Sugar Refining Co. Ltd*. Sydney: Angus and Robertson.

Mackson, J. 1983. 'Trash retention: dollars in your pocket?'. *Australian Canegrower*, December, 22–4.

Matthews, A. and Makepeace, P. 1981. 'A new slant on soil erosion control'. *The Cane Growers Quarterly Bulletin*, 45 (No. 2): 43–7.

Middleton, B. 1970. *Burrum Shire Handbook*. Brisbane: Queensland Department of Primary Industries.

Middleton, B. and Barnes, L. 1974. *Douglas Shire Handbook*. Brisbane: Queensland Department of Primary Industries.

McTainish, G. and Leys, J. 1993. 'Soil erosion by wind' in McTainsh, G. and Boughton, W. (eds), *Land Degradation Processes in Australia*, pp. 188–233, Melbourne: Longman Cheshire.

Mechanization Committee of the South African Sugar Association. 1949. *Mechanisation on South African Farms. Report No. 2*. Supplement to *The South African Sugar Journal*, January 1949. 18 pp. booklet.

Moller, R. 1958. 'Soil conservation in the Childers Area'. *The Cane Growers Quarterly Bulletin*, 21 (3): 93–6.

Moller, R. 1959. 'Soil conservation in the Isis Area'. *Proceedings of the Queensland Society of Sugar Cane Technologists*, 89–91.

Morgan, R. 1979. *Soil Erosion*. London: Longman Group Limited.

Ormrod, Richard. 1979. 'The evolution of soil management practices in early Jamaican sugar planting'. *Journal of Historical Geography*, 5, 2: 157–70.

Penrose, B. 1998. 'Medical experts and occupational illness: Weil's disease in North Queensland, 1933–1936'. *Labour History*, 75: 125–43.

Pick, Jock. 1942. *Australia's Dying Heart: Soil Erosion in the Inland*. Melbourne: Melbourne University Press, 1942.

Pink, H. 1975. 'Protecting the soil in new cane areas'. *Producers' Review*, 65 (12): 19–22.

Prove, Brian 1991. 'A study of the hydrological and erosional processes under sugar cane culture on the Wet Tropics coast of northeastern Australia'. Unpublished PhD thesis, James Cook University.

Prove, B., Truong, P. and Evans, D. 1986. 'Strategies for controlling caneland erosion in the Wet Tropical Coast of Queensland'. *Proceedings of the Australian Society of Sugar Cane Technologists*. 8: 77–84.

Prove, B. and Hicks, W. 1991. 'Soil and nutrient movements from rural lands of North Queensland'. In David Yellowlees (ed.), *Land Use Patterns and Nutrient Loading of the Great Barrier Reef*, pp. 67–76. Townsville: Sir George Fisher Centre for Tropical Marine Studies, James Cook University.

Prove, B., Doogan, V. and Truong, P. 1995. 'Nature and magnitude of soil erosion in sugarcane land on the wet tropical coast of north-eastern Queensland'. *Australian Journal of Experimental Agriculture*. 35: 641–9.

Quabba, Ray 2000. 'Achieving ESD as a cane grower'. In P. Hale, A. Petrie, D. Moloney and P. Sattler (eds), *Management for Sustainable Ecosystems*, pp. 159–62. Brisbane: Centre for Conservation Biology, The University of Queensland.

Ratcliffe, F. 1938. *Flying Fox and Drifting Sand: The Adventures of a Biologist in Australia*. London: Chatto and Windus.

Richardson, Bonham. *The Caribbean in the Wider World, 1492–1992. A Regional Geography*. Cambridge: Cambridge University Press, 1992.

Ridge, D., Hurney, A. and Chandler, K. 1979. 'Trash disposal after green cane harvesting'. *Proceedings of the Australian Society of Sugar Cane Technologists*, 1: 89–92.

Robinson, Guy 1995. 'Deregulation and restructuring in the Australian cane sugar industry'. *Australian Geographical Studies*, 33 (2): 212–27.

Rosser, J. 1961. 'Mackay growers use contour banks to control erosion'. *The Cane Growers' Quarterly Bulletin*, 25 (1): 13–16 & 21–3.

Sallaway, M. 1979a 'Trash retention as a soil conservation technique'. *Proceedings of the Australian Society of Sugar Cane Technologists*, 1: 133–7.

Sallaway, M. 1979b. 'Soil erosion studies in the Mackay district'. *Proceedings of the Australian Society of Sugar Cane Technologists*, 1: 125–32.

Sallaway, M. 1981. 'Soil erosion processes in Mackay Canelands'. Unpublished Master of Philosophy thesis, Griffith University, Brisbane.

Saunders, Kay. 1982. *Workers in Bondage. The Origins and Basis of Unfree Labour in Queensland 1824–1916*. Brisbane: The University of Queensland Press.

See, J. and Crouch, H. 1963. 'Mechanisation of Sugar Cane Harvesting'. *Australian Department of Labour and National Service Personnel Practice Bulletin*, 40–48.

Seton, D. 1975. *Hinchinbrook Shire Handbook*. Brisbane: Queensland Department of Primary Industries.

Shlomowitz, Ralph. 1979. 'The search for institutional equilibrium in Queensland's sugar industry 1884–1913'. *Australian Economic History Review*, 19 (2): 91–122.

Sloan, W. 1947. 'Some aspects of the problem of soil erosion control in Queensland cane fields'. *The Cane Growers' Quarterly Bulletin*, 10 (4): 155–61.

Skerman, P., Fisher, A. and Lloyd, P. 1988. *Guiding Queensland Agriculture 1887–1987*. Brisbane: Queensland Department of Primary Industries.

Smith, N. 1955. 'Combatting the soil erosion problem in the Childers area'. *The Cane Growers' Quarterly Bulletin*, 18 (3): 103–4.

Soil Conservation Authority, Victoria. 1953. *The Soil Conservation Authority, Victoria, Australia: A Brief History of Victorian Erosion Control*. Melbourne: Victorian Soil Conservation Authority.

Sypkens, E. 1970. 'Are short rows necessary in contoured canefields'. *The Cane Growers' Quarterly Bulletin*, 23 (4): 137–41.

Stephens, C. 1945. *The Nature and Incidence of Soil Erosion on the Sugar Cane Fields of Queensland*. Divisional Report No. 20 of 1945, Division of Soils, Council for Scientific and Industrial Research.

Taylor, A. 1957. 'Attempts at soil conservation on South Johnstone farms'. *Proceedings of the Queensland Society of Sugar Cane Technologists*, 149–53.

Vallance, L. 1947. 'A soil erosion control experiment in the Isis district'. *The Cane Growers' Quarterly Bulletin*, 10 (3): 118–28.

Vallance, L. 1950. 'The History of Sugar Soils Investigations and Agricultural Research'. In Queensland Bureau of Sugar Experiment Stations, *Fifty Years of Scientific Progress: A Historical Review of the Half Century Since the Foundation of the Bureau of Sugar Experiment Stations*, pp. 21–37. Brisbane: Queensland Government Printer.

Veurman, J. 1975. 'Contours protect threatened soil'. *Producers' Review*, 65 (4): 94–5.

Peter Griggs

Veurman, J. 1977. 'Soil conservation in the central cane district'. *Queensland Agricultural District*, 103 (6): 579–87.

Watts, D. 1987. *The West Indies: Patterns of Development, Culture and Environmental Change since 1492*. Cambridge: Cambridge University Press.

Williams, Michael. 1974. *The Making of the South Australian Landscape*. London: Acdemic Press.

Willis, G. 1972. *The Harvesting and Transport of Sugar Cane in Australia*. Department of Geography Monograph Series, No. 3. Townsville: James Cook University of North Queensland.

Wilson, T., Wissemann, A. and Dwyer, G. 1982. *Report on the Soil Conservation Practices and Related Attitudes of Innisfail Canegrowers*. Project report No. Q082015. Brisbane: Queensland Department of Primary Industries.

Wood, Andrew 1991. 'Management of crop residues following green harvesting of sugarcane in north Queensland'. *Soil and Tillage Research*, 20 (1): 69–85.

Woods, E., Cox, P. and Norrish, S. 1997. 'Doing things differently: the R, D & E revolution?' In Keating, Brian and Wilson, J. (eds), *Intensive Sugarcane Production: Meeting the Challenges Beyond 2000*, pp. 469–90. Wallingford, UK: CAB International.

Wright, J. 1970. 'Is soil erosion a problem on the farm?'. *The Cane Growers' Quarterly Bulletin*, 23 (4): 134–6.

Yeomans, Percival. 1965. *Water for Every Farm*. Sydney: Murray Books.

Did they Really Hate Trees? Attitudes of Farmers, Tourists and Naturalists towards Nature in the Rainforests of Eastern Australia

Warwick Frost

THE FARMER AND THE SCIENTIST

Between 1929 and 1931, Francis Ratcliffe, a young English biologist with the Australian Council for Scientific and Industrial Research, was engaged in a study of fruit bats (or flying foxes) in sub-tropical Australia. Orchardists had complained that the bats were destroying their crops. Ratcliffe's tasks were to gain information about their numbers and life-cycle and recommend what action should be taken to control them.[1]

As the first scientist to systematically study the fruit bats, Ratcliffe needed to observe them in a range of locations and to obtain specimens. The gregarious bats tended to gather in semi-permanent colonies, known as *camps*. To find these camps Ratcliffe turned to local farmers for help.

Ratcliffe and a friend visited a rainforest on Tamborine Mountain, about 100 kilometres south of Brisbane. Their guide was a farmer's wife, Mrs Curtis, aged somewhere in her twenties,

> My friend had been a naturalist since his schooldays, and I had spent a year in the bush as a professional biologist: but in her company we could only listen and learn. She had spent her whole life on the mountain, and knew it as a man knows his own golf course. She knew the habits of every bird and beast that lived there, and where the rare ferns and orchids could be found. The bird kind, I think, were her favourites. She had made friends with them, studied them, photographed them, her infinite patience outweighing the deficiencies of her apparatus.[2]

Ratcliffe wished to shoot some fruit bats as specimens. Mrs Curtis excused herself and walked some distance away. She disapproved of the shooting of wildlife. She was upset and in turn Ratcliffe became upset and embarrassed that he had so disturbed her tranquil piece of rainforest.

CLEARING AUSTRALIA'S RAINFORESTS

Tamborine Mountain was a part of the long thin line of rainforests which ran down Australia's east coast for 4,000 kilometres or so from Cape York to southern

Tasmania.[3] Their clearance, primarily for farming, began in the late eighteenth century, but peaked between the 1870s and the 1930s. While it is not possible to exactly say how much rainforest was cleared, as official statistics were not collected, a reasonable estimate is two to five million hectares, but it could easily be argued as higher.[4] In some instances, such as the sub-tropical Big Scrub and the mixed eucalypt-temperate rainforest Great Forest of South Gippsland, the rainforest was nearly completely cleared. The farmers were *selectors*, family farmers buying uncleared land from the government. They were attracted to the high rainfall of these forests, which ranged upwards from 1,000 millimetres (40 inches) per year. However to utilise that rainfall for agricultural purposes, the farmers had to clear the immense forests quickly. This was done through a process of cutting, drying and burning, universally known by settlers as *The Burn*.[5]

Clearing the tall dense forests was hard, dirty, dangerous work. One selector described cutting rainforest as, 'no matter where or how you hit anything it invariably falls on top of you, and every damn thing has spikes on it'.[6] Another recorded how 'thousands of little hooks and needle-points wait for the slightest contact of the unwary, tearing the clothes to ribbons, and lacerating hands and arms'.[7] Deaths and injuries resulted from falling trees and limbs and axes skidding and slipping.[8] Picking up half-burnt logs to restack and burn again, 'was hard, rough work, and only strong men could stand it; the charcoal on the logs when wet would wear the skin off the hands until they bled'.[9] The farmers were frustrated by their underestimation of the task they had undertaken, their chronic lack of capital and the rapidity of forest regrowth. Clearing the forests was hard, dangerous and frustrating. Many gave up and abandoned their farms.[10]

HATRED, FEAR AND ALIENATION: THE FARMERS' PERSPECTIVE

Settlers' feelings of hatred, fear and alienation towards the Australian environment have been amongst the major themes of Australian history. Given the difficulties and expense involved in clearing the rainforests, it is not surprising that such feelings have been painted as being at their strongest amongst the farmers in the rainforests of the east coast. Four major works are considered here as examples of how farmers' attitudes towards rainforests (and more generally the Australian environment) have been represented.

In 1930 Keith Hancock wrote his history/commentary *Australia*. Provocatively he labelled the settlers as invaders and declared that, 'the invaders hated trees ... [and] the greed of the pioneers caused them to devastate hundreds of thousands of acres of forest-land'.[11] Furthermore, 'to the early settlers, the Bush was an unfriendly wilderness', especially the coastal rainforests. The invaders wanted open spaces, 'they were overjoyed when, pushing beyond the dense mountain forests of the belt of heavy rainfall, they found more manageable, more familiar country, "like a park and grounds laid out"".[12]

A generation later, A.J. Marshall continued in this vein. Writing generally of Australia, he argued that, 'the bush, to our great-grandfathers, was the enemy: it brooded sombrely outside their brave and often pathetic little attempts at civilization … It, not they, was alien'.[13]

This notion of a hatred of trees arising from fear and alienation was taken up by Geoffrey Bolton in *Spoils and Spoilers* (1981), Australia's first environmental history textbook. In a chapter titled 'They hated trees', Bolton wrote,

> The first European arrivals in New South Wales may well have been oppressed by what they saw as the vastness of its forests. Many of the earliest drawings of the Australian bush tend to exaggerate the size of the trees and to dwarf the human figures among them, and this probably reflects the way the newcomers felt about their new surroundings.[14]

Finally, George Seddon argued that Australians developed a double standard towards rainforests. On the one hand they saw them as 'satanic' and 'full of menace', provoking an 'attitude of dread'. On the other hand they greatly appreciated them in botanic gardens and as street trees, but only 'drained of menace by their new context, safely suburban'.[15]

In addition, Australian historiography has been greatly influenced by accounts of farmers' experiences overseas, especially in similar regions of recent European settlement. Particularly influential was Roderick Nash's *Wilderness and the American Mind* (1967). Nash argued that the American pioneers transferred Old World attitudes to the forests of the Atlantic coast. For those on the frontier, 'constant exposure to wilderness gave rise to fear and hatred on the part of those who had to fight it for survival and success'.[16] Furthermore,

> Wilderness not only frustrated the pioneers physically but also acquired significance as a dark and sinister symbol. They shared the long Western tradition of imagining wild country as a moral vacuum, a cursed and chronic wasteland.[17]

NATURALISTS AND ARTISTS

In recent years, there has been growing interest in those naturalists and artists who enjoyed and valued the Australian environment, particularly the rainforests. However, this is no major revisionism. Rather, these groups have been characterised as only a minority counterpart to the majority's perspective of the rainforests as alienating and intimidating. The views of these few groups are comfortably familiar to a modern environmentally minded audience, but they have also been used to remind us that until quite recently, these were unusual attitudes and that most Australians hated the rainforests.

Natural history was one of the great genteel pastimes of the nineteenth century. Indeed it was popular amongst some of the officers of the early convict fleets.[18] It seemingly developed almost exclusively amongst urban professional elites, such as lawyers, schoolteachers and doctors.[19] Most of its enthusiasts were amateurs, its pursuit a respectable leisure pastime.[20] Their obsession was possession, they were a new type of *hunters and collectors*.[21] Most importantly, their numbers were very small.

Rainforests were a favourite topic of many landscape artists in the nineteenth century.[22] Prominent rainforest painters included Conrad Martens, Augustus Earle, John Skinner Prout, Eugene von Guerard, Nicholas Chevalier, Louis Buvelot, Isaac Whitehead and (in the early twentieth century) Arthur Streeton. In addition there was a group of prolific rainforest photographers, including Richard Daintree, J.W.Lindt and Nicholas Caire.[23]

These works both fed off and stimulated further the fascination for rainforests. Von Guerard's *Ferntree Gully in the Dandenong Ranges* (1857) attracted tremendous interest and appreciation. Furthermore it encouraged large numbers to visit the site which it depicted, making Ferntree Gully an early tourist attraction and in time a National Park.[24] A conservation ethic was also promoted through the paintings of Arthur Streeton in the 1920s and 1930s. Streeton's work contained strong warnings of deforestation, as in his bluntly titled *Gippsland Forests for Paper Pulp* and *Our Vanishing Forests* and his apocalyptic *Silvan Dam and Donna Buang – AD 2000*.[25]

TOURISTS AND THE CREATION OF NATIONAL PARKS

Naturalists and artists may be construed as a small and exclusive urban elite. However, rainforests also attracted a far wider audience. While in the nineteenth century, rainforest loving artists ensured that Australia's 'museums and art galleries are full of ferns', broad public tastes also meant that 'colonial buildings, from butchers' shops to ballrooms, were equally full of ferns'.[26] This fascination with ferns and their rainforest environment came from two sources. The first was an English craze for ferns (labelled *pteridomania*), transported to Australia as part of the colonists' *cultural baggage*.[27] The second was that rainforests represented coolness, shade and moisture, desirable qualities in a hot and dry climate.

This passion for rainforests is best seen in public tree plantings. Melbourne's Royal Botanic Gardens are dominated by nineteenth century rainforest plantings, including 600 palms, 110 tropical figs, 192 lillypillies and 164 rainforest conifers. In contrast there are only 75 elms, 100 oaks and 81 true pines.[28] Regional towns like Bendigo boasted public ferneries.[29] As early as 1866 the Adelaide Botanic Garden planted an avenue of over twenty Moreton Bay Figs. In the 1920s the Perth seaside suburb of Cottesloe planted hundreds of Norfolk Island Pines as street trees, creating a look which soon spread to most seaside areas. Bunya Pines and Moreton Bay Figs flourished in the manicured gardens of wealthy squatters in Victoria's Western District, an excellent example of Victorian era conspicuous consumption, given their high water requirements.

One better than bringing the rainforest to the city was for visitors to travel to and experience the rainforests in their natural setting. The geography of Australia's east coast made such tourism easily possible, even inevitable. All the capital cities of the east coast were very close to mountains and rainforests. Melbourne, for example, was only 40 kilometres from the rainforest gullies of the Dandenong Ranges.

Popular attractions for tourists included waterfalls, fern-gullies and individual large trees. Typical activities included picnicking, walks and games. Fern-collecting was a popular, though destructive activity.[30] The extension of railways into mountains increased the volume and pressure of tourism. A railway excursion to the forests of Mirboo North (Victoria) in 1908 by 100 American sailors from the visiting Great White Fleet was particularly illustrative of the widespread appeal of rainforests.[31] Demand was especially high for Sunday excursion trains, which allowed ordinary wage-earners the chance to see the rainforests.[32]

The flow of tourists into the rainforests remained strong through to the 1920s and 1930s. Stays lengthened. Resorts with guest houses or holiday homes boomed. Famous holiday home users in the Dandenongs ranged from the Prime Minister Billy Hughes to the gangster 'Squizzy' Taylor. New trends amongst tourists in the 1920s and 1930s included the use of motor vehicles to reach rainforest attractions and the popular craze for *bushwalking*.[33]

Whilst tourists could be destructive in rainforests,[34] they also contributed to their preservation. This was particularly so through the creation of National Parks to cater for their demands and protect the natural environment.[35] Many early National Parks were small patches of rainforest greatly popular amongst visitors, such as Witches' Falls at Mount Tamborine, Queensland's first National Park established in 1908.[36]

DID THEY REALLY HATE TREES?

The literature provides a convenient division. Farmers, pioneers and settlers, it was argued, were alienated and intimidated by the dense dark rainforests and accordingly hated them and cut them down wherever they could. In contrast, there were small urban groups (naturalists, painters, tourists) who appreciated the rainforests, but were generally powerless to protect them. Such a simple model is attractive as it can easily be compared to today.

However, there are those who disagree. The agricultural scientists Neil Barr and John Cary in *Greening a Brown Land* (1992) argued that, 'there has not been a total hatred of the trees as some commentators would have us believe', rather there is 'an historical tradition of native and exotic tree appreciation on Australian farms'.[37] Speaking on National Tree Day 2001, Ian Donges, President of the National Farmers' Federation declared, 'farmers deserve a solid pat on the back for their commitment to the environment and myths that they rape and pillage the land should not be nurtured'.[38]

Is the division between the farmers on the one hand and the naturalists, artists and tourists on the other just too black and white? Were there not certain common values which they all shared? Were not farmers sometimes tourists? Were there not farmers who were naturalists (as John Muir was in the USA)? Ian Tyrrell has recently explored the environmental values of Californian farmers.[39] There needs to be a similar exploration of the attitudes towards nature of Australian farmers. This paper considers the evidence of some farmers (such as Mrs Curtis) who did enjoy and appreciate their rainforest surrounds.

FIRST IMPRESSIONS IN THE 1820S AND 1830S

The 1820s and 1830s saw a number of detailed accounts of visits to rainforests. Rather than being intimidated or frightened by the dense forests, these writers found them beautiful and invigorating. On his first visit to the Illawarra region of NSW around 1827, Alexander Harris told how he and his guide had to leave open ground and cut through the dense brush. Writing for the English reader, Harris could have been excused for spicing up his adventure by highlighting his fears. Instead he described 'the most novel and beautiful scenery', and throughout his account betrayed no notion of fear amongst him or his companions.[40] Harris often referred to the forest as 'gloomy', but he explicitly clarified that as meaning that light levels were lower under the canopy, not that it was depressing or scary. Indeed, he observed that while the constant shade left the cedar-cutters pallid they were generally very healthy.[41]

Peter Cunningham, another visitor to Illawarra in the 1820s, described the forests as 'beautiful, fertile and romantic'.[42] In 1826, James Atkinson described the rainforests of NSW as 'magnificent', the sassafras tree as 'elegant', the tree fern as 'very beautiful' and that there were 'many other beautiful plants and trees'.[43] In the 1830s James Backhouse described north-west Tasmania as 'one of the most magnificent of forests', the crowns of the eucalypts as 'elegant' and the tree ferns as 'splendid'.[44]

These were the comments of some of the earliest visitors to the rainforests. Their accounts were intended for publication in England and they were clearly striving to convey the immensity and exoticness of the forests to an unfamiliar English audience. Did they represent a view commonly held by European settlers in Australia at that time? Certainly these four authors gave no indication that their views were contrary or unusual.

A LOVE OF NATURE (SOUTH GIPPSLAND)

In the 1870s and 1880s South Gippsland in Victoria was flooded by settlers. A Mrs Williams was initially distraught, but quickly came to love her forest surroundings,

> Oh! how I used to love the early mornings, when everything awoke to new life; I would just stand and feast on the beauty and glory of it all. There was a spot down by the river which I never tired of looking at, the tall tree ferns, with their graceful spreading plumes, the bracken, the swordgrass, clematis, maiden-hair fern, and Xmas [sic] trees, etc., made a picture impossible for me to describe.[45]

Her neighbour, Miss Elms, came to be housekeeper for her bachelor brother. Travelling to his selection for the first time, 'I was charmed by the beauty of the surrounding scrub and the song of the birds, especially the beautiful clear note of the lyre bird, which I had not heard before.' She made a garden around their log hut, using flowering plants transplanted from the nearby bush, such as fire-weed and supplejack, as ornamental flowers. She recorded that some visitors were highly amused to see what they thought of as being weeds cultivated. It is notable that the only exotics she recorded as in her garden were of practical value, such as parsnips and hops.[46]

In summer Miss Elms 'used to love riding along the pretty tracks looking like beautiful avenues with the supplejacks' lovely blossoms wreathing and festooning the trees'. A favourite spot was 'a glorious mass of tree ferns and blackwoods in a gully that we admired very much, and which my brother tried to reserve as a beauty spot, but the ruthless fires swept through it all when burning other scrub'.[47]

Niels Petersen recalled her childhood in the 1880s at Poowong East in South Gippsland,

> On a bright, sunny morning the forest was not lonely, all around was the song of the brightly coloured birds ... There were some very beautiful trees in the forest, such as the Pittosporum, Blanket Wood, Hazel, Wattle, Mintwood (or Christmas Tree) the Musk with its beautiful scent and white and yellow dotted flowers, and sometimes a tree fern was seen with a green creeper growing over it, the latter having lovely snow white bell shaped flowers.[48]

Charles Barrett, a noted naturalist and journalist, came often to Poowong to observe the Giant Earthworm and the Lyrebird and on an unsuccessful search for Leadbeater's Possum. He recorded his guide as Lou Cook, a local dairy farmer. Indeed Barrett, who rarely named his guides, gave much of the credit for their work to Cook.[49] Other nearby farmers whose writings qualified them as naturalists were T.J. Coverdale and F.P. Elms.[50]

Two small rainforest gullies in South Gippsland were reserved as National Parks. They were Bulga (1904) and Tarra Valley (1909). Both were initially used by local farmers as picnic sites. In sparsely populated frontier settlements such sites were especially important for the community to get together. The reservation of these National Parks was due to the efforts of the local shire council, particularly the Shire Engineer, Frank Corrigan. As well as walking tracks and picnic facilities, in 1938 the council constructed a suspension footbridge over the gully at Bulga.[51]

THE ATHERTON TABLELAND

The Atherton Tableland in Northern Queensland was settled from the 1880s onwards. One selector, Charles Bryde, was like Mrs Williams in that his initial dismay turned to wonder and joy. He wrote that,

> The spirit of the romance of pioneering took possession of us. We were the only inhabitants of a new-found beautiful world; we were shipwrecked on an unspoiled pre-Adamite island; we were well – just a couple of enthusiastic bush-lovers, with some ability to appreciate the beauty of old mother Nature.[52]

As in South Gippsland, local settlers tried to save small portions of the rainforests from clearance, and in time some were developed as much loved beauty spots and picnic grounds. One particular noteworthy battle concerned Lakes Barrine and Eacham. In 1912 Charles Bryde was thankful to be one of a party of twenty settlers who took a much needed break from clearing to enjoy a Christmas picnic at Lake Barrine.[53] Earlier six 40 acre freehold lots had been released around Lake Eacham and the shores partly cleared. A proposal for a similar release around Lake Barrine galvanised locals into action. The 1897 Royal Commission on Land Settlement heard pleas from John Byers (Crown Land Ranger), William Kelly (storekeeper and farmer) and Robert Ringrose (Secretary, Herberton Chamber of Commerce) for the conservation of the lakes. Ringrose exclaimed that the untouched Barrine was 'a most wonderful lake', which 'has such a quiet peacefulness buried in the scrub'.[54] As a result Barrine was saved and the freehold land around Eacham resumed.

GREEN MOUNTAINS

In 1911 five O'Reilly brothers and their three cousins selected land on the Lamington Plateau in Southern Queensland. The Plateau was actually the remains of a giant *caldera* (or volcanic cone). Its rich soil and high rainfall supported lush sub-tropical rainforest, with some temperate rainforest on the highest parts. However, its forbidding steeply cliffed sides had caused settlement to by-pass the Plateau. Only a shortage of unsettled land had convinced the Queensland Lands Department to open it for selection.

The commencement of clearing caused an outcry. The area had been proposed as a National Park in 1896, by Robert Collins, a wealthy farmer and Member of the Queensland Parliament. Collins' interest in National Parks stemmed from a world tour he had undertaken in 1878. He was particularly impressed by Yellowstone National Park and Yosemite, and he believed that parts of Queensland should be similarly protected.[55] Combining lobbying by the Queensland Royal Society and guided tours for influential persons (including Queensland Governor Lord Lamington and visiting writer Arthur Conan Doyle), Collins and his supporters advanced the case for National Parks.[56] Following Collins' death in 1913 leadership of this movement passed to Romeo Lahey, the son of a wealthy timber family.[57] Grabbing the opportunity of a close state election, Lahey also enlisted the patriotic zeal for World War One, arguing that, 'I want to get away to Europe to my next duty, but cannot leave this one unfulfilled'.[58] Lahey's campaign struck a chord with the public, for, 'Queensland, after an orgy of [rainforest] destruction, had reached a stage where it could well begin to think of conservation'.[59] Public support was demonstrated by a petition of 521 voters from mainly rural southern Queensland, who called on the Queensland Government to declare a National Park.[60] Ironically, the O'Reillys climbed the plateau by following tracks made by advocates of a National Park.[61]

In 1915 a National Park was declared and all available Crown land on the Plateau was withdrawn from selection. Plans for the government to build a road were shelved, the O'Reillys eventually making their own. However, they remained alone and isolated, without the schools or dairy factories which other settlers would have brought. There were lengthy discussions regarding their farms being resumed and added to the National Park.[62] However, while the Government offered a land exchange, the O'Reillys countered that they also wanted compensation of 6,000 pounds for seven years of clearing and other improvements.[63] Wishing neither to pay that amount of cash, nor to cause alarm by forcing the O'Reillys out, the Government adopted a waiting strategy, confident that the O'Reillys could not make farming pay on the isolated Plateau.

However, the O'Reillys did not fail. On the doorstep of the new National Park, they found a steady stream of visitors wishing to purchase supplies and guiding services from them. Lodging was provided at first in 'various slab humpies'.[64] These visitors were typical urban naturalists, such as Archibald Meston, a Queensland journalist whom they guided in 1918.[65] The O'Reillys found themselves suited to such work. Their farmer father had been a keen naturalist – 'never once did he return [from droving] without his saddle bags stuffed with seeds, cuttings and roots'[66] – and the boys were keenly interested in birds and had excelled at Nature Studies at school.[67]

Gradually the O'Reillys began to see the potential of mixing tourism with farming. The claim that they chose the Lamington Plateau because its scenery excelled the Blue Mountains[68] needs to be approached with caution, given it was made over 30 years later. However, by 1919 the idea was established. Putting the case for their remaining to the Minister of Lands, they argued, 'the few hundred acres of [our] settlement will prove an asset to the National Park in its development as a health and pleasure resort'.[69] In turn their critics argued that the O'Reillys were hanging on with the view to 'subdivide and sell as building lots for summer cottages'.[70]

While the government delayed its decision one of the brothers, Mick O'Reilly, a returned serviceman, was appointed Working Overseer (effectively Park Ranger) of the National Park. The other brothers were appointed Honorary Rangers.[71] Eventually, as it became clear that the Government would not resume the land, the O'Reillys moved further into tourism, in 1926 opening a guest house under the name 'Green Mountains' to service the growing number of visitors to the National Park.[72] Their venture was an immediate success, 'the people who had for many years enjoyed the shelter of our humpy were only too glad to recommend us to their friends ... [and it] put the plateau within the reach of many more people'.[73] A tremendous boom at the time in nature-based tourism, especially in mountains and forests, ensured their popularity.[74]

One early guest was Charles Chauvel. Significantly, he was the son of dairy farmers in nearby Fassifern, which had been settled and cleared of rainforests in the 1880s. However, Chauvel chose film-making over farming and he used the Lamington Plateau for three films. The first was *Heritage* (1934), in which he filmed a pioneer's bullock dray struggling up the impossibly steep road to the Plateau.[75] Chauvel returned a decade later to use the rainforest as a substitute for the jungles of New Guinea in his war film *The Rats of Tobruk*.[76]

On this visit Chauvel read the newly published *Green Mountains* by Bernard O'Reilly, a younger brother of the five original O'Reillys. Bernard had achieved fame in 1937 by rescuing the survivors of an aeroplane which had crashed on the Lamington Plateau. Public interest in the rescue led to him writing of it and the broader story of the O'Reillys. Chauvel bought the film rights and made *The Sons of Matthew* (1949), a fictionalised account of the O'Reillys efforts at farming the rainforests.[77] It is important to note that Chauvel did not demonise the rainforest. Instead the pioneers are shown as being enchanted with their new environment.

CONCLUSION

Farming in the rainforests of eastern Australia was extremely difficult. To the settlers the rainforests were exotic, unlike anything they had experienced in Britain or the wheatbelt of the inland plains of Australia. The rainforests were difficult, dangerous and expensive to clear. The settlers consistently underestimated the magnitude of the task, only to realise their mistake when well-committed. To farm successfully, to recoup their investment and to create a home for their families, the farmers had to remove the rainforest. Farming, whether dairying or cropping, could not take place *in* the rainforest. Rainforest was an obstacle to farming.

Many farmers, perhaps the majority, demonised that obstacle. For them, rainforest became an enemy, hated, intimidating, frightening. However, not *all* farmers shared these feelings. Some found the rainforests beautiful, magical and enchanting. They still had to clear large sections in order to succeed as farmers, but many tried to preserve parts of the rainforests.

This article provides a number of instances of farmers who loved the rainforest. It is perhaps open to criticism in that it is only a small sample and the cases are drawn from a limited range of rainforest regions. However, it must be understood that there were many significant rainforest regions (for example the Big Scrub of NSW) where there were few accounts left by farmers. The regions covered in this article tend to be those where we have the most number of accounts. Analysis is also limited by the large numbers of farmers who did not write down their stories, or if they did, then did not record their feelings towards nature. An interesting example of this is provided by Mrs Williams of South Gippsland. She recorded her attitudes in detail, while her husband also wrote his account, which was longer, but did not discuss his feelings.[78] Unfortunately we cannot tell whether or not he shared his wife's feelings. In addition, in looking at farmers' writings we must always be wary, measuring any complaints against the possibility of Tuchman's Law, 'a greater hazard, built into the very nature of recorded history, is overload of the negative'.[79]

It is impossible to tell whether or not this small sample is typical of the majority of farmers, simply because the majority left no relevant records. However, it is particularly telling that 521 voters, most of whom were farmers, signed a petition in favour of the Lamington National Park.

Bearing these limitations in mind some interesting patterns emerge. The observations range from the 1820s to the 1940s, from Northern Queensland 3,000 kilometres south to Gippsland. They were made by both men and women. Only one was made by someone (the former sailor Bryde) who clearly had no previous farming experience. On the other hand the O'Reillys and Kelly in Queensland and the Williams and Petersens in South Gippsland were farmers

Warwick Frost

before coming to the rainforests. All the evidence points to the farmers referred to here as hard-working and pragmatic.

These accounts blur the distinctions between farmers on the one hand and naturalists and tourists on the other. The dairy farmer Lou Cook worked on three projects with the distinguished naturalist Charles Barrett and Barrett gave him equal credit. A group of 53 South Gippsland farmers called their reminisces *The Land of the Lyrebird*, while Niels Petersen titled hers *Close to Nature's Heart*. Mrs Curtis volunteered to work with the scientist Ratcliffe. The O'Reillys worked with a range of scientists and once it seemed their title was clear, built a guesthouse for tourists. A group of leading citizens on the Atherton Tableland asked a Royal Commission into Land Settlement for assistance for agriculture, a government-built railway *and* the preservation of the rainforest around two lakes. At Tarra Valley, Bulga and Tamborine Mountain the locals campaigned for their picnic spots to be protected and promoted as National Parks.

Over the last two decades, there has been a great deal of research into Australia's environmental history. Much of this research has attempted to make sense of Australia's past in order to understand current environmental issues. However, with greater research, the past is being seen to be increasingly complex. A simple division between uncaring farmers and concerned urbanites is no more tenable for the past than it is today.

NOTES

1. In the end Ratcliffe argued that any widespread eradication campaign 'would be a waste of both time and money' (Ratcliffe 1938, 5). The fruit bats are a good example of how European settlement sometimes encouraged native plants and animals (Frost 1998). Generally found in sub-tropical Australia, since 1983 they have settled in Melbourne's Royal Botanic Gardens. By the late 1990s their numbers had grown to over 8,000 and they were causing tremendous damage to many trees in the Gardens. A cull in 2001 has sparked much debate. Ratcliffe went on to be one of the founders of the Australian Conservation Foundation.

2. Ratcliffe, 1938: 19–20.

3. Definitions of rainforest have been the subject of much debate, especially in Australia. The term rainforest was coined by Andreas Schimper in 1898 to cover what we now regard as tropical rainforest. Essentially it is an Eurocentric term, describing non-European forests which have characteristics different from European forests. Where the term has been applied to non-tropical forests it has led to controversy. The debate is not just about scientific nomenclature, for the term rainforest has acquired significant cultural and economic meaning. In Australia the debate has centered on whether or not dense wet eucalypt forest with rainforest understories and gullies can be defined as rainforest. For the purpose of this article I have used the broader definition, including most notably the forests of south Gippsland. For a more detailed consideration of the problems of defining rainforests see Figgis (1989), Cameron (1992) and Frost (2001).

4. For example, Figgis (1989: 28) argued that nearly eight million hectares have been cleared since European settlement.

5. Frost, 1997 and 1998.

6. Bryde, 1921: 59.

7. Sorenson, 1911: 187.

8. Bryde, 1921: 108–9; South Gippsland, 1972: 108–110 and 351; O'Reilly, 1944: 77–8.

9. South Gippsland, 1972: 283.

10. Frost, 1998: 140.

11. Hancock, 1930: 33–4.

12. Hancock, 1930: 56. Despite his great influence, some inconsistencies in Hancock have been commented upon. McIntyre (2001: 49) noted that having labelled settlers as invaders and raised issues of environmental degradation, Hancock stopped this discussion abruptly with, 'and yet, if a balance could be struck, it would probably be reckoned that alien men and animals and vegetation have enriched the soil more than they have impoverished it' (Hancock, 1930: 35)

13. Marshall, 1966: 2.

14. Bolton, 1992: 37.

15. Seddon, 1997: 91–2. The notion of settlers' fears was occasionally reversed (perhaps even parodied) by environmentally sympathetic contemporaries. For example, in 1883 La Meslee warned, 'there are parts of the bush ... which are really terrifying. There are areas, miles and miles in extent, where not a leaf or a piece of living bark is to be seen on the trees ... nothing can give any idea of the infinitely sad and desolate air of these dead forests ... it is a vision of Ezekiel: and the forest resembles a multitude of skeletons raising their long, fleshless arms to the sky' (La Meslee, 1973: 30).

16. Nash, 1982: 43.

17. Nash, 1982: 24.

18. Bonyhady, 2000: 14–39.

19. Powell, 1976; Stevens, 1991; Finney, 1993; Griffiths, 1996.

20. Ratcliffe's companion on Mount Tamborine was an amateur naturalist.

21. Griffiths, 1996.

22. Ritchie, 1989.

23. The Daintree River and Rainforest of north Queensland was named after the photographer/geologist.

24. Bonyhady, 2000: 105–7.

25. Bonyhady, 1993: 9–12. Ironically Streeton made his name as one of the Heidelberg School, which can be viewed as a reaction against the realist landscapes of the late nineteenth century. The political statements incorporated in his rainforest paintings have often been overlooked and many were dismissed as just ordinary rural landscapes.

26. Bonyhady, 2000: 102–3.

27. Ritchie, 1989: 133; Bonyhady, 2000: 103–4.

28. Seddon, 1997: 90–1.

29. Ritchie, 1989: 138.

30. Ritchie, 1989: 135–41.

31. Murphy, 1994: 107–9. In 1869 the officers from a visiting English naval fleet had visited the Fern Tree Gully in the Dandenongs (Ritchie, 1989: 141).

32. Bonyhady, 2000: 116.

33. Frost, 2000: 40–2.

34. Ritchie, 1989: 135; Bonyhady, 2000: 116.

35. There is an important ongoing debate over whether National Parks are primarily to protect the environment or to provide leisure and recreational experiences for visitors. Irrespective of that current debate, most national parks established in Australia before World War Two were primarily created in response to tourist demand.

36. For this and others see Anderson, 2000; Hall, 2000: 32–4 and Ritchie, 1989: 138–50.

37. Barr and Cary, 1992: 82.
38. National Farmers' Federation, 2001.
39. Tyrrell, 1999.
40. Harris, 1964: 25.
41. Harris, 1964: 45.
42. Cunningham, 1966: 48.
43. Atkinson, 1975: 3.
44. Backhouse, 1967: 111.
45. South Gippsland, 1972: 351.
46. South Gippsland, 1972: 367–8.
47. South Gippsland, 1972: 370.
48. Petersen, 1951: 18–20.
49. Barrett, 1940: 23 and 37.
50. South Gippsland, 1972: 20–8, 34–8 and 103–5.
51. Anderson, 2000: 70–2. The suspension bridge was probably the first elevated viewing structure for a rainforest anywhere in the world. It is also notable that both National Parks were given Aboriginal names.
52. Bryde, 1921: 102.
53. Bryde, 1921: 83–4.
54. Queensland Royal Commission, 1897: Q. 4147–50, 4247 and 4301–6.
55. Groom, 1949: 58–63.
56. A strategy identical to that used by John Muir with Theodore Roosevelt and Milo Dunphy with Prince Charles and NSW Premiers Neville Wran and Bob Carr.
57. Hutton and Connors, 1999: 33–4, 62.
58. quoted in Hutton and Connors, 1999: 34.
59. O'Reilly, 1944: 87.
60. Groom, 1949: 81–2.
61. Groom, 1949: 85.
62. O'Reilly, 1944: 87.
63. Correspondence between O'Reillys and Lands Department, 14 April 1919 and 22 April 1919, *O'Reilly Papers*.
64. O'Reilly, 1944: 116.
65. O'Reilly, 1944: 19–20. See O'Reilly, 1944: 116, for a list of scientists who visited the Plateau. One not mentioned was Francis Ratcliffe who visited in the early 1930s.
66. O'Reilly, 1944: 52.
67. O'Reilly, 1944: 61–4.
68. O'Reilly, 1944: 70.
69. O'Reilly brothers to Minister of Lands, 8 April 1919, *O'Reilly Papers*.
70. Land agent's memo., 10 April 1919, *O'Reilly Papers*.
71. O'Reilly 1944: 111–2, 115.
72. The name was chosen to both distinguish from, but also suggest something similar to the Blue Mountains.
73. O'Reilly, 1944: 133–4.
74. Frost, 2000.
75. The road (built by the O'Reillys in 1915) was effectively a 'stand-in' for one in the Blue Mountains.
76. Chauvel, 1973: 9–12, 67–8, 97.

77. Chauvel, 1973: 97–9.
78. South Gippsland, 1972: 289–95, 348–51.
79. Tuchman, 1989: xx.

REFERENCES

Anderson, Esther 2000. *Victoria's National Parks: A Centenary History.* Melbourne: Parks Victoria.

Atkinson, James 1975. *An Account of the State of Agriculture and Grazing in New South Wales.* Sydney: Sydney University Press. 1st pub. 1826.

Backhouse, James 1967. *A Narrative of a Visit to the Australian Colonies.* New York: Johnson. 1st. pub. 1843.

Barr, Neil and Cary, John 1992. *Greening a Brown Land: The Australian Search for Sustainable Land Use.* Melbourne: MacMillan.

Barrett, Charles c.1940. *On the Wallaby: Quest and Adventure in Many Lands.* Melbourne: Robertson and Mullens.

Bolton, Geoffrey 1992. *Spoils and Spoilers: A History of Australians Shaping their Environment.* Sydney: Allen and Unwin. 1st pub. 1981.

Bonyhady, Tim 1993. 'A different Streeton'. *Art Monthly Australia*, 61: 8–12.

Bonyhady, Tim 2000. *The Colonial Earth.* Melbourne: Miegunyah.

Bryde, C.W. 1921. *From Chart House to Bush Hut: Being a Record of a Sailor's Seven Years in the Queensland Bush.* Melbourne: Champion.

Cameron, D. 1992. 'A portrait of Victoria's rainforests: distribution, diversity and definition'. In P. Gell and D. Mercer (eds), *Victoria's Rainforests: Perspectives in Definition, Classification and Management.* Melbourne: Monash publications in geography.

Chauvel, Elsa 1973. *My Life with Charles Chauvel.* Sydney: Shakespeare Head Press.

Cunningham, Peter 1966. *Two Years in New South Wales.* Sydney: Angus and Robertson, 1st pub. 1827.

Figgis, Penny 1989. *Rainforests of Australia.* Sydney: Ure Smith.

Finney, Colin 1993. *Paradise Revealed: Natural History in Nineteenth-century Australia.* Melbourne: Museum of Victoria.

Frost, Warwick 1997. 'Farmers, government and the environment: the settlement of Australia's "wet frontier", 1870–1920'. *Australian Economic History Review*, 37(1): 19–38.

Frost, Warwick 1998. 'European farming, Australian pests: agricultural settlement and environmental disruption in Australia, 1800–1920'. *Environment and History*, 4(2): 129–44.

Frost, Warwick 2000. 'Nature-based tourism in the 1920s and 1930s'. In Ewen Michael (ed.), *Proceedings of the Tenth Australian Tourism and Hospitality Research Conference.* Melbourne: La Trobe University, pp. 39–46.

Frost, Warwick 2001. 'Rainforests'. In David B. Weaver (ed.), *The Encyclopedia of Ecotourism.* Oxford: CABI, pp. 193–204.

Griffiths, Tom 1996. *Hunters and Collectors: The Antiquarian Imagination in Australia.* Cambridge: Cambridge University Press.

Groom, Arthur 1949. *One Mountain after Another.* Sydney: Angus and Robertson.

Hall, C. Michael 2000. 'Tourism and the establishment of National Parks in Australia'. In Richard W. Butler and Stephen W. Boyd (eds), *Tourism and National Parks: Issues and Implications.* Chichester UK: Wiley, pp. 29–38.

Hancock, W. Keith 1930. *Australia.* London: Ernest Benn.

Harris, Alexander 1964. *Settlers and Convicts: Or Recollections of Sixteen Years' Labour in the Australian Backwoods.* Melbourne: Melbourne University Press. 1st pub. 1847.

Hutton, Drew and Connors, Libby 1999. *A History of the Australian Environment Movement*. Cambridge: Cambridge University Press.

La Meslee, Edmond Marin 1973. *The New Australia 1883*. London and Melbourne: Heinemann. 1st pub. 1883.

McIntyre, Stuart 2001. '"Full of hits and misses": a reappraisal of Hancock's *Australia*'. In D.A. Low (ed.), *Keith Hancock: The Legacies of an Historian*. Melbourne: Melbourne University Press, pp. 33–57.

Marshall, A.J. (ed.) 1966. *The Great Extermination: A Guide to Anglo-Australian Cupidity, Wickedness and Waste*. London and Melbourne: Heinemann.

Murphy, John 1994. *On the Ridge: The Shire of Mirboo, 1894–1994*. Sydney: Allen and Unwin.

Nash, Roderick 1982. *Wilderness and the American Mind*. New Haven and London: Yale University Press. 1st pub. 1967.

National Farmers' Federation 2001. *Environmental protectors – not pillagers*. www.nff.org.au/nr01/91. Accessed 13 August 2001.

O'Reilly, Bernard c.1944. *Green Mountains*. Brisbane: Smith and Paterson.

O'Reilly papers. Letters and clippings held at O'Reilly's Mountain Resort, Queensland.

Powell, J.M. 1976. *Environmental Management in Australia, 1788–1914: Guardians, Improvers and Profit*. Melbourne: Oxford University Press.

Petersen, N.A. 1951. *Close to Nature's Heart: A Tribute to the Pioneers of the Danish Settlement, East Poowong 1877*. East Poowong: author.

Queensland Royal Commission on Land Settlement 1897. 'Minutes'. *Votes and proceedings 1897 (3)*.

Ratcliffe, Francis 1938. *Flying Fox and Drifting Sand: The Adventures of a Biologist in Australia*. London: Chatto and Windus.

Ritchie, Rod 1989. *Seeing the Rainforests in 19th Century Australia*. Sydney: Rainforest Publishing.

Seddon, George 1997. *Landprints: Reflections on Place and Landscape*. Cambridge: Cambridge University Press.

Sorenson, Edward 1911. *Life in the Australian Backblocks*. London: Whitcomb and Tombs.

South Gippsland Pioneers' Association 1972. *The Land of the Lyre Bird: A Story of Early Settlement in the Great Forest of South Gippsland*. Korumburra: Shire of Korumburra. 1st pub. 1920.

Stevens, P. 1991. 'Plants, forests and wealth: vegetation conservation in Queensland, 1870–1900'. In B.J. Dalton (ed.), *Peripheral Visions: Essays on Australian Regional and Local History*. Townsville: James Cook University.

Tuchman, Barbara 1989. *A Distant Mirror: The Calamitous 14th Century*. London: Pan Macmillan. 1st pub. 1978.

Tyrrell, Ian 1999. *True Gardens of the Gods: California-Australian Environmental Reform, 1860–1930*. Berkeley: University of California Press.

'Folk-Ecology' in the Australian Alps:
Forest Cattlemen and the Royal Commissions of 1939 and 1946

Chris Soeterboek

...the landscape did not lie about like a shattered watch, its pieces inert and scattered, but like a deeply traumatised yet still living entity that somehow continued to function and that masked, often for decades, the full extent of its damages and infections...The European era had to end, if only from its own exhaustion and excess.

Stephen J. Pyne, 1991[1]

What manner of people caused this destruction? They were not greedy and ignorant as is too often stated; many of them had a background of hundreds of years of good British farming... but it was beyond human achievement to assess Australia correctly. It was more a new planet than a new continent

Eric Rolls, 1997 [2]

INTRODUCTION

These quotations frame different aspects of the environmental history of Australia. Stephen Pyne paints a picture of the destruction caused by alien farming practices being forced onto an unsuited Australian landscape. This is the picture of the Australian bush commonly painted by environmental historians: a burnt and eroded victim of rapacious European settlers and land-management, of ignorance and unsustainability and of cataclysmic change. In contrast, Eric Rolls makes the important point that the farmers, graziers and bushmen were not deliberately destructive or villainous. They observed their environment and developed and adapted practical knowledge to help them to exploit it most effectively. In settler societies around the world, the earliest European invaders pushed beyond the bounds of 'civilisation' and were forced to come to terms with new and different landscapes from which they tried to live and profit. In most of these societies, settlement outpaced scientific understandings of the new environments so settlers created their own, often complex, bodies of knowledge about local landscapes based on observation of the land, its effect on the condition of their animals and their traditions of the past.[3] This essay is a case study of these understandings,

Environment and History **14** (2008): 241–263.

the 'folk-ecology' that forest cattlemen of the Australian Alps developed for the rugged and isolated landscape in which they lived.

The Australian Alps in the south eastern corner of Australia constitute the continent's major alpine environment. While comprising only 0.3 per cent of the land mass, their ecological and cultural significance far exceeds this small figure. Below the snow line, the mountains are covered by wet-sclerophyll forests dominated by the highly valued eucalyptus alpine ash (*Eucalyptus delegatensis*) and mountain ash (*Eucalyptus regnans*). These forests stretch south and west onto the foothills of the Alps in Victoria, clothing the catchments of the main Victorian rivers so important for urban water supply and agricultural irrigation. Rich summer grasslands, valuable (and highly flammable) forest timbers, and the proximity to Melbourne's water supply have combined to create a European history where pastoralists, foresters, timber-workers, and water supply engineers have competed for control of the land.[4]

Folk-ecology or folk-biology are terms more commonly used by anthropologists describing small-scale subsistence societies:[5] one does not frequently

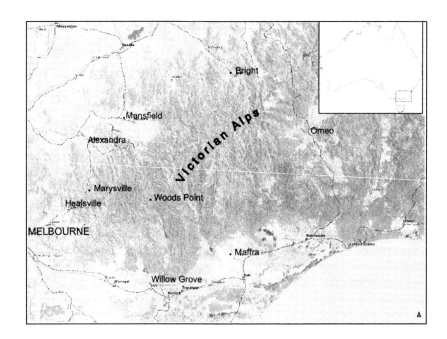

FIGURE 1. The Victorian Alps showing forested areas. Adapted from Victorian Department of Sustainability and Environment, <http://www.dpi.vic.gov.au/dpi/vro/vrosite.nsf/pages/os_dse_mapshare>.

encounter them in historical discourse. However, historian Thomas Dunlap has looked at the folk-biology of settler societies in Australasia and North America and in his book on history of the mountain ash forests of Victoria, Tom Griffiths also commented on the 'folk reality' revealed in a 1939 Royal Commission. Similarly, Stephen Pyne's exhaustive work on the fire history of Australia described the folk-*practices* of graziers.[6]

I want to use the concept of folk-ecology as a way of examining how rural land-managers saw their local environment, why they used methods now seen as destructive, and why mountain graziers were (and still are) in conflict with proponents of scientific land management. There has been a tendency among environmental historians to judge past farmers, graziers and bushmen harshly, but such judgements can only be valid in the context of contemporary knowledge of problems, causes and practical remedies.[7] This essay aims to map out this contemporary knowledge in some detail. There was a belief, among bushmen, that fire was an inevitable part of the Australian environment and this justified their burning practices. The origins of this view can be traced back to bushmen's perceptions of 'the early days' of European settlement. There were contemporary warnings that cattle grazing and fire were degrading the alpine environment but these came from ecologists and foresters. The grazier's world-view made it difficult for them to accept they had caused this damage to the environment and for those who had 'direct experience' of the land to accept the 'theoretical' and 'impractical' ideas about the environment put forward by scientists.[8]

'A full scale enquiry into Australian bush culture'

Black Friday was 13 January 1939 when the bush-fires ravaging South Eastern Australia culminated in a day of devastation and loss of life.[9] After the fires had cooled, recriminations began. Foresters blamed graziers for illegally burning-off in dangerous conditions and graziers accused the Victorian Forests Commission of not carrying out enough protective burns. The newspapers, depending on their orientation, cited the criminal irresponsibility of graziers or the professional incompetence of the Forests Commission.[10] In this context a Royal Commission was launched into 'the Causes of and Measures Taken to Prevent the Bush Fires of January 1939'. The Commissioner was Judge Leonard B. Stretton, who would go on to conduct two more Royal Commissions into bushfire related issues. The transcripts of evidence from this commission provide a wide-ranging review of attitudes towards fire in Australian rural communities; Griffiths described it as 'nothing less than a full scale enquiry into Australian bush culture'.[11] Within its pages the tensions that existed between millers and timber-workers, between foresters and graziers, and between city and bush are played out. This Commission questioned the sustainability of Australian fire practices and has been seen as a watershed in the environmental history of Australia.[12]

Chris Soeterboek

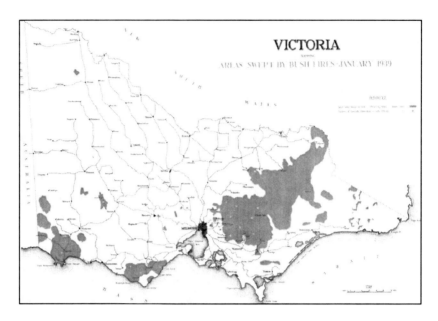

FIGURE 2. Areas Affected by the 1939 Black Friday Bushfires. Victorian Department of Sustainability and Environment, 'Black Friday 1939 Victorian Bushfires' <http://www. dse.vic.gov.au/DSE/nrenfoe.nsf/LinkView/C4BCA40C95A4C061CA256D9600144420D-8AC9C23269FA53B4CA256DAB0027ECC4 >.

The issues raised in the 1939 report were taken up again by Stretton in the 1946 Royal Commission 'to enquire into forest grazing'. It was established in an international context of growing concern over soil conservation and land degradation that had developed from an awareness of the massive ecological damage caused by poor farming practices, most spectacularly in the 'Dust Bowl' of the USA and the dry north-west of Victoria.[13] This Royal Commission was given wider terms of reference and looked at rural people's attitudes towards their local landscapes and the signs of environmental damage that were appearing around them. Stretton's report became an influential document for post-war utilitarian conservation groups.[14]

Both these sets of transcripts provide an insight into graziers' folk-practices and understandings that would otherwise come to us through the critical lens of foresters' reports.[15] Graziers gave reasons for their fire practices to the Royal Commissions and explained, or explained away, the emerging evidence of land degradation. Although they give voice to a group largely absent from the historical record, the Royal Commissions, like any investigations with political origins, were subject to powerful vested interests. Stretton was well aware of this; in his 1939 report he admitted: 'the truth was hard to find...Much of the evidence was coloured by self interest. Much of it was quite false. Little of it was wholly truthful'.[16] Graziers feared they could be held responsible for Black Friday so emphasised the protective aspects of their fire practices and were reluctant to admit their use of fire to improve pasture. The 1946 Royal Commission was less politically charged and those who gave evidence were more willing to openly discuss their land-management practices, although the Commission was plagued by rumour that its recommendations would deny pastoralists their grazing licences. As a result, some graziers attempted to present their land-management practices in conservationist terms.[17]

The 'bushman' of the high country

From very early in the European history of Australia, graziers were attracted to the high-rainfall grasslands of the Australian Alps and by the early nineteenth century had occupied the open woodlands and treeless plains with huge but isolated snow leases where they practised transhumance cattle grazing. Gradually, as more accessible land was subdivided for closer settlement, these cattlemen relied more upon the high plains and as the dependence on this landscape increased, so did their use of fire to promote grass growth and clear the forest understorey. In time, the landscape of the Australian Alps and the challenges it posed to graziers became central to their identity as 'bushmen' and their long occupation on grazing leases made them think they had a right to use the land.[18] Bushmen had the most intimate work relationship with the alpine environments. They generally lived and worked away from settlements and considered themselves practical men with a wide range of experience and knowledge of their occupation and environment. They often had long family associations with these districts, their parents or grandparents being the some of the earliest pioneers. For those who gave evidence to the Commissions, 'bushman' was also a label given to the most experienced and was a title held with pride. Representatives of the Forests Commission of Victoria, the Melbourne Metropolitan Board of Works (MMBW), ecologists and scientists also gave evidence to the Royal Commissions. They tended to be proponents of modern, 'enlightened', utilitarian land-management, based on scientific principles, and so were generally allied against graziers' traditional practices and ways of thinking. Despite this, they bickered frequently with each other over their differing views of how much management should be applied to forests. For example, the MMBW strongly advocated total exclusion of fire and 'natural' succession of forests, mainly because of its concern for water quality. However, they were similar in their official conception of the 'natural' condition of the Australian environment and their adherence to the influential ecological succession theories of the American prairie ecologist Frederic Clements.[19] These factors distinguished them from folk-ecologists.[20]

FIRE

Burning-off irreversibly changed the spread, density and species composition of plant communities in forest and alpine areas.[21] The problems of fire control, fire-weeds, and fuel reduction were central concerns of the Royal Commissions in both 1939 and 1946. The majority of evidence given to the Commissions was about fire practices and the conflicts between foresters and graziers centred on their vastly different views of how fire could be managed. Graziers saw it as inevitable and necessary, while foresters saw it as human-induced and preventable. Bushmen conceptualised the period of early European settlement as a time that established a template of an 'original', 'natural' or 'normal' condition for the environment, which in turn strongly influenced how they judged the conditions of their own time. By contrast, foresters invoked 'science' rather than 'history' as the authority for their expertise.

Fire's place in the environment

Graziers distinguished between two types of fire; uncontrollable bushfire and their own 'burning off'. This dualistic view can be seen in the seemingly contradictory comment by the grazier John Findlay: 'it has never had a fire in it until now... I burned it whenever it would burn'.[22] Findlay distinguished between the uncontrollable bushfire that threatened European settlement and the burning he carried out to prevent it. He also illustrates the ambivalence many rural people felt towards fire. On the one hand, it was their most powerful tool, something they believed they could control and something that allowed Europeans to better exploit their environment. On the other hand, fire was an uncontrollable force that destroyed their fences, killed their flocks and herds and denuded their pasture.

Many graziers also saw fire as an inevitable part of Australian life. Harry Treasure, an 'old bushman' from a famous grazing family, stated: 'I firmly believe that it would be impossible to absolutely prevent fires altogether'.[23] Almost all the high country graziers who gave evidence to the Royal Commissions shared his view. They believed forests would inevitably have fires, either caused by humans or by lightning or other 'natural' causes. This assumption encouraged a blasé attitude to the existence of fire in the landscape: their familiarity with it bred complacency. John Findlay summed up this fatalistic view: 'you have to control nature and man both, and I do not see how it is possible to do it'.[24]

Fire practices

Burning practices also show most sharply the divisions between the folk-ecology of the bushmen and scientific-forestry. The reasons graziers gave for their burning practices fall into three categories; burning for protection from bushfires, burning to improve pasture and burning to clear land. Graziers believed a forest left in a 'dirty filthy condition' was a menace to the surrounding country that would eventually ignite in a bushfire. In the context of the Black Friday bushfires, graziers emphasised the protective role of their fires. Justifications for protective burning were based on the belief that fire was an inevitable threat in Australian life. John Bussel stated, 'It would be better if burning could be abandoned altogether, but I think that is impossible. A light burning here and there, now and then cleans up a certain amount of rubbish and prevents a heavy fire like that of 1939'.[25] Black Friday was frequently invoked as a warning of what would occur if forests were left unburnt. The blame for this was often placed squarely onto the Forests Commission, whose practice at the time was not to burn forests. They were condemned for 'refusing to burn and clear up the floor of the forest'.[26] Graziers believed burning was the only way to reduce the ferocity of bushfires, a view James Treasure expressed most clearly: 'We have to adopt the method of immunisation with the menace itself.'[27]

In reality, the major reason graziers burnt the forests was to encourage new grass growth, called 'sweet feed' or 'green pick', which flourished in the nutrient-rich ash after a fire.[28] Some brave individuals admitted to this practice, such as John Findlay who burned 'to get good grass', and George Purvis who made it plain, 'I never made any secret of the fact that we burn our leased land in order to get good feed for cattle.'[29] The final reason for burning was to make the land more open and suitable for grazing. Again, George Purvis claimed, 'if you take over a forest block, there is only one way to clear it, and that is with fire and the axe'.[30] Similarly, Andrew Finn, a lifelong stockman who gave evidence in Corryong, claimed that on the alpine plains in the Mt Jagungal area of New South Wales, fire was explicitly used to clear up scrub and convert long unpalatable grass to 'excellent grass about 6 inches high'.[31]

The early days of European settlement

These views and practices were based on assumptions about what the Australian environment was like in the 'early days'. Many of the graziers who justified and defended their use of fire based this on the example set by the pioneers of their district. The environmental conditions inherited from pioneering practices were often seen as the 'natural' and 'normal' state of the environment. Reliance on 'traditional knowledge' was common among graziers. A forester in the 1930s had complained that 'old ideas and practices are accepted without question', and the former grazier Percy Weston confidently cited the 'information at my disposal – that is, the opinions of the oldest residents in my district'.[32] Fire was seen by graziers as a primordial component of the Australian environment because it had been used by early pioneers and even earlier by Aboriginal people.

Foresters and scientists at this time tended to see fire as something introduced to the environment and therefore outside it.

The reference point for the 'early days' described by graziers differed depending on the location of settlement. Some places, like Omeo, had been occupied by graziers for more than a hundred years, while other areas deep in the rugged forests had been settled in the memory of older bushmen. Despite this, there were remarkable similarities in how graziers thought their predecessors had managed the land. A characteristic account was given by grazier Neil Gow of Wandiligong near Bright. His story began with prospectors working in the hills between the 1860s and 1890s who lit fires frequently, 'anywhere it would burn'.[33] In other areas, these early settlers were timber-splitters or bullockies. Edward Leeder from Marysville recalled timber-splitters lighting fires in the mountains to protect their timber palings, he said 'they burned patches of scrub… but would not do the forest much harm'.[34] The condition of the country at this time was often described in positive terms. Ovens Valley grazier William Blair remembered, '50 years ago when it had the appearance of a park and most of it was grassy… there was plenty of grass'.[35] The good conditions were nearly always linked to the burning regime these pioneers carried out. Reginald Barnewall from Alexandra described the early settlers 'course of systematic burning …The fires were light and the country was clear'.[36] John Findlay, the self-declared 'first white man' in the forests around Rubicon, claimed that cattle runs were always 'burnt whenever they would burn' and 'the floor of the forest was as clean as a whistle'.[37] When asked whether the early forests were 'in a natural condition', Edward Leeder answered 'Yes, the ridges had been burned, the splitters kept them clean'.[38] The success of this early burning regime was usually attributed to the expertise of bushmen. Clarence Poole from Ten Mile near Woods Point summed this up clearly: 'their properties were periodically burned off and the fire was never destructive to the forest or other properties, because those men knew when to burn'.[39]

The final stage in this story of the 'early days' was the suppression of fire in the forests, a move that resulted in the accumulation of scrub, leading to destructive bushfires. Blame for this was placed on the much despised Forests Commission, who had limited the extent of forest burning soon after their formation in 1918.[40] Edward Leeder in 1939 declared: '[the bushfires] killed big stretches of mountain ash which were beautiful'.[41] Such professed concern for the safety of the forests may have been influenced by some graziers' desire to portray themselves as friends of the forests during the politically charged 1939 Royal Commission. However, the pervasiveness of this concern suggests a genuine appreciation of the forests destroyed on Black Friday.[42] The Forests Commission's practice of limiting burning, a 'primitive' method of forestry, was seen by some graziers as the end of an early 'golden' period of thick grass, open woodlands and frequent burning-off.[43] Such views of the 'early days' became the template against which the present was judged: the present was plagued by rabbits, thick scrub and destructive bushfires.

Aboriginal burning

Griffiths has commented that the language used by forest workers in the 1939 Royal Commission: '"burning to clean up the country" — was uncannily like that of Aboriginal people'.[44] Some graziers drew on the connection between Aboriginal burning and their own practices to defend their use of fire. In 1939, the bushman Peter O'Mara argued the thick undergrowth of ash forests 'did not exist in the graziers' time or the aboriginals' time' because 'they kept the floor of the forest clear'. He claimed Aboriginal man did this 'so that he could spear kangaroo and wallaby'.[45] Alfred Saxton, a former sawmiller and farmer from north Gippsland, showed a similar understanding of Aboriginal fire use: 'I believe aboriginals fired periodically to get feed to bring game'. He suggested 'if a tribe of aboriginals had been let loose in that forest and had carried on in their old ways, they would have preserved that forest'.[46] In 1946, William Kelly, a farmer and Shire President from Maffra, argued 'fires were continuous in those days' because of Aboriginal occupation.[47] These examples were unique in the evidence given and the impression from the transcript pages is that those in the courtroom were surprised by the examples and did not take them seriously. After O'Mara's comment, Stretton quipped: 'you would be in favour of taking the Forests Commission out of the forest and putting in a tribe of blackfellows to look after it?'[48]

The surprising thing about these extracts is that they foreshadow ideas about 'the sophisticated environmental stewardship of Aboriginal people' described in Rhys Jones' famous study of 'firestick farming' thirty years later in 1969.[49] In these extracts there seems to be a clear understanding of what constituted Aboriginal burning that is remarkably close to current understandings.[50] Aboriginal burning was used to prove a point. The question of what occurred 'before the white man' recurred frequently in the Commission's evidence because the fire regime of this period was sought as a possible solution to Victoria's destructive forest fires. Scientific resource-managers ignored or downplayed Aboriginal occupation and confidently referred to this period as the 'original' state of the forest to which they aimed to return their forests by excluding fire.[51] In the face of this, graziers wanted to provide an ancient precedent for their burning regime, presenting a picture of 'a big fire sweeping the country from end to end before the white man went there'.[52] Human use of fire, they argued, was part of the 'natural' and 'normal' state of the environment. This connection was explicitly made by Percy Weston: 'in the days preceding settlement by the white man, there existed a natural balance, which the aborigine, who is acknowledged to have possessed a "fire conscience" which shames the present generation, played no small part in preserving'.[53] Aboriginal burning, it was argued, had protected the forests for hundreds of years. In essence, the Aboriginal burning example was used to justifying the practices of the graziers giving evidence to the Royal Commission.

On the surface, the burning regimes of Aboriginal people and graziers appeared to be remarkably similar. Both aimed to promote landscapes favourable for production of food and both burned grasses to promote green pick for animals (be they marsupials or sheep and cattle). European burning – especially by nomadic splitters, prospectors and bullockies – may have had a similar environmental result to Aboriginal mosaic burning. The difference lay in the use of the land that was burned. Aboriginal people burned to distinguish between different native flora and conformed to the natural rhythms of Australia. Graziers' burning was confined to specific places and times, determined by introduced species that did not conform to Australian conditions.[54]

On this matter, the views of 'educated' men on the Commission differed greatly from those of the bushmen. In the scholarly opinion of the time, Aboriginal people were depicted as part of pristine nature and assumed to have had little environmental impact.[55] In contrast, graziers saw Aboriginal people as having significant but not destructive impacts on the environment. This difference can be seen most sharply in an exchange between the farmer William Kelly and the forestry representative Geoffrey Dyer. Dyer attempted to lead Kelly into admitting aboriginal people were too few in number to have much environmental effect and that 'the black man was far more interested than the white settler in conserving the natural resources of the country'. Kelly disagreed on all of these points arguing aboriginal people 'would probably burn more than we do now' and that 'the black man would burn without any thought of the future'.[56] Dyer presented a picture of the 'noble savage' while Kelly showed a more realistic appreciation of human impact on the environment. Despite such testimony, Aboriginal people did not easily fit into the idea of a scientific wilderness and Stretton ignored them in his reports.

Scientific wilderness

The question of what fire regime had existed 'before the white man' shows most sharply the difference between folk and scientific perspectives. Scientific forestry tended to downplay human disturbance, a perspective that could be described as scientific wilderness. The evidence given to the 1939 Commission by the Chief Inspector of Forests for the Commonwealth of Australia, Charles Edward Lane-Poole, was the most authoritative account of this perspective. Lane-Poole, the quintessential imperial forester in training and temperament, concluded 'fires in the blackman's country were very small in comparison with those of our day' and he argued fires were less frequent and of low intensity during that time.[57] He was typical of many of his scientifically trained colleagues in his adherence to Clementsian ecological succession theory. They believed that if fire was excluded from forests they would 'succeed' to a 'climax' fire-proof state of 'thinner shrubs as the tree canopy increases'.[58] Lane-Poole assured the Commission 'we can prevent fires anywhere ... it is simply wrong to regard fires as inevitable'.[59] The 'natural' or 'original' state of the forests was assumed to be almost without fire, and mainly without humans. The significant influence of ecological theories on these 'expert' witnesses is consistent with the greater status and authority that was accorded to ecological conservation at the time in Australia and internationally.[60] Maisie Fawcett, an ecologist from the Botany School at the University of Melbourne had been sponsored by the Soil Conservation Board to study the erosion and ecology of the alpine catchment for the economically important Murray River.[61] Her evidence and that of the scientifically trained foresters ultimately carried more weight with the Royal Commission than the bushmen. Stretton's 1939 report stated: 'When the early settlers came to what is now this State, they found for the greater part a clean forest. Apparently, for many years before their arrival, the forest had not been scourged by fire. They were in their natural state... But the white man introduced fire to the forests.'[62] Griffiths has persuasively concluded that Stretton outlined an historical and ecological vision of pre-*European* Australian nature that was 'stable and relatively unmodified by humans', a view that was common to his generation.[63]

This view of bushfire as an alien, unnatural element in the environment was also widespread at this time. In popular culture, bushfire was commonly depicted as a military struggle between humans and fire. Editorials from *The Argus* described the 'onslaughts of the firefighters' who were 'those great hearted men who cheerfully battle with the elements' and when 'fire threatens fare forth to fight it'. In these accounts, fire was presented as an avoidable tragedy and hope was always expressed fires would be prevented in the future.[64] The general view of fire was that of the senior foresters, a dangerous alien element in the environment. Pyne suggests that such ideas had become deeply ingrained in English thinking, originating in eighteenth century changes to English farming that rejected the use of fire as dangerous, chaotic and uncivilised.[65]

SEEING HUMAN IMPACTS

Graziers were generally reluctant to admit environmental changes caused by their land-management were destructive. This was based on their understanding of the dynamics of the environment, especially the notion of a fire cycle.

Chris Soeterboek

The fire cycle

In his report from the 1939 Royal Commission, Stretton gave a clear description of the problem with European burning practices in south-eastern Australian forests:

> The white man introduced fire into the forests. They burned the floor to promote the growth of grass and to clear it of scrub which had grown where, for whatever reason, the balance of nature had broken down. The fire stimulated grass growth; but it encouraged scrub growth far more. Thus was begun a cycle of destruction which can not be arrested in our day. The scrub grew and flourished, fire was used to clear it, the scrub grew faster and thicker, bushfires, caused by the careless or designing hand of man, ravaged the forests; the canopy was impaired, more scrub grew and prospered, and again the cleansing agent, fire, was used.[66]

Stretton clearly saw the cause and effect of the fire cycle based on the evidence he had heard from foresters, timber workers, farmers, bushmen and graziers. Graziers and bushmen were more reluctant to see changes in the environment as part of this sort of fire cycle.

In the 'early days' of settlement, the environment was said to have had 'the appearance of a park'; the floor of the forest was 'as clean as a whistle' and open enough that 'one could drive a mob of sheep anywhere'.[67] This later changed to a thick, scrubby and dirtier condition of which graziers frequently complained.[68] The change was noticed by most graziers because it affected the livestock-carrying capacity of their land. William McCoy, a grazier from Ensay, contrasted the '3000 head of cattle' capacity in 1907 with the 'little more than 500 or 600 cattle' in 1939.[69] A more sophisticated observation was made by Harry Treasure. After the bushfires, he said, 'there will be more grass than we know what to do with' which would last 'for one season, but each year the undergrowth will grow up more' until 'in a few years later it will make an enormous litter on the ground and grazing will be practically nil'.[70] There was a divergence of views on the cause of this 'dirty' condition. John Langtree explained that 'a slow burn will not bring up more undergrowth, but a heavy burn will'.[71] This distinction was also shown in the comments of John Findlay that, 'the fire of the graziers is not the fire that kills trees'[72] and of Buxton grazier Maurice Keppell who blamed 'heavy fire' for the originally 'almost park-like' landscape to have 'thickened up considerably'.[73]

The closest most graziers came to Stretton's notion of a fire cycle was an admission that poor practices by inexperienced graziers may have caused damage. Most of the graziers who gave evidence considered themselves experienced and conscientious bushmen who would use fire wisely and knew the country intimately. John Cameron from Mansfield claimed to 'never have known a grazier who would light a dangerous fire' and Andrew Finn, a stockman, emphasised that 'judgement was used with the burning'.[74] Correct burning according to Weston ensured a harmless 'slow' or 'light' fire when done at the correct time of year.

He explained: 'If you burn during the spring and summer you attack the natural grass covering and promote the growth of bracken and scrub. If you burn in the autumn you kill the growth of bracken and scrub.'[75] Stretton was influenced by this distinction in his 1946 report where he commented, 'the permitted grazier of long standing behaves somewhat better than the newcomer'.[76]

Finding other causes

Graziers and bushmen suggested other causes for the observed changes. The link between rabbits and flammability of environments was a common one among those who gave evidence. Arthur Pearson, a stock agent from Omeo, blamed rabbits as 'the first factor' while 'bushfires formed the final factor'.[77] Similarly, Percy Weston claimed 'there are two main causes – the misuse of fire, and over-grazing, for which the rabbit is totally to blame'.[78] Another response was to deny that change was human induced at all and claim the effects of the fire cycle were 'natural'. William Blair, an Ovens Valley grazier, claimed the drying out and less snow on alpine grass-plains (usually ascribed to fire) was rather 'the seasonal cycle'. He claimed he could not see any ecological change other than 'seasonal conditions'. A similar comment was made by Whitfield grazier Herbert Swinburne who simply gave the cause as 'nature'.[79]

Unlike the foresters and Stretton, bushmen regarded a frequently burnt environment as the 'natural' state and did not see humans as agents of destruction. This is illustrated by grazier Michael McNamara's statement that 'many years ago, *before this "fire business" was started*, it used to be burnt according to our ideas'.[80] The benevolent environment of the 'early days' was an example from which to take methods that could 'bring the country back to normal'.[81] In contrast to this, many graziers blamed destructive bushfires on the abandonment of these traditional burning practices by the Forests Commission.

Folk understandings of human impacts were also based on bushmen's own observation and casual experimentation. A good example of this was Percy Weston's explanation of bracken fire ecology: 'If you burn in the autumn you kill the growth of bracken and scrub. Their growth only takes place during the summer months. They are dormant for the six months during the winter, as far as I can see. When that growth receives a set back, the grass covering, without any competition, has a chance to make a headway.'[82] Comments on the fire ecology of various plant species were common throughout the Royal Commission transcripts. Anthony Mangan observed 'after a hot fire seedlings germinate' while Fredrick Barton observed 'fires help the bracken'. Charles Lumsden clearly articulated the close relationship between autumn burning, soil moisture, ground covering and rainfall runoff in very similar terms to Weston.[83]

Graziers tended to downplay or even deny signs of degradation in their local environment or they ascribed causes for them that were outside their own environmental footprint. What Stretton described as a fire cycle – a destructive downward spiral – was for them not a cycle at all but rather the *abandonment* by inexperienced foresters and government officials of sound burning-practices in favour of 'locking up' the forests. Environmental problems were not the long-term effects of inappropriate and excessive exploitation, but were aberrations, caused by rabbits and inexperienced men, that could be overcome by returning to practices associated with the 'early days'. In this context, Griffiths' claim that the Black Friday bushfires were 'a culmination of a century of white settlement and environmental practice' gains additional resonance.[84]

FOLK VERSUS SCIENCE: FIRE ECOLOGY

The regeneration of mountain ash forests requires a hot fire that kills mature trees and causes their seeds to be released to germinate in the rich and insect free ash bed on the forest floor. David Ashton, an ecologist who devoted his life to studying these trees, described this process as a 'miracle of timing'; if the forests were burned too soon (up to 15–20 years old) the trees died without producing seeds. If the forest was without fire for hundreds of years it was overtaken by cool temperate rainforest.[85]

The Royal Commissions paid particular attention to the effect of fire on mountain ash forests because this was a valuable timber, but also one that posed great fire danger by growing thickly in often rugged terrain. The transcripts of evidence show a rich body of folk-understanding of the relationship between mountain ash and fire; one that did not accept the authority of science. Edward Leeder observed it was 'killed by intense fire' but also observed that 'after smaller fires ... the mountain ash would always come up'.[86] Peter O'Meara, a bushman and timber-getter, claimed 'they are delicate timbers, and will not stand the slightest fire. Those timbers will not stand trampling or being knocked about'.[87] The vulnerable period of young mountain ash trees when the entire tree population would be wiped out if burnt by fire was also understood by many bushmen. John Findlay described the situation after a fierce fire: 'I saw saplings 10 to 12 ft high, so that shows it will grow again. However, if the fire had gone through it before the trees were matured enough to drop seeds, there would have been no young forest'.[88] According to this folk-knowledge, bushmen advocated light burning of mature ash forests claiming these fires did not damage the trees.[89] There were also some who advocated not burning ash forests at all and instead burning the surrounding less valuable (usually messmate, *Eucalyptus obliqua*) forests.[90] These examples show a practical understanding of the effects of fire on forests, one that utilised knowledge about forest ecology to harvest timber and burn the forests to allow grazing.[91] A forester observed of cattlemen that 'although they know the practical side of the job from A to Z ... there is no attempt to probe into things and find out the why and wherefore'.[92] Similarly, a passing comment by Maisie Fawcett about a plant species that 'is one of the few alpine plants to which the cattlemen have paid sufficient attention to give it a name' also shows an assumption that people who used their environments tended to only be interested in useful plants and landscapes.[93]

The transcripts reveal understandings as much ecological as utilitarian. The anthropologist Scott Atran has also observed that for many societies in close contact with the natural environment general folk-understanding extends beyond useful plants and animals.[94] There are examples in the Royal Commission transcripts of bushmen who provided sophisticated explanations of the how and why of ecological relationships. A good example is from Alfred Saxton, who we have already heard from regarding Aboriginal burning. He commented:

> There is no question about it, the eucalypt is a fire plant. It comes from the seed. It is a query even among forestry officers and bushmen as to actually whether the seed floats down after the fire or whether it is there beforehand. Generally the bushman's idea is that it is also in the ground, and that is my idea.[95]

This comment hints at an ecological understanding. Saxton's comments suggest there was a body of knowledge and discussion among bushmen about the fire-ecology of the eucalypt. This was not just confined to how quickly it could be burned or how soon it could be harvested. Other bushmen observed the role of fire in the germination of eucalyptus seeds, and others again gave sophisticated explanations of the relationship between bracken, fire and soil erosion.[96]

Alfred Saxton went on from his explanation of eucalypts to tell a story that gives an indication of how folk-ecological knowledge was acquired:

> a settler that I knew was walking over a paddock that had been burnt years before. He noticed a square iron plate lying on the ground. He did not know how it got there, but he picked it up and threw it aside. The space where it had been was left uncovered. There had been a good fire over the ground. Some months afterwards that square came up in native oats. You may get a fire that brings up the native oats. We imagine that when there is a certain amount of dampness in the ground, when a fire goes over it, it brings up the native oats. That would have to be burnt if you wanted a growth of forest or for seedlings to come up later.[97]

This story shows most clearly the basis of folk-ecology in common day-to-day observation. Casual experimentation like this was also practised by Percy Weston who told the Commission 'I have carried out tests in the last two years just to see how vulnerable a young tree is to fire'.[98] Weston and Saxton were perhaps unique in the sophistication and interest they showed in studying their environment, but both were old experienced bushmen whose common-sense observations were respected among the rural community.

The most significant difference between forestry and folk views of mountain ash fire-ecology was the reluctance of officials to see any role for fire in the process. Rather, they emphasised the danger posed to mountain ash by fire. The fierce opponent of burning Alexander Kelso declared 'even a small fire does destroy the mountain ash sooner or later'.[99] However, there were also some foresters, most often in the field like Adrian Beetham, who claimed it would be desirable to use 'light fires' to keep down scrub so long as it did not harm timber.[100] Pyne has described this apparently contradictory practice as something most foresters considered 'wrong and dangerous' but temporarily necessary in Australian conditions. Tensions that had existed between practical and 'educated' foresters in the early days of Australian institutional forestry from the 1910s to 1920s still resonated in the 1930s and 1940s.[101] In fact, the Forests Commission representatives on the 1939 Royal Commission were frequently concerned that some junior or field foresters might contradict official Forests Commission policy.[102] The Commission's reluctance to tolerate fire was based on its adherence to the Clementsian ecological succession model which argued that plant communities would succeed to a stable 'climax community', becoming increasingly stable with each stage.[103] Lane-Poole claimed that fire would restart the succession therefore preventing a stable community.[104] These strong scientific frameworks set limits on how foresters, even traditional 'practical' ones, could construct theories about the ecology of their local landscapes. They relied less upon their own observations about the local environment because they were trained by an institution with a systematic and organised school of thought.[105]

Authoritative Forms of Knowledge

To understand why differences existed between the observations made by scientific and folk forms of knowledge it is worth referring to Atran's general observations of the relationship between science and common-sense beliefs. Atran distinguished common sense, which he defined as a human's 'spontaneous apprehension of living kinds', from speculative thought systems (which could include scientific or religious knowledge) that he saw as not 'spontaneously elaborated' and not always in apparent accordance with common sense.[106] Dunlap applied this theory further suggesting the observation inherent in scientific thought may not be visible to non-specialists and may even appear wrong.[107] The distinction provides a plausible model for the differences between folk-ecology and scientific forestry. It also provides a starting point from which to analyse the views bushmen held towards scientific knowledge and it provides the beginning of an explanation why science was not considered authoritative by most bushmen.

Alfred Saxton's explanation of fire-ecology suggested 'It is a query even among forestry officers and bushmen as to actually whether the seed floats down after the fire or whether it is there beforehand'.[108] Saxton did not see science as an authoritative explanation of the environment. He viewed it as on-par with, or even inferior to, the folk-ecology theory. When Maisie Fawcett gave evidence to the Royal Commission in Omeo 1946 this scepticism was evident. During her evidence on the early stages of erosion and its relationship to fire, she was interrupted by an objection from one of the cattlemen present.[109] Those who gave evidence, following her, disputed almost all of her claims and the grazier John Gibson made a pointed observation that the 'so called experts' didn't know the country as well as he did.[110] In the view of many bushmen, scientific knowledge was the theoretical and impractical basis of foresters' poor fire management. Edmund Cornwall, a bushman from Noojee, criticised '*impractical* forest men' who copied their practices and views 'from a conference on *theoretical* forestry in America' while Alfred Webb, a dairy farmer from Willow Grove, complained of foresters that 'they will not know the conditions when they are educated in a college'.[111]

Atran's conclusions provide an insight into this scepticism of science. He observed that lay people only accept modifications of folk-biological knowledge if the scientific alternative 'proves compatible with everyday common sense realism'.[112] Dunlap suggested the people who had direct experience with their environments would be less likely to accept the explanations of scientists.[113] Bushmen placed the highest value on 'practical common sense; the highest praise graziers paid to each other was to label them a practical bushman. John Cameron when asked if he agreed with the preceding evidence of the graziers William Lovick and John Bostock, answered he did, simply because they were '*practical* bushmen'.[114] Graziers and bushmen rejected scientific explanations from elsewhere in favour of their home-grown folk-ecology.

CONCLUSION

The folk-ecology of cattlemen in the Australian Alps was based on a particular conception of the 'early days' of pioneer settlement in Victoria. Fire practices and ecological knowledge were drawn from a period they considered 'natural' and 'normal', an imagined landscape of open forest and pastures, which fire had an important ecological role in maintaining. This knowledge challenged the evidence of environmental degradation and the remedies suggested by ecologists. In contrast, scientific forest management at this time imagined a 'natural' environment, which largely excluded fire and which regarded it as an introduced and destructive element in the landscape. In hindsight, these folk-understandings of the role of fire are in many ways closer to current scientific understandings of Australian fire-ecology. It is significant that Stretton was sufficiently influenced by folk-ecology to advocate a new direction in Australian fire management: one that accepted the presence of fire in the Australian landscape and committed itself to its use in fire management.

NOTES

1. Stephen J. Pyne, *Burning Bush* (New York: Henry Holt and Coy, 1991), 276.
2. Eric C. Rolls, 'The Nature of Australia,' in *Ecology and Empire*, ed. Tom Griffiths and Libby Robin (Melbourne: Melbourne University Press, 1997), 40.
3. Thomas R. Dunlap, *Nature and the English Diaspora* (Cambridge: Cambridge University Press, 1999), 24.
4. Tom Griffiths, *Forests of Ash* (Melbourne: Cambridge University Press, 2001).
5. Douglas L. Medin and Scott Atran, 'Introduction', in *Folkbiology*, ed. Douglas L. Medin and Scott Atran (Cambridge, Mass.: The MIT Press, 1999), 5.
6. Dunlap, *Nature and the English Diaspora*, 23–7, 139–40, Griffiths, *Forests of Ash*, 137; Pyne, *Burning Bush*, ch. 12.
7. John Bradsen, 'Soil Conservation: History, Law and Learning', in *Environmental History and Policy, Still Settling Australia*, ed. Stephen Dovers (Oxford: Oxford University Press, 2000), 273–4.
8. Dunlap, *Nature and the English Diaspora*, 140.
9. See Paul Collins, *Burn: The Epic Story of Bushfire in Australia* (Sydney: Allen & Unwin, 2006), ch. 1; Griffiths, *Forests of Ash*, ch. 10; W.S. Noble, *Ordeal by Fire: The Week the State Burned Up* (Melbourne: Jenkin Buxton, 1977); Pyne, *Burning Bush*, 309–14.
10. E.g. *The Age* newspaper (editorial 12 Jan 1939, 10) labelled some graziers as 'potential murderers' while Stretton on 28 February 1939 threatened journalists of *The Sun* newspaper with expulsion from proceedings if they continued to vilify the Forests Commission. Leonard Stretton, Comments, Transcript of evidence, 'Royal Commission into the Causes of and Measures Taken to Prevent the Bush Fires of January 1939', (1939), 1102.
11. Griffiths, *Forests of Ash*, 140.
12. Ibid., viii; Pyne, *Burning Bush*, 312, 325.
13. Bradsen, 'Soil Conservation: History, Law and Learning', 278.
14. For example, its conclusions on soil erosion were taken up by the 'Save the Forests Campaign', C.E. Isaac, *An Inseparable Trinity* (Melbourne, 1950).
15. For example B.U. Byles said of the graziers that 'their stock of fundamental knowledge about the grass lands from which they get their living is practically nil', B.U. Byles, 'A Reconnaissance of the Mountainous Part of the River Murray Catchment in New South Wales', *Commonwealth Forestry Bureau, Bulletin* 13 (1932): 26. D.H. Thompson commented that farmers had 'an astonishingly limited knowledge of the bush around them', quoted in Griffiths, *Forests of Ash*, 137.
16. Leonard E. B. Stretton, 'Report of the Royal Commission into the Causes of and Measures Taken to Prevent the Bush Fires of January 1939', (1939), 7.
17. Leonard Stretton, Comments, Transcript of evidence, 'Royal Commission into Forest Grazing', (1946), 221. The 1946 Royal Commission was also partially initiated because of pressure from the Save the Forests Campaign which at times was highly critical of forest grazing practice. See: C.E. Isaac, *An Inseparable Trinity*; for criticism: Alfred Douglas Hardy, Transcript of evidence (1946), 313–16.
18. Peter Brian Cabena, 'Grazing the High Country: An Historical and Political Geography of High Country Grazing in Victoria, 1835–1935' (unpublished MA thesis, University of Melbourne, 1980), 112; Griffiths, *Forests of Ash*, 33.
19. Michael G. Barbour, 'Ecological Fragmentation in the Fifties', in *Uncommon Ground*, ed. William Cronon (New York: W. W. Norton & Company, 1995); Donald Worster, *Dust Bowl: The Southern Plains in the 1930s* (New York: Oxford University Press, 1979), 201.

20. See: Reginald Edward Torbet (MMBW), Transcript of evidence (1939), 2340–5; Charles Lane-Poole (FC), Transcript of evidence (1939); Peter Gell, 'Forest Ecology and Ecologists', in *Created Landscapes, Historians and the Environment*, ed. Don Garden (Carlton: The History Institute, 1992), 90–92, Griffiths, *Forests of Ash*, 140.

21. Pyne, *Burning Bush*, 199–209, 212–15.

22. John Findlay, Transcript of evidence (1939), 500.

23. Harry Treasure, Transcript of evidence (1946), 2.

24. John Findlay, Transcript of evidence (1939), 502; Stephen Pyne placed emphasised this fatalism: Pyne, *Burning Bush*, 204, 248.

25. John Henry Bussel, Transcript of evidence (1946), 232.

26. A motion passed by the Graziers Association in Mansfield, 'News', *Mansfield Courier*, 17 Feb 1939.

27. Letter from J. Treasure, Transcript of evidence (1946), 17.

28. Pyne, *Burning Bush*, 212–15.

29. John Findlay (Rubicon), Transcript of evidence (1939), 509; George Purvis (Willow Grove), Transcript of evidence (1939), 989; by 1946 there was more willingness to explain the reasons for burning: see Hedley Gordon Stoney (Mansfield), Transcript of evidence (1946), 270.

30. George Purvis, Transcript of evidence (1939), 1006; see also Griffiths, *Forests of Ash*, 35-6.

31. Andrew Burnett Finn, Transcript of evidence (1946), 168; Pyne treats these practices more generally: Pyne, *Burning Bush*, 213.

32. Byles, 'A Reconnaissance of the Mountainous Part of the River Murray Catchment in New South Wales', 26.; Percy George Weston, Transcript of evidence (1939), 1437; see also D.H. Thompson's observation of 'a tradition of fire', 'Forests Fire Prevention and Control in the Cann Valley Forest District', Diploma of Forestry Thesis (Victoria), (1957), 1.

33. Neil Gow, Transcript of evidence (1946), 212–13.

34. Edward Leeder, Transcript of evidence (1939), 408; see also, Joseph Smedley, Transcript of evidence (1939), 411; Thomas William Irvine, Transcript of evidence (1939), 250–1.

35. William Blair, Transcript of evidence (1946), 205.

36. Reginald Barnewall, Transcript of evidence (1946), 284.

37. John Findlay, Transcript of evidence (1939), 507, 499.

38. Edward Leeder, Transcript of evidence (1939), 407.

39. Clarence George Henry Poole, Transcript of evidence (1939), 775.

40. Griffiths, *Forests of Ash*, 44.

41. Edward Leeder, Transcript of evidence (1939), 402.

42. For example, John Findlay, Transcript of evidence (1939), 498; William Francis Lovick, Transcript of evidence (1939), 699; Similar sentiments from an earlier time are common in the collection of settlers accounts: *The Land of the Lyre Bird – A Story of Early Settlement in the Great Forest of South Gippsland*, (Drouin, 1998), 269; pointed out in Griffiths, *Forests of Ash*, 35.

43. On forestry, see Pyne, *Burning Bush*, 21–22.

44. Griffiths, *Forests of Ash*, 140.

45. Peter O'Mara, Transcript of evidence (1939), 1130.

46. Alfred Saxton, Transcript of evidence (1939), 1047.

47. William Kelly, Transcript of evidence (1946), 383.

48. Peter O'Mara, Transcript of evidence (1939), 1133.

49. Rhys Jones, 'Fire-stick farming', Australian Natural History 16 (1969); Tom Griffiths, 'Judge Stretton's Fires of Conscience', *Gippsland Heritage Journal* 26 (2002): 17.

50. A.M. Gill, R.H. Groves and I.R. Noble, *Fire and the Australian Biota* (Canberra: Australian Academy of Science, 1981), ch. 3; Pyne, *Burning Bush*, ch. 6, 8. There remains significant debate as to the *extent* of aboriginal use of fire. See D.M.J.S. Bowman, 'Tansley Review No. 101: The Impact of Aboriginal Landscape Burning on the Australian Biota', *New Phytologist* 140 (1998).

51. E.g. Barber, Transcript of evidence (1939), 1450; Kelso, Transcript of evidence (1939), 1455.

52. Leslie George Gambold, Transcript of evidence (1946), 239.

53. Percy George Weston, Transcript of evidence (1939), 1437–8.

54. Pyne, *Burning Bush*, 148, 140, 184.

55. Griffiths, 'Judge Stretton's Fires of Conscience', 14, 17

56. William Kelly, Transcript of evidence (1946), 383–4.

57. Lane-Poole, Transcript of evidence (1939), 2375, 3276–7.

58. Lane-Poole, Transcript of evidence (1939), 2383–4; Griffiths, *Forests of Ash*, 141, Griffiths, 'Judge Stretton's Fires of Conscience,' 14.

59. Lane-Poole, Transcript of evidence (1939), 2329–30.

60. Libby Robin, 'Radical Ecology and Conservation Science', *Environment and History* 4, 2 (1998): 191–208, doi: 10.3197/096734098779555691.

61. Linden Gillbank, 'Scientific Exploration of the Botanical Heritage of Victoria's Alps', in *Cultural Heritage of the Australian Alps*, ed. Babette Scougall (Canberra: Australian Alps Liaison Committee, 1992), 224.; Stella Grace Maisie Fawcett, Transcript of evidence (1946), 127

62. Leonard E.B. Stretton 'Report of the Royal Commission', (1939), 11.

63. Griffiths, 'Judge Stretton's Fires of Conscience', 17.

64. *Argus*, 11 January 1939, 3, 10; see also, *The Age*, 16 January 1939, 12; *The Argus*, 20 January 1939, 2; editorials, *The Sun News Pictorial*, 16 and 21 January 1939; see also Pyne, *Burning Bush*, 254, 248.

65. Stephen J. Pyne, 'Frontiers of Fire,' in *Ecology and Empire*, ed. Tom Griffiths and Libby Robin (Melbourne: Melbourne University Press, 1997), 21–2. Arguably, little has changed.

66. Leonard E. B. Stretton, 'Report of the Royal Commission', (1939), 11; an almost identical description of this cycle was again made in Leonard E.B. Stretton 'Report of the Royal Commission to inquire into Forest Grazing', (1946), 18; on the fire cycle see also: Pyne, *Burning Bush*, 132–3, 204.

67. Examples from Bright: William Blair, Transcript of evidence (1946), 205; from Rubicon: John Findlay, Transcript of evidence (1939), 499; from Thornton: Reginald John Barnewall, Transcript of evidence (1946), 286.

68. Eg: Fredrick Alexander Ross, Transcript of evidence (1946), 248; Herbert John Robert Swinburne, Transcript of evidence (1946), 236.

69. William Douglas McCoy, Transcript of evidence (1939), 1206.

70. Harry Lewis Treasure, Transcript of evidence (1939), 1181.

71. John Samuel Langtree, Transcript of evidence (1939), 1204.

72. John Findlay, Transcript of evidence (1946), 289.

73. Maurice Francis Keppell, Transcript of evidence (1946), 294.

74. John Alexander Cameron, Transcript of evidence (1946), 276; Andrew Burnett Finn, Transcript of evidence (1946), 167-8.

75. Percy George Weston, Transcript of evidence (1939), 1450–1450A, see also William Francis Lovick, Transcript of evidence (1946), 263; Peter O'Meara, Transcript of evidence (1946), 328

76. Leonard E. B. Stretton 'Report of the Royal Commission', (1946), 15. Even the fiercely anti-grazing forester Alfred Douglas Hardy conceded 'most graziers probably value the forests', Transcript of evidence (1946), 307.

77. Arthur Mervyn Pearson, Transcript of evidence (1946), 114.

78. Percy George Weston, Transcript of evidence (1939), 1438.

79. William Francis Blair, Transcript of evidence (1946), 205, 209; Herbert John Robert Swinburne, Transcript of evidence (1946), 236. See also Leonard E. B. Stretton 'Report of the Royal Commission', (1946), 18; Byles, 'A Reconnaissance of the Mountainous Part of the River Murray Catchment in New South Wales,' 19; Pyne, Burning Bush, 213. Note: there is possibly much that is accurate in Blair's description. Ruth Lawrence has identified changes in the moisture of the environment of the Bogong High Plains, with a drier period from the 1890s to the mid 1940s that corresponds with the most intense fire period. Ruth Lawrence, 'Environmental Changes on the Bogong High Plains, 1850s to 1980s', in Australian Environmental History: Essays and Cases, ed. Stephen Dovers (Melbourne: Oxford University Press, 1994), 187.

80. Michael McNamara (from Omeo), emphasis mine. Transcript of evidence (1946), 150.

81. Reginald John Barnewall, Transcript of evidence (1946), 285.

82. Percy George Weston, Transcript of evidence (1939), 1450–1450A.

83. Anthony Mangan, Transcript of evidence (1946), 203; Fredrick John Barton, Transcript of evidence (1946), 302; see also, William Henry Whitehead, Transcript of evidence (1946), 180; Charles Gordon Lumdsen Transcript of evidence (1946), 220.

84. Griffiths, Forests of Ash, 135.

85. David Ashton, 'Fire in tall open forests', in Gill, Groves and Noble, Fire and the Australian Biota, 348–62.

86. Edward Leeder, Transcript of evidence (1939), 405.

87. Peter O'Meara, Transcript of evidence (1946), 327.

88. John Findlay, Transcript of evidence (1939), 500.

89. Joseph Smedley, Transcript of evidence (1939), 411; Edward Leeder, Transcript of evidence (1939), 402; Thomas Wilmont, Transcript of evidence (1939), 527; William Lovick, Transcript of evidence (1939), 696.

90. George Clifton Purvis, Transcript of evidence (1939), 1007; Alfred Saxton, Transcript of evidence (1939), 1042; Carl Lamont Wraith, Transcript of evidence (1946), 194.

91. Many of the bushmen had been timber-getters, sawmillers, or would harvest the timber that existed on their own land.

92. Byles, 'A Reconnaissance of the Mountainous Part of the River Murray Catchment in New South Wales', 26.

93. Stella Grace Maisie Carr, 'Upper Hume Catchment. Ecological Report 1940–47' (Melbourne University, c.1947), 10.

94. Scott Atran, Cognitive Foundations of Natural History (Cambridge/Paris: Cambridge University Press/Editions de la Maison des sciences de l'homme, 1990), 216.

95. Alfred Saxton, Transcript of evidence (1939), 1048.

96. For fire germination of seeds see: Evan Evans, Transcript of evidence (1946), 224; Anthony Mangan, Transcript of evidence (1946), 203; Percy Weston, Transcript of evidence (1939), 1450–1450A.

97. Alfred Saxton, Transcript of evidence (1939), 1049.

98. Percy George Weston, Transcript of evidence (1939), 1458b.

99. Alexander Edward Kelso (MMBW), Transcript of evidence (1939), 110.

100. Adrian Herbert Armstrong Beetham, Transcript of evidence (1939), 641–2.

101. Pyne, Burning Bush, 252–4, 256, 250.

102. FC representatives Alfred Oscar Lawrence and E.H.E. Barber expressed this throughout the 1939 Royal Commission. eg: Stretton warned the FC about coaching and placing hand-picked witness before the Commission. Transcript of evidence (1939), 353; also, Stretton commented that many young foresters were cautious in their evidence for fear of harming their career chances in the FC. Leonard E. B. Stretton 'Report of the Royal Commission', (1939), 7.

103. Griffiths, *Forests of Ash*, 141; Griffiths, 'Judge Stretton's Fires of Conscience,' 14.

104. Charles Lane-Poole, Transcript of evidence (1939), 2383–4.

105. On the effect of institutional scientific thinking on environmental understanding, see James C. Scott, *Seeing Like a State* (New Haven: Yale University Press, 1998), ch.8.

106. Atran, *Cognitive Foundations of Natural History*, 263–4.

107. Dunlap, *Nature and the English Diaspora*, 139.

108. Alfred Saxton, Transcript of evidence (1939), 1048.

109. This was indicated by a brief recorded comment or reprimand in the transcripts (which I cannot recall). Stella Grace Maisie Fawcett, Transcript of evidence (1946), 124–31.

110. Patrick John Kelly, Transcript of evidence (1946), 144; Michael McNamara, Transcript of evidence (1946), 150; John Douglas Gibson, Transcript of evidence (1946), 147, 154, 156 (Fawcett comment).

111. Edmund Cornwall , Transcript of evidence (1939), 1139, emphasis mine; Alfred Joseph Webb, Transcript of evidence (1939), 1025; see also Alfred Saxton, Transcript of evidence (1939), 1042; Allam Murray Dobson, Transcript of evidence (1939), 518–20; and Claude Staff who complained about foresters who were too young, too inexperienced or who followed impractical regulations too closely, Transcript of evidence (1939), 1031–2 .

112. Atran, *Cognitive Foundations of Natural History*, 7.

113. Dunlap, *Nature and the English Diaspora*, 140.

114. John Alexander Cameron, Transcript of evidence (1939), 701; similar examples see: Clarence George Henry Poole, Transcript of evidence (1939), 773; William George Reed, Transcript of evidence (1939), 711.

INDUSTRIAL

'Wilderness to Orchard': The Export Apple Industry in Nelson, New Zealand 1908–1940

Michael Roche

The forum on environmental history in *Pacific Historical Review* in 2001 reveals US environmental history in the early twenty-first century as less interdisciplinary and more fully incorporated into the domain of history than in earlier years. Nevertheless, two insights from the *Pacific Historical Review* forum papers are, first, the observation that environmental history notwithstanding the recent interest in global environmental history, must 'be based firmly on the local'[1] and, second, evidence that a rapprochement has been reached between Worster, who essentialised environmental history as the study of capitalist agriculture, and Rosen and Tarr, who asserted the case for urban environmental history.[2] Cronon separately argued that,

> we need to embrace the full continuum of a natural landscape that is also cultural, in which the city, the suburb, the pastoral, and the wild each has its proper place, which we permit ourselves to celebrate without needlessly denigrating the others.[3]

He eventually settles on the garden as an appropriate site for understanding people's place in nature and the meaning of sustainability. A further gendered layering of meaning has been added by Norwood who accentuates feminine nature and a masculine gardener.[4]

Commercial apple growing in New Zealand differs from Cronon's garden. Fruit consumption increased in industrialising nations from the late nineteenth century. The demand was incipiently global even in the early twentieth century, when New Zealand apples were sold in the UK competing with US, Canadian and Australian fruit. Orchardists had to confront a variety of old and new pests and diseases in order to maintain production for local and British markets. Howard Seftel uses the phrase 'fruit pests in the entrepreneurial garden',[5] deftly linking gardens and orchards to encapsulate the Californian experience, and stresses that 'fruit growers regarded pest control as absolutely essential to the long-term success of their enterprise'.[6] Increasing insecticide usage was a response to commercial monocultures and international trade.

To look at orcharding in New Zealand is to recover rural narratives overshadowed by the dominant sheep and dairy industries. In terms of environmental history, it provides an opportunity to uncover the particular values and attitudes invested in apple growing as a lifestyle and the orchard as an ideal relationship with nature. The inconsistencies in the depiction of orcharding as an Arcadian

Environment and History **9** (2003): 435–50

lifestyle are simultaneously apparent in the amount of spraying that was required to produce the quantity and quality of apples destined for export.

Apples arrived in New Zealand with European contact. They were grown for local consumption as fresh fruit, for cooking and for cider. A local commercial industry was quick to develop, shaped by soils, climate, the perishability of fresh fruit and the proximity of urban markets.[7] From 1910 Nelson was transformed into an export-oriented apple growing region.

THE ROLE OF THE STATE IN ENCOURAGING FRUIT GROWING

State involvement in fruit growing rested on legislation for the control of pests and diseases, the creation of a Biology and Pomology Division within the Department of Agriculture, and provision of export incentives. The *Codlin Moth Act, 1884* provided for the proclamation of districts within which orchardists had to make declarations about levels of infection and contribute to a fund to pay for orchard inspection. Infected trees were to be bandaged to capture the caterpillars, and infected fruit destroyed. The *Orchard and Garden Diseases Act, 1896* extended the earlier legislation by requiring the destruction of diseased trees and fruit and more closely regulating the importation of plants and fruit. A revised 1903 Act included in its definition of disease Phylloxera, which was devastating the grape industry as well as San Jose Scale, Mediterranean or Western Australian Fruit Fly and Queensland Fruit Fly. Except for Phylloxera and Australian Fruit Fly all of these could infect apples, as could all those listed in a second schedule including American Blight (later known as Woolly Blight), Apple Scab, Codlin Moth, Mussel or Oyster Scale and Red Mite. These checks to the spread of introduced pests and diseases, though rudimentary, encouraged additional investment in the industry from the 1890s, although with commercialisation and new inexperienced growers the regulations were on occasions disregarded.[8]

THE NELSON ORCHARD BOOM, 1910–1915

A closely subdivided landscape coupled with high annual sunshine levels (2500 hours or more), reliable rainfall (800 to 1200 mm) and a long frost-free season (late October to mid April), led to commercial orcharding in the 1860s.[9]

Successful apple export trials began in 1908, stimulated by a government export price guarantee of 1d per lb, when local firm E. Buxton & Co., copying Tasmanian methods of packing, dispatched 1236 cases of apples to the UK.[10] Consignments of 4499 cases to the UK in March 1910 and subsequent shipments to Montevideo in 1912 were important triggers to the expansion of the New

Zealand industry. In 1914 the government even sent Nelson grower G. C. Tacon to Argentina to report on the prospects for apple exports.[11]

Land in Nelson was available for orchards. The Moutere Hills area was originally cleared of bush and grassed by settlers in the 1870s. Because of a trace element deficiency farming had failed except for some rough sheep grazing and a few orchards. Land development syndicates in the 1910s subdivided, planted

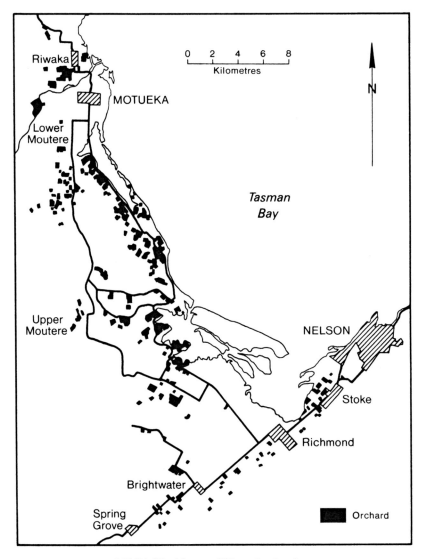

MAP 1. The Moutere Hills orchard region.

and sold orchard blocks in the area. Centrally identified with the promotion of export apple growing in Nelson was Arthur McKee (1863–1943) who migrated to New Zealand from Liverpool in 1890. Although not the first commercial orchardist in the Moutere Hills, McKee was a tireless promoter of apple growing in the district. He was convinced that New Zealand should follow Tasmania in exporting apples, and unsuccessfully lobbied politicians for a state-assisted community fruit growing export scheme. In 1910 his earlier publishing and real estate experience came together in his promotion of the Moutere Hills area for apple growing for export. He acquired a 300 acre block in 1911 and later 2000 acres, which he named 'Tasman', after the Dutch navigator Abel Tasman who visited this coast in 1642. Here his company Tasman Fruitlands Ltd. developed small orchard blocks for sale. A promotional coup for McKee was the sale of a block to T. W. Kirk, head of the Biology Division of the Department of Agriculture.

McKee's newspaper advertisements in December 1910 promoted Tasman as the 'New Fruit Colony'. More impressive was his lavishly illustrated pamphlet *Apples for Export*, successive editions of which extolled the economic merits of apple growing and labelled Nelson as 'The Fruit-Growers' Paradise'. McKee contrasted city versus country life:

> The city man who desires to change his mode of life for the better should study this fruitgrowing proposition. There is nothing surer in farming, nothing more interesting nor so easily learned ...Tasman is the type of country that appeals to the practical fruit grower. While the sunbathed slopes are all that can be desired for apple and pear cultivation the deep sandy loams of the flats – rich drained swamp land – will give to perfection stone fruit, hops, raspberries, cereals &c.[12]

The booklet *Sunny Nelson* also proclaimed a transition from 'wilderness to orchard'.[13] The soils and climate of the district were claimed to be ideal for apple growing enabling 'trees to make such a robust and solid growth that insect pests and disease have no terror for the orchardist of Nelson Province'.[14] *The Colonist*, a local newspaper, observed in 1915 of McKee's scheme that:

> The manuka-clad hills had been cleared, ploughed and planted in orchard. Smiling homesteads have arisen, and the district is carrying a population of nearly two hundred persons, where formerly only one or two families resided. The transformation has been rapid, but judging by the satisfactory growth of the trees, the district in the near future promises to be a thriving and prosperous one.[15]

Similar sentiments were expressed in the *New Zealand Stock and Station Journal*, which, reporting on the rapid expansion of the industry, asked 'What impulse, what fairy's wand, has conjured up this miracle – the conversion of once-barren and despised hills into potential gold mine?'[16] Fruit growers from elsewhere in New Zealand were more circumspect, one commenting that, 'there has been an overeagerness to utilise every acre, and some of it is certainly too steep for the purpose'.[17]

'Wilderness to Orchard'

FIGURE 1. The pictorial record of the Nelson apple industry is replete with images of young women posed with trees in blossom, with ripe fruit and actually picking fruit. Harvesting was, however, typically undertaken by men, women and children, though sorting and packing fruit was typically women's work, while pruning and spraying were carried out by men. (Photo: Nelson Provinvcial Museum)

Michael Roche

McKee's 'Fruit Colony' was to be a rural Arcadia; 'there would be all the delights of the country combined with the principal conveniences of the city, without the drawbacks inseparable from city life'.[18] Other companies joined McKee to plant the Moutere Hills. From 4,125 acres in 1910–11 the orchard area in Waimea County expanded to 10,081 acres by 1917–18. The more unsuitable areas were subsequently abandoned and the acreage reduced to 5,080 acres by 1928–9.[19]

Some came to regret their involvement. A Wellington schoolteacher, Mary Piggford, who had purchased a 10-acre lot in the Tasman West subdivision in 1915, brought a court action against McKee's company in 1922, alleging misrepresentation about the suitability of the land for apple growing, the profit levels, and the claim that orchard blocks would support a comfortable lifestyle. The Supreme Court found for the appellant, leading McKee to the Court of Appeal, where his case was successfully put by a leading barrister in what was described as a 'characteristically bizarre judgement' on the part of Chief Justice Stout.[20]

SCIENCE AND APPLE GROWING

Nelson, although without a university, was the site of the privately endowed Cawthron Institute, established in 1919 to undertake agricultural research.[21] Cawthron scientists carried out research vital to the local apple industry. This enabled growers to overcome the limitations of the Moutere Hill soils, which were found to be low in phosphoric acid and required liming and green manures, without which tree growth suffered and yields were low.[22] Their efforts were also linked to imperial science: the Empire Marketing Board provided a grant as part of a much larger biological control project.[23] Scientific research locally carried out in Nelson was thus partly driven by imperial concerns about the quality of local UK fruit and for securing of empire-wide supplies. New Zealand government and industry funding enabled a research orchard, agitated for since 1915, to be established in 1930 under Department of Scientific and Industrial Research (DSIR) control at Appleby in Nelson.[24] R. G. Hatton of the UK East Malling Research Station produced a Fruit Research Scheme for DSIR in 1931.[25] Later initiatives included frost control work, and cool storage experiments.[26]

COMBATING PESTS AND DISEASE BY SPRAYING

Chewing insects such as Codlin Moth were controlled by spraying the foliage and fruit with a coating of poison that was ingested. Arsenate of lead was replacing other sprays for this purpose in the 1920s. Sucking insects (e.g. San Jose Scale), which inserted a proboscis into the tree leaf or fruit were unaffected by a poisonous coating on the foliage. They were destroyed by using chemicals that had a solvent action, known as contact sprays. Fungoid diseases were countered by a film spray such as Bordeaux Mix so that spores settling on the foliage were destroyed.[27] From the 1920s orchards and sprays developed in conjunction with each other. As such the orchard was a blending of nature and rural economy where sprays were essential to the defence of the natural equilibrium of the orchard.

Orchard work from the 1910s included increasing amounts of spraying. McKee's orchard as an idealised amalgam of family, economic unit and nature was from the first fraught with contradictions. *The Colonist* in 1910 contained advertisements for insecticides, fungicides, arsenate of lead (3 January) The latter was used to control Codlin Moth (5 June) and at the meeting of a local fruit growers association members listened to papers on Bitter Pit and Bordeaux Mixture, and on Woolly Aphid. Department of Agriculture advice on spraying with lime and sulphur to combat aphid and various soap sprays to counter apple sucker (*Psylla mali*) were also reported (2 August). Nelson firm Tasker and Levien advertised that it could supply growers with an array of chemicals and pumps (1 August). An article in *The Colonist* in 1915 is typical of advice presented to orchardists:

> The devout preacher says 'let us pray'. The successful orchard says 'Let us spray'. We need always to pray, but proper and thorough spraying will help the prayer immeasurably in the production of first class fruit ... spraying is one of the great things which mark the difference between success and failure in the apple business.[28]

Harry Everett, a pioneering Nelson orchardist, later recalled:

> Our chief troubles were Mussel Scale, Red Spider, Codlin Moth, and some San Jose Scale which it seems we had for years without discovering it. For Scale we used lime, salt and sulphur and crude black oil which was sometimes difficult to emulsify: sometimes a few trees got all the oil and others got all the water. In such event the oil certainly controlled the Scale, an in many cases the tree, too.[29]

Spraying trials were undertaken by individual growers at their own expense on various orchards in Nelson in order to identify what chemicals were reliable locally. These experiments resulted in Bordeaux Mixture being replaced by lime sulphur for Black Spot.[30]

Unseasonally wet or warm weather further disrupted the spray regimes. Not only was Lime Sulphur more effective than Bordeaux Mixture but it was also much easier to make up and to use. By 1924, however, Bordeaux Mixture was again favoured to combat Black Spot.[31] Underneath McKee's Arcadian vision lay agrarian capitalism. As one grower observed 'it is essential nowadays to do everything possible with the idea of saving labour, and the details attached to Bordeaux mixture would not be missed if they could be avoided'.[32]

Michael Roche

Fireblight (*Erwinia amylovora*), a significant bacterial threat to the apple industry in New Zealand in the 1920s, never established itself in the Nelson district.[33] The major problems faced by Nelson growers were Black Spot [or Apple Scab], Woolly Aphis, Mealy Bug, and Codlin Moth. Growers depended on both export sales and spraying, as the Nelson Correspondent for the *New Zealand Fruitgrower and Apiarist* observed:

> Spraying has naturally taken up the greater part of the time, as everyone is anxious to secure as much first grade fruit as possible … many growers who in the past were more or less proud of the fact that they could dispose of Black Spot and other fruit showing blemishes, are now coming around to the opinion that the way to deal with low grade fruit is not to grow it.[34]

Spray systems developed from simple pump and bucket apparatus of the 1890s to motorised pumping systems towed by horses, in use until the early 1930s, and tractor-towed sprayers from the 1920s. Other innovations included agitators to prevent the chemicals from settling, higher water pressures, and from the 1940s dispersal with fan powered units. Some Nelson growers from the mid 1920s also experimented with fixed spray systems. The advantages included savings on horses and labour and the ability to spray when ground conditions were wet.[35] Initial problems with pipe systems of spraying related to uneven spray mixes. Some innovations such as dry powder chemical application proved to be dead-ends of limited success.

REGULATION SPRAYS AND BIOLOGICAL CONTROL

Spraying was essentially unregulated until 1913 when the Department of Agriculture issued certificates of competency in spraying and pruning. Subsequently there were few restrictions on the use of orchard sprays. Voluntary certification operated from 1937 but compulsory registration of orchard sprays was not required in New Zealand until 1959.

By the early 1900s Nelson apple growers had available to them several sprays developed in France, for example, Lime Sulphur in 1851, Bordeaux Mixture in 1882, Burgundy Mix in 1887 and others of US origin, notably spray oils and Lead Arsenate, the first use of which dates to 1880 and 1892 respectively.[36] Some of these chemical treatments had been further adapted for use in New Zealand conditions. The Department of Agriculture also undertook research into orchard diseases and pests and advised growers through a network of extension officers.

Of interest is the Department of Agriculture's stance on spraying versus biological control: Codlin Moth parasites had been imported from the United States and although it was considered too early to pronounce them an unqualified success, hopes were high in 1908.[37] Simultaneously some extension staff

FIGURE 2. Spraying fruit trees was part of the male orchardists' work routine. Note the absence of any protective clothing. The dangers of spraying to consumers and the orchardists were down played even in the aftermath of the UK arsenate scare of 1926. (Photo: Nelson Provinvcial Museum)

endorsed the use of sprays to combat Codlin Moth. By the 1920s however, most growers were depending on arsenate of lead sprays to counter Codlin Moth.

The way in which spraying was incorporated into the orchard arcadia was summed up by the *New Zealand Farmer Stock and Station Journal's* Nelson Correspondent in September 1917:

> All the countryside is busy with the spray pump, and the smell of lime and sulphur is with us and in us. It is the prevailing smell throughout the district at the present time, a smell that one has to get used to like, but a healthy one withal. Add to this the magnificent view by night of the Aurora Australis, and who will say that Nelson is not the best fruit growing district in New Zealand.[38]

Some growers were quite sanguine about their longer-term prospects while they had to 'wage war against all the pests that the apple is heir to'.[39] The very label 'pests' legitimates the use of chemicals for their destruction. In the longer term growers were increasingly locked into a spray regime because export markets could be lost if fruit was pest-ridden. In 1921 Australia banned New Zealand apples because of Fireblight. Argentina was similarly closed to New Zealand fruit, while France proposed to ban imports from countries with San Jose Scale.[40]

There were, however, two competing streams within the research community, a group who strongly supported biological controls and another who championed the role of chemical sprays. The first was represented by DSIR scientist G. C. Cunningham, whose monumental *Fungous Diseases of Fruit-Trees in New Zealand* was notable for its description of the botany of fruit trees, their diseases and spray remedies; the latter by Cawthron entomologist R. J. Tillyard, whose investigations resulted in the introduction of a parasite (*Aphelinus mali*) which successfully countered the Woolly Aphis problem. This saved the apple industry by enabling more of the favoured export varieties to be established. The parasite was eventually successfully released in other apple growing districts. However, reflecting on US research, Tillyard did sound a cautious note about achieving complete success through biological control.[41]

THE FIRST SPRAY CRISIS

The first significant market reaction to arsenate sprays occurred in 1926 when arsenate deposits on imported apples exceeded that on local UK apples. The residue problem was eventually linked to late spraying to deal with a serious Codlin Moth outbreak amongst two apple varieties in selected US growing districts. The UK was supplied largely from Canada, Australia and the USA, nevertheless the scare had implications for New Zealand growers.[42] Tillyard, usually the detached scientist, expressed the view that the fruit trade henceforth be conducted within the British Empire, urging closure of the UK market to American fruit 'so as to prevent them dumping either cool-stored or arsenicated products on to our legitimate market and so ruining it'.[43]

The Department of Agriculture from the 1910s widely published the results of chemical spray tests for orchards. This included a lengthy series of articles on different classes of sprays prepared by Cunningham in 1932–3.[44] DSIR and Cawthron research on biological control was continued into the 1920s. Cunningham however, was 'scathing about biological control'.[45] In reporting progress in orchard disease control he identified some areas where local practice was diverging from that of the US and Europe.[46] For instance, Lime Sulphur sprays (for diseases such as Black Spot) had by 1930 again replaced the traditional Bordeaux Mixture, which tended to scorch fruit and foliage. Lime Sulphur was also believed to have some desirable insecticidal properties.

Cunningham's research was intended to foreshadow regulations for the certification of chemicals under the *Fungicides and Insecticides Act, 1927*.[47] This act was a response to local New Zealand grower dissatisfaction with the variability of chemicals, both within and between brands.[48] The legislation lacked detailed standards, which were left to be established by Order-in-Council in the future. Given the lack of scientific analysis, this was not an unreasonable option.

DSIR tests in 1931 confirmed orchardists' opinion about their ineffectiveness and variable quality of available chemicals.[49] In 1937 the Plant Diseases Division (PDD) of the Plant Research Bureau of DSIR assumed responsibility for the certification of sprays and fumigants (collectively termed therapeutants).

Subsequently the PDD introduced its certification system, inviting manufacturers to send samples for analysis and field trials. The first certification list in 1937 specified four acid lead arsenates, one colloidal sulphur, one nicotine sulphate, three polysulphite sulphurs and seven spraying oils. With the exception of two of the spraying oils, the copper sulphate and three of the lime sulphurs, the rest of the suppliers were from the UK (four), USA (four), and Australia (five). They included Cooper, McDougall & Robertson, makers of the 'Arsinette' brand. This company, founded in 1843, had branches in Australia, USA, Uruguay, Argentina and South Africa. Approved red and white spraying oils included the Avon brand of the Tide Water Associated Oil Co of San Francisco. By 1944 some 36 products were listed.[50]

Copper sulphate for Bordeaux Mixture was initially sourced from the UK, although some was later made locally from scrap copper. Some 133 tons of copper sulphate valued at £3,911 and other spray oils, arsenites, and nicotine sulphates valued £20,917 were imported in the period 1926 to 1930.[51] Orchard chemicals were supplied by British, Australian and (after World War One) by American firms, another aspect of global reach into the Nelson region. Arthur McKee, however, was importing orchard sprays from 1910. Difficulties in securing supplies prompted him to form in 1931 a family-run private company, Fruitgrowers' Chemical Co., to manufacture lime sulphur at Mapua in Nelson. It later expanded production to include spraying oils and sulphur compounds.[52] By 1941 they were producing a certified 'Consul 40', a colloidal sulphur, 'Polysol', a polysulphide sulphur, 'Sprayol White' and 'Sprayol Red' spraying oils.

Government scientists played a dual role in extending knowledge about orchard pests and disease as well as developing competing chemical and biological controls and actually developing regulatory systems. Although biological control advocates enjoyed some successes, powerful establishment scientists such as Cunningham were disparaging and championed chemical solutions to the pest problem.

CONCLUSION

Grown under quite specific environmental conditions, fresh fruits are seasonal and perishable. Apples among fresh fruits are comparatively durable, but to ship them half way around the world and through the tropics to provide a counter-seasonal supply for UK markets awaited the application of quite sophisticated cool storage technology. Fruit buyers demanded higher quality fruit of specific

Michael Roche

varieties making spraying a necessity for growers which was reinforced by the threat of transporting pests and diseases on infected fruit. Possibly the spray regime was more intensive in New Zealand than in the UK, for exporters could not risk dispatching poor quality shipments. Scientific co-operation was also global in terms of British funding of orchard research in New Zealand. UK and US chemical companies and their Australian subsidiaries also offered competing spray products to Nelson growers.

The orchard is worthy of the status that Cronon ascribes to the garden as a site for better understanding people and/in nature. The New Zealand orchard lay between the city or town and the pastoral economy based on sheep for meat and wool and the dairy industry. It was a smaller scale enterprise than dairy farming, it evoked different images of nature for owners than farmland hewn from forest land a generation earlier or the tussock grasslands converted to pasture. The orchard as constructed by McKee placed people in a 'nature' that was a combination of metabolic and social processes. McKee's orchard lifestyle, where a family by honest toil could build a home and business by harvesting nature's bounty, was superior to that of the city. Ever the entrepreneur, he always saw orcharding as providing better financial returns than conventional pastoral agriculture. The

FIGURE 3. The actual experience of orchard establishment on the Moutere Hills was somewhat at odds with the promotional material. This family is shown picking fruit in what is still a comparatively barren landscape. (Photo: Nelson Provinvcial Museum)

contradiction between the orchard as idealised lifestyle in a beneficent natural setting while at the same time maintaining it by applying more and more sprays to keep waves of pests and diseases at bay went unnoticed.

Initially the pest problems were played down. Yet spraying was essential to the maintenance of the orchard as an 'island' of nature. It was less a case of an imperfect nature improved by human intervention than an invading nature that had to be held at bay with chemical sprays. British economic entomologists of the time viewed spraying as a means of restoring an equilibrium in nature. Some Nelson entomologists experimented with biological predators, but for other scientists and growers the spray regime remained the only means of combating successive waves of pests and disease, a threat to the apple crop as their economic livelihood and the orchard as their home in nature. From a point in the mid 1920s biological control lost out to the chemical sprays, the latter being a much more powerful alliance of scientists, industry growers, and regulators.

This discussion of the early years of export apple growing perhaps unfortunately reinforces environmental history as essentially rural. A fuller examination of apple consumption would however bring a domestic and overseas urban population into focus. Changing varietal preferences and fruit quality expectations from afar increasingly shaped the spraying regime in Nelson orchards. Apple growing in Nelson provides an example of the ways in which the local and the global in environmental history are simultaneously and mutually constituted.

NOTES

1. J. Donald Hughes, 'Global Dimensions of Environmental History', *Pacific Historical Review* 70 (2001): 98. Yet the peripheral status of environmental history is maintained by Ted Steinberg, 'Down to Earth: Nature, Agency, and Power in History', *American Historical Review* 107 (2002): 798–820. This essay only became available after this paper was drafted.

2. Donald Worster, 'Transformation of the Earth Toward an Agroecological Perspective in History', *The Journal of American History* 76 (1990): 1087–1110; and Christine Rosen and Joel Tarr, 'The Importance of an Urban Perspective in Environmental History' *Journal of Urban History* 20 (1994): 299–310.

3. William Cronon, 'The Trouble with Wilderness; or Getting Back to the Wrong Nature', *Environmental History* 1 (1996): 24.

4. Vera Norwood, 'Disturbed Landscape/Disturbing Processes: Environmental History for the Twenty-first Century', *Pacific Historical Review* 70 (2001): 77–89.

5. Howard Seftel, 'Government Regulation and the Rise of the Californian Fruit Industry: The Entrepreneurial Attack on Fruit Pests, 1820–1920', *Business History Review* 59 (1985): 375.

6. Ibid., 373. See also James Whorton, *Before Silent Spring, Pesticides and Public Health in pre-DDT America* (Princeton NJ: Princeton University Press, 1974); John Perkins, *Insects, Experts, and the Insecticide Crisis* (New York: Plenum, 1982); and John Clark, 'Bugs in the system: Insects, Agricultural Science, and Professional Aspirations in Britain, 1890–1920', *Agricultural History*, 75 (2001): 83–114.

Michael Roche

7. Norman Adamson, 'History of Fruitgrowing in Nelson', *New Zealand Journal of Agriculture* 98 (1959): 70–81, 167–76; Frank Bailey, 'Development of Fruit Growing in the Auckland District', *New Zealand Journal of Agriculture* 96, (1958): 583–91; William Kemp, 'History of Fruit Growing in Central Otago', *New Zealand Journal of Agriculture* 90 (1955): 169–85; Rose Mannering, *100 Harvests, a History of Fruitgrowing in Hawkes Bay* (Wellington: PSL Press, 1999); G. Ward, *Early Fruit Growing in Canterbury, New Zealand* (Christchurch: Spotted Shag Press, 1995); and B. Wells, *The Fruits of Labour: A History of the Moutere Hills Area Served by the Port of Mapua* (Nelson: General Printing Services, 1990).

8. 'Orchard Disease Controls', *New Zealand Farmer and Stock and Station Journal*, April 1916: 563.

9. James McAloon, *Nelson: A Regional History* (Whatamanga Bay: Cape Catley, 1997) and Adamson, 'History of Fruitgrowing in Nelson', 71.

10. Adamson, 'History of Fruitgrowing in Nelson', 73.

11. George Tacon, *Fruit Markets and Fruitgrowing in South America*, Industries and Commerce Bulletin 53 (new series) (Wellington: Industries and Commerce, 1915).

12. Arthur McKee, *Apples for Export* (Nelson: Riwaka, 1910).

13. Leo Fanning, *Sunny Nelson* (Wellington: Evening Post, 1915), 40.

14. Fanning, *Sunny Nelson*, 43.

15. 'Colonist Summary for Abroad', *The Colonist*, March 1915 (no pagination).

16. 'The Romance of the Moutere Hills, The Beginnings of a Great Industry', *New Zealand Farmer Stock and Station Journal,* July 1916: 1005.

17. George Alderton, 'Round the Dominion', *New Zealand Farmer Stock and Station Journal,* April 1916: 570.

18. 'Co-operation in Production', *New Zealand Farmer Stock and Station Journal,* June 1916: 828. In fact poor roading was a significant handicap to the area for many years, though the fruit was exported by water.

19. *New Zealand Statistics*, 1907–1917, *New Zealand Agricultural Statistics* 1929.

20. James McAloon, *Nelson: A Regional History*, 158.

21. David Miller, *Thomas Cawthron and the Cawthron Institute*, (Nelson: Cawthorn Institute, 1963).

22. Theodore Rigg, *A Lecture on Soil Survey in its Relationship to the Nelson District with Particular Reference Moutere Hills Soils* (Nelson: Nelson Evening Mail, 1920) 12

23. Ross Galbreath, *Making Science Work. DSIR for New Zealand* (Wellington: Victoria University Press, 1998).

24. R. Hallam, A Fruit Research Scheme for New Zealand, 1931. File SIR (Scientific and Industrial Research) 1/14 Archives New Zealand, Wellington.

25. G. Hopkirk, *Appleby Research Orchard 1930–1980*, DSIR Information Series (Wellington: DSIR, 1980). The local MP Harry Atmore was heavily involved in the political manoeuvring to establish this research orchard.

26. John Atkinson, 'Science and the Fruit Industry', in Francis Callaghan (ed.), *Science in New Zealand* (Wellington: Government Printer, 1957), 164–75.

27. John Sinclair, *Fruit Growing in New Zealand* (Auckland: Whitcombe and Tombs, 1921).

28. J. Reynolds, 'The Benefits of Spraying', *The Colonist*, 27 April 1915.

29. Harry Everett, '50 Years of Orcharding, Mr Harry B. Everett Retires', *The Orchardist of New Zealand*, November 1957: 5–7.

30. Our Correspondent, 'Orchard News from the Nelson District', *New Zealand Farmer Stock and Station Journal*, June 1916: 862.

31. Notes from Nelson', *The New Zealand Fruitgrower and Apiarist*, 16 September 1924: 700.

32. Nelson Correspondent, 'Orchard News from the Nelson District', *New Zealand Farmer Stock and Station Journal*, December 1915: 1661.

33. Alfred Cockayne, 'The Fire Blight Menace', *New Zealand Farmer Stock and Station Journal*, July 1921: 858–9; and Kathleen Curtis, 'Fire Blight: A Survey of Current Knowledge and Recent Advances', *Supplement to the Orchardist of New Zealand*, 1 June 1934: 3–8.

34. Our Correspondent, 'Notes from Nelson', *The New Zealand Fruitgrower and Apiarist*, 16 November 1923: 70.

35. 'Notes from Nelson', *The New Zealand Fruitgrower and Apiarist*, 16 September 1924: 700 and Nelson Orchardist, 'An Improved Method of Orchard Spraying', *The New Zealand Fruitgrower and Apiarist*, 16 October 1925: 26.

36. Gordon Cunningham, 'Orchard Sprays in New Zealand. 1: The Sulphur Series'. *New Zealand Journal of Agriculture*, 44 (1932): 177–87; Gordon Cunningham, 'Certification of Sprays', *The Orchardist of New Zealand*, 2 October 1935: 257–8. For the earlier industrial use of arsenate see P. Bartrip 'How Green was my Valance? Environmental Arsenic Poisoning in the Victorian Domestic Ideal', *The English Historical Review*, 109 (1994): 891–913.

37. New Zealand Department of Agriculture, *Annual Report* (Wellington: Government Printer, 1908), 132.

38. Our Correspondent, 'Orchard News from the Nelson District', *New Zealand Farmer Stock and Station Journal*, September 1917: 1056.

39. Our Correspndent, 'Moutere-Motueka Notes', *New Zealand Farmer Stock and Station Journal*, July 1917: 814.

40. 'Fruit Industry', *New Zealand Smallholder* July 1932: 583. See also Keith Sinclair, 'Fruit Fly, Fireblight, and Powdery Scab: Australia–New Zealand Trade Relations, 1919–1931', *Journal of Commonwealth and Imperial History*, 1 (1972): 27–48.

41. Robin Tillyard, 'Position of the Fruit-Growing Industry in NZ', *The Orchardist of New Zealand*, 1 March 1928: 5.

42. T. Parker, 'Insecticidal Sprays: "Arsenic Scare" and its Lessons', *New Zealand Fruitgrower and Apiarist*, 6 June 1926: 489–90; and R. L. Andrew, 'Arsenic in New-Zealand-Grown Apples', *New Zealand Journal of Science and Technology*, 9 (1927): 206–9.

43. Tillyard, 'Position of the Fruit-Growing Industry': 5.

44. These were reprinted in consolidated form as Gordon Cunningham, *Plant Protection by the Aid of Therapeutants*. (Dunedin: John McIndoe, 1935).

45. Galbreath, *Making Science Work*, 89

46. Gordon Cunningham, 'Five Years' Progress in Orchard Disease Control', *The Orchardist of New Zealand*, 1 September 1930: 9–12.

47. Cunningham, 'Orchard Sprays in New Zealand'.

48. *New Zealand Parliamentary Debates* 214 (1927): 219–20.

49. Gordon Cunningham, 'Certification of Therapeutants', Supplement to *The Orchardist of New Zealand*, 1 July 1937: 1–2; and Gordon Cunningham, 'Certification of Sprays', *The Orchardist of New Zealand*, 2 October 1939: 257–8.

50. Gordon Cunningham, 'Certification of Therapeutants', Supplement to *The Orchardist of New Zealand*, 1 Februrary 1944: 1–2.

51. A. Mouat and A. Rodda, 'Exports and Imports', in Department of Agriculture, *Farming in New Zealand* (Wellington: Department of Agriculture, 1950) 273–327.

52. Cunningham, 'Orchard Sprays in New Zealand', 185.

'Potatoes Made of Oil': Eugene and Howard Odum and the Origins and Limits of American Agroecology

Mark Glen Madison

INTRODUCTION

> This is a sad hoax, for industrial man no longer eats potatoes made from solar energy; now he eats potatoes partly made of oil.

Howard T. Odum, *Environment, Power, and Society*, 1971[1]

For the brothers Eugene and Howard Odum, understanding the relationship between humans and their immediate environment was something of a family tradition. Their father, Howard Washington Odum, was a leading American sociologist in the 1930s and 1940s, whose works on Southern regionalism sought to explain the environmental, racial, and cultural factors that made the South unique.[2] One of Howard W. Odum's primary concerns was the 'achievement lag', by which he meant that 'man has too often failed to apply his technical skills to prevent the social problems that have been created by the rapid expansion in technology'.[3] The father's interests seemed initially lost on the sons, as they went off to study ornithology and biogeochemistry; however, over time their work betrayed a continuing Odum tradition in its concern about the predicament of American agriculture. Agriculture struck the sons as a field that could be both explained and improved by applying the new methodology of 'systems ecology' (a term coined by Eugene) to overcome some of its technical problems. The Odums' attempt to understand the agroecosystem was reminiscent of their father's earlier attempts to understand how humans and the environment interact and, in doing so, improve the situation for both human and natural systems. A social role for the scientist in American society was ultimately the most important Odum family legacy.

The eldest brother, Eugene Odum (b. 1913), was initially trained in ornithology under Victor Shelford at the University of Illinois.[4] After receiving his doctorate in 1939, Eugene joined the faculty of the University of Georgia in 1940 where he remained for the rest of his career. His younger brother, Howard, was moving towards ecology via a similarly circuitous route, gaining a doctorate in biogeochemistry from Yale in 1951 and obtaining a post at the University of Florida at Gainesville. The two brothers saw their careers intersect in 1954 when

Environment and History **3** (1997): 209-38

both were hired by the Atomic Energy Commission (AEC) to study a coral reef at the Eniwetok atoll atomic test bomb site.[5]

Eugene's credentials as an ecologist at this point were the more impressive, as he had already published the first edition of his *Fundamentals of Ecology* (1953), the first textbook to be organised around A.G. Tansley's 1935 concept of the 'ecosystem'. Eugene had also been doing ecological field research for the AEC on the succession and productivity of abandoned farmland near the Savannah River nuclear facility. Howard, meanwhile, was busy studying fresh water springs in Florida. Neither ecologist had any particular background in coral reefs, but the 1950s was an important period of federal largesse as regards ecological programmes. Both ecologists had experience with federal funding and this was ultimately the experience that mattered most.[6] The six weeks spent at the Eniwetok Atoll were to have two important effects on the brothers. First, it was to link the brothers inextricably in the public mind as sharing a common paradigm of systems ecology. This was not an inaccurate perception since Howard was to contribute the chapters on energy in Eugene's textbook and both were fond of quoting and using each other's work in an almost symbiotic manner. The other result of the Eniwetok study was to convince the Odums that energy was the means to unlock the secrets of any ecosystem.

While at Eniwetok the Odums studied the entire reef as a system to determine its energy budget.[7] Strikingly, the results of the Odums' study seemed to show that most of the energy in a coral reef ecosystem was used to sustain the system. Energy for production (or photosynthesis) was nearly equalled by the energy respired – leading to their interpretation of a coral reef as a steady-state system. In the years that followed, the coral reef system was to remain an exemplar to the Odums of a mature ecosystem as a self-regulating, self-maintaining, steady-state system. As Howard went on to study the Puerto Rican rainforests, while Eugene studied marshes and woodlands, their ecosystem data confirmed their belief that conditions of stability were characteristic for all mature ecosystems.

In part, this concept was reminiscent of Clementsian succession where the climax community was the end of succession, thereafter maintaining a relatively steady state, barring some disaster such as fire or the mouldboard plough. The Odums shared with Frederic Clements a belief in evolution at the level of a system and a modified dynamics of successional stages culminating in a climax community, which the Odums defined as a 'steady-state' and self-maintaining condition.[8] However, the Odums' analysis differed in two important ways. First, the Odums always regarded their focus of analysis as arbitrarily determined by the ecologist. As Eugene liked to note in his textbooks, the ecosystem under study could range from a puddle to the entire biosphere depending on an ecologist's interests. For the Odums, all human systems also fell under the domain of the systems ecologist, a far cry from Clements' description of naturally occurring and recognisable plant communities. Second, the mode of analysis for the Odums was energy, not a flora or typological species as it had been for Clements. For the Odums, energy was the proper way to evaluate and analyse the ecosystem unit and, as a tool, ecoenergetics (the flow of energy through a system) allowed a meaningful comparison among units – something that had not been particularly easy to achieve with Clementsian communities. Most importantly, energy had a real meaning for human ecosystems and therefore provided an inroad for proactive ecologists, such as the Odums, to begin an analysis of human ecosystems along ecological lines. The Odums made this connection explicit in the introduction to their early Eniwetok coral reef study.

> Perhaps in the structure of organisation of this relatively isolated system man can learn about optima for utilising sunlight and raw materials, for mankind's great civilisation is not in steady state and its relation with nature seems to fluctuate erratically and dangerously.[9]

Moving beyond Clements was in keeping with the Odums' belief that previous attempts to study the agroecosystem were less than scientifically rigorous. The most famous attempts to study the agroecosystem ecologically had previously occurred within the Soil Conservation Service (SCS), a branch of government well-acquainted with the elder Odum's sociological work. Eugene, in his textbook and articles, was forever using the vast archive of SCS photos to demonstrate good and poor land-use practices and he even included a description of the agency's 'land-use maps' in all of his textbooks. These land-use maps were developed during the New Deal as a farm 'blueprint' so that SCS technicians could implement soil and water conservation projects.[10] Based on Clementsian ideas of agriculture as a disclimax and devoted to technological and engineering methods to protect the soil, the SCS offered few means of comparing various agricultural systems. Lacking energy analysis and any attempt to study the cycling and nutrient system on a farm, the SCS was hopelessly behind the atomic ecology of the 1950s, which was actively employing radioactive tracers to study the various cycles in every conceivable system. Even worse, the SCS had firmly tied itself during World War II to increasing production and a series of engineering projects – including reclaiming wetlands and straightening rivers – without any proper means of evaluating the environmental consequences.[11] Always focused on the key inputs of soil and water conservation, the SCS largely transformed itself into a narrowly technical 'agricultural corps of engineers' in the aftermath of World War II. Eugene's final verdict on the SCS was that it had become 'increasingly bureaucratic' and 'less responsive to the real needs' of American agriculture.[12]

The other important source of science on the farm lay in the state agricultural colleges and local extension agents. The extension system was rather single-mindedly committed to higher yields, even while crop surpluses were once again becoming a threat to the farm economy in the 1950s. Eugene dismissed extension agents as 'technicians who had great skill, but no understanding', while Howard described them as having 'forgotten how to farm without poisons' and who 'must go back to school as soon as the agricultural schools put courses in lower energy farming back into the curriculum.'[13] The extension service was described as moving in the opposite direction of more ecologically-based farming.

The final constituency with an interest in the agroecosystem arose in the post-World War II era, as increasing numbers of amateur alternative agriculturists took a non-traditional view of agriculture. Agrarian romantics – such as Louis Bromfield and the 'The Friends of the Land' – all sought to create some sort of 'balance' between humans and nature, while remaining tied to SCS ideas of how that balance might be achieved through soil conservation.[14] This was at odds with the Odums' claim that the human and natural systems did not need to be 'balanced', but rather the human ecosystems needed to more closely 'resemble' their natural counterparts.[15] The other important alternative agriculturists were the organic farmers, who in the 1940s began an American movement to bring agriculture more into line with what they saw as natural processes.[16] The organics eschewed chemicals and based their holy grail of the compost pile on the natural process of humus creation on a forest floor. In addition, the organics were intrigued by new developments in American environmental sciences (such as the early years of agroecology) which promised to validate their practices. Yet the organics could never completely shake off the aura of being health eccentrics. Their personal commitment to farming with nature arose from pseudo-medical beliefs about the concentration of proteins in organic vegetables and the supposed health benefits of uncooked edibles. They frequently came across as one step removed from herbalists and hence were not seen as an effective spokesgroup or source of support for the new agroecology and were instead often spoken of condescendingly by practicing ecologists.[17]

All three agricultural interest groups sought the professed goal of the Odums, to stabilise and improve American agriculture. Yet the Odums felt each of these groups was flawed in their outlook. The SCS was primarily concerned about preserving soil and water on farm lands, believing that the conservation of these two most precious resources would trickle down to the preservation of American agriculture. The land-grant institutions, by the same standard, looked to increasing production and farm income as a panacea, while remaining relatively indifferent to the transformation of farmers and their farms into agricultural workers on agricultural factories. Finally, the alternative agriculturists sought personal redemption by a return to the land, while remaining largely suspicious of many scientists, blaming them for most modern agricultural problems. The Odums felt that all three groups had focused on only one segment of the problem and it was up to ecologists to focus on the big picture. This may explain the Odums' dismissive treatment of their predecessors.

The three agricultural interests dismissed by the Odums – the SCS, land-grant institutions, and the alternative agriculturists – had all produced precursors to agroecology which were ignored by the Odums. The SCS, for example, may have been the first organisation to present the term 'agroecology' to the American public.[18] In a 1938 article in the agency's official publication, *Soil Conservation*, Basil Bensin described 'agroecology' as the 'basic science of soil conservation.'[19] Bensin traced the term back to the Czechoslovak Botanical Society in 1928 and described the new science as emphasising the 'relationships between species and types of crop plants and their environment'.[20] Bensin's breakdown of the three main components of agroecology would not be out of place in a current agroecology text, as he emphasised: '(1) crop plants and their regional types, (2) regional environment as it affects crops, and (3) culture as a dynamic factor in agroecology, comprising the agrotechnique of the region'.[21] Bensin's primary omission was in any concept of 'energy' informing his study of agroecology. If the Odums ignored this early pioneer in bringing agroecology to America, so did the SCS, since there were no future references to agroecology until the 1970s and Bensin's methodology seems to have had little effect on SCS policies up to the present day. Still, the disappearance of this early article from traditional histories of agroecology is odd.

Likewise, buried in slightly less obscurity, there was an early article from an important land-grant university scientist, Alfred Transeau, about 'The Accumulation of Energy by Plants'.[22] This work seems to be a direct predecessor of the Odums' agroecology, as Transeau sought to calculate specifically the flow of energy in corn to determine whether agricultural crops might provide a future renewable energy source to replace limited supplies of coal, petroleum and natural gas.[23] Transeau concluded that corn was only able to convert about 1.6.% of available energy into usable energy (i.e., kernels) and hence crops were an unlikely substitute for traditional energy resources.[24] Although this particular line of reasoning may seem quaint in lieu of later attempts to limit the flow of fossil fuels *into* agriculture, the study's broader energy concerns, and attempts to understand agronomy via energy conversions, are reminiscent of present studies in agroecology. This makes the study's omission by the Odums all the more perplexing.[25]

Finally, the omission of all domestic organic farm work by the Odums seems unwarranted. The types of Asian and Indian agriculture lauded by Howard Odum in his comparative studies had already been transplanted on to a number of American organic farms through the information spread by Sir Albert Howard and J.I. Rodale.[26] While the organics continued to cover new agroecological advances in their periodicals, the organic farms in America remained an understudied resource for agroecologists, who persisted in looking to the developing world for their examples.

In contrast to these early examples of agroecological thought arising in fields the Odums dismissed, the Odums presented a history almost exclusively devoted to a pantheon of ecologists. In Eugene's attempt to describe the history of energy flow studies (and by extension agroecological studies) not a single non-ecologist appears, and the list is heavily weighted towards the Odums' work and the traditional heroes of American ecology (e.g., Stephen Forbes, Charles Elton, Raymond Lindeman).[27] The result of this revisionist history, written from the ecologist's point of view, was the omission of credit for early pioneering work in federal agricultural programmes, land-grant colleges, and alternative agricultural communities. The new field of agroecology was to be defined along narrowly scientific lines and to exclude all other inputs as extraneous. The Odums were certain that the reform of American agriculture would arise from the ecological community. In Eugene's words:

If biologists do not rise to the challenge, who will advise on the management of man's environment – the technicians who have great skill, but no understanding, or the politicians who have neither?[28]

This attempt narrowly to define agroecology as a subdiscipline of ecology was to have ramifications for the development of the discipline in the 1970s and 1980s, when agroecologists sought to broaden their appeal. For the Odums, however, the key role of agroecology was management and previous attempts at agricultural management had removed themselves from the equation by virtue of their past failures.

The fact that agriculture needed managing seemed obvious to the Odums upon even the most superficial inspection. The return of some smaller dust bowls in the 1950s, the continuing decline of the farm population, the repeated agricultural booms and busts, and the pollutants arising from farmlands all pointed to a system dangerously unstable and unsustainable – the exact opposite attributes of a stable and self-maintaining mature ecosystem. If agriculture, with its unique status between the natural and the human environment, could not be effectively managed by ecologists, there was little hope for a meaningful role for ecologists in the management of other human systems. The Odums, continuing in their father's tradition, determined that a scientist's ultimate goal must lie in the human realm. Howard visualised the ecologist's role as wielding a 'macroscope.'[29] The goal of the ecologist was to look at all his data through a macroscope that would 'eliminate the details' of the tangled web of life and allow a simple diagram to be created (see Figure 1). This was all in keeping with the work of most ecologists in this period. But the final step in both Odums' conception of ecology involved the ecologist explicating the principles of the diagram and then 'managing with actions' the human system based on these ecological laws. Both ecologists were overcome with an excess of optimism in assuming ecologists would follow this banner. Howard predicted that in the

Mark Glen Madison

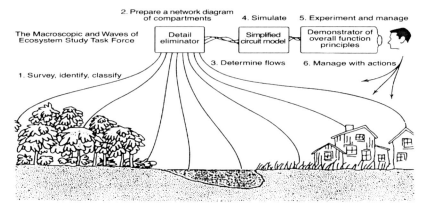

FIGURE 1. Cartoon of the 'macroscopic view' in which the detail eliminator simplifies by grouping parts into components of similar function. Once ths simplified model has been created it is the ecologist's role to 'manage' both the human and natural systems portrayed in the cartoon. (Howard Odum 1971: 10)

future schoolchildren would be taught the 'three E's' ('energy, environment, and economics') and even helped to prepare a prototype of just such a textbook.[30] Eugene, meanwhile, predicted that ecology would become 'the link between the natural and the social sciences' and produced a textbook to show how this could be achieved.[31] The path was laid for the reworking of American agriculture – the most obviously awry human ecosystem as regards natural ecosystem laws. Farm management was to be moved into the hands of the ecologists, while the farmers themselves were to be reduced to variables in an ecological diagram. With a working theory, a faith in the ecologist's role in human systems, and an important problem to tackle, the stage was now set for the Odums' analysis of American agriculture.

'EXPLANATION, PREDICTION AND CONTROL'

> The diagram of energy flow might be referred to by some as an 'Odum' device…
>
> Eugene Odum, 'Energy Flow in Ecosystems: A Historical Review', 1968.[32]

The first thing the Odums required, to create a meaningful comparison of human and natural ecosystems, was a means to reduce both to a common language – in this case, it was energy, their preferred ecosystem language since the groundbreaking Eniwetok study. Ecoenergetics was to be the great unification model for human and natural systems. Initially a wheat field, a natural grassland, or a forest shared little resemblance to one another. The difficulty in dealing with human environments had led Clements to dismiss all agricultural systems as 'disclimaxes'.[33] This sense that human environments were beyond the pale of ecological investigation continued well into the 1950s. However, the Odums felt that by reducing all systems to energy and, then, by further reducing this energy analysis to a simple model, meaningful comparisons could be made. The details of both a natural and a human ecosystem could be overly distracting, what was needed were simple models to act as 'detail eliminators' that would 'extend the capacity to see the wholes and parts simultaneously'.[34] In Howard's words the energy diagrams would serve the three roles of 'explanation, prediction and control'.[35] By extending the explanatory weight given to their ecoenergetic models, the Odums made them the key to their own (and future) agroecological comparisons. They also risked allowing their models to develop an explanatory power greater than the facts might bear. The Odums were to transform their energy diagrams from a merely visual representation of energy transformations in an ecosystem, into a heuristic device for discerning new connections, deriving new theories, and making long-term predictions.

In the collaboration between the two brothers, it was Howard who was the more gifted modeller. Since his dissertation on 'The Biogeochemistry of Strontium', Howard had always taken a keen interest in physical ecology. Well-versed in both biochemistry and biophysics, it was natural that Howard's interests drifted towards energy – a common denominator in most chemical and physical experiments. Energy circuits had been common in the physical sciences for decades but they were virtually unknown in the biological sciences, particularly ecology which still had a strong bent towards natural history and applied botany while the Odums were being trained. Howard's spiritual mentor in unifying the physical and the biological was Alfred Lotka, a physical chemist whose 1925 work, *The Elements of Physical Biology*, predicted that all of biology could be reduced to matter and energy exchanges. Although Lotka did not mention how these exchanges should be depicted, some type of energy diagram was a reasonable way to present these exchanges for a visual field such as ecology. Howard's first diagrams of ecoenergetics in the early 1950s showed the various mass and energy amounts as they decreased from one trophic level to the next, while much of the energy was dispersed in heat (see Figure 2). These diagrams were effective visual aids and largely based on his study of Florida springs and estuaries. These early diagrams, according to Joel Hagen, continue to be published in textbooks up to the present.[36] However, these large masses gradually shrinking into higher trophic levels tended to confuse issues of energy and matter exchange and Howard, at an early stage in his diagramming, sought to move to the next level – energy circuit diagrams.

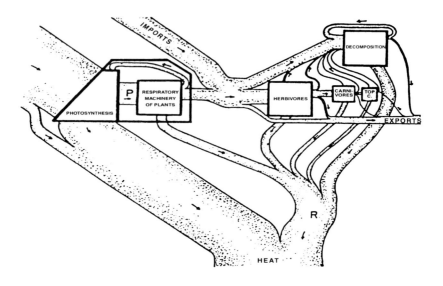

FIGURE 2. Energy flow diagram for an ecosystem. The diagram represents data obtained in 1956 from Howard's study of warm mineral springs in Florida. The various trophic levels are labeled in the boxes and the amount of available enegry declines at each stage (represented by changing width of energy bands) as respired energy goes to the 'heat sink' as a result of energy transformation. This particular diagram was reprinted in a number of ecology texts in the 1960s and 1970s. (Howard Odum 1960: 1)

Energy circuits were a traditional way for physical scientists to demonstrate the transfer of at least one form of energy – electricity. Although Howard's early energy circuit diagrams, presented in 1959 before the Ecological Society of America, did not look especially auspicious to the attendees (see Figure 3), they had several things in their favour as far as the Odums were concerned.[37] First, Howard quickly abandoned the strictly analogous energy circuit diagrams of the electrical engineer for a more symbolic energy language he developed himself which combined some of the best design elements of his trophic diagrams – directional arrows, discrete stages, ease of visual analysis – with the traditional language used in electrical engineering. The other advantage, not immediately realised with the earliest energy circuit diagrams, was that it provided a gateway for the addition of both positive and negative feedback loops.

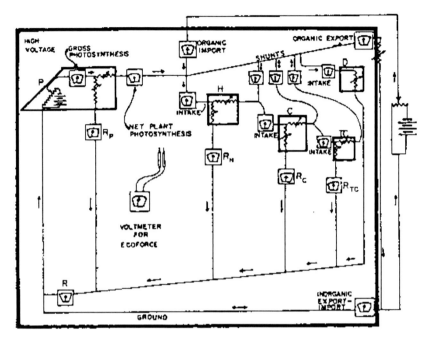

FIGURE 3. Howard's first presentation of a electrical analogue circuit. In this diagram of a steady state ecosystem, the flow of electrons corresponds to the flow of carbon. These strictly analogous diagrams were poorly received by ecologists and by the late 1960s Howard had replaced them with a more symbolic representation of energy flows through a system. (Howard Odum 1960: 4)

The promise of cybernetics in World War II had yet to fully manifest itself in ecology, prior to the Odums.[38] Physiology had incorporated homeostasis into its discipline and computing had adapted many cybernetic concepts, but the transition to ecology remained elusive. Howard's energy circuit diagrams, however, offered a way into cybernetics as his representations turned back upon themselves through feedback loops in a visually impressive and immediately apparent manner. These modeling devices quickly came to dominate a fair bit of the literature on ecosystems, in part, through the Odums' dominance in the field of American ecology in the 1950s and 1960s. Eugene's textbook, *Fundamentals of Ecology*, went though three editions (1953, 1959, 1971) and this definitive work in the 1950s and 1960s was filled with Howard's views on energy and with both brothers' energy diagrams, which derived from Howard's new energy models. Furthermore, the two brothers helped to train a new generation of college instructors through an advanced course in ecosystems biology at the Woods Hole Marine Biological Laboratory from 1957-1961. Together, the brothers enjoyed a quick trajectory of fame, winning three high prizes in ecology from three nations in three different decades as they shared the Mercer Prize in 1956 from the Ecological Society of America, the French Prix de l'Institute de la Vie in 1975, and the Crafoord Prize in 1987 from the Swedish Academy of Science. Eugene also went on to assume the presidency of the Ecological Society of America from 1964-1965. In the late 1950s and early 1960s the two ecologists were well-poised to introduce a new field of study in agroecology, enjoying secure research institutions, graduate students, and renown.

The agroecosystem was an overriding concern for both brothers when it came to trying out their new modeling system of energy. In many ways the American farm became the exemplar for the Odums' system of energy modelling for human ecosystems, just as the coral reef had served as the exemplar for natural ecosystems. Eugene's revolutionary 1964 article 'The New Ecology' had as its first illustration an SCS photo of a healthy farm, and he included lengthy sections in all his textbooks elucidating the basics of the agroecosystem. Howard, likewise, tended to introduce his energy diagrams using the agroecosystem in both popular books (e.g., *Energy Basis for Man and Nature*, 1976) and scientific texts (e.g., *Systems Ecology*, 1983). For both ecologists the agroecosystem was to be an important test case for the modelling and predictive powers of the new energy flow diagrams.

Howard first unveiled his new symbolic method of following energy flow in an article prepared for the Panel on the World Food Supply. In the published proceedings of the Presidential Commission, *The World Food Problem* (1967), Howard introduced his new energy diagrams specifically as a means to compare different agricultural systems around the world (see Figure 4). In the opening sentence of the article Howard immodestly declared that: 'The problem of world food production and the population explosion is one of system design.'[39] The rest of the article proceeded to lay out energy flow diagrams as a means of comparing different agricultural systems and evaluating them.

For the models of both natural and human ecosystems to be of any predictive or theoretical value they had to follow certain laws. Early on, Howard described the three ecoenergetic laws which would play a role in interpreting all such diagrams.[40] The first law was the familiar one of conservation of energy, which required that in each diagram all energy be accounted for somewhere in the model. The second law was a variation on the entropy law, which required the necessary degradation of energy at each stage of transformation, usually into a heat sink (or respiration). The final law was variously described by both Odums as the Darwin-Lotka law or 'the maximum power principle'; this law stated that natural selection selected for maximum effectiveness in the use of available energy resources. This third law was variously interpreted as working at the system level and the individual level, but in effect almost all the Odums' examples derived from the level of systems (i.e., rainforests, lakes, estuaries,

'Potatoes Made of Oil'

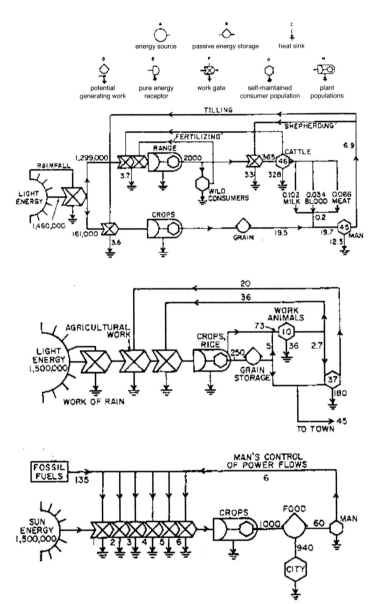

FIGURE 4. Energy circuit diagrams for: tribal cattle system in Uganda (top); unsubsidised monsoon agriculture in India (middle); and fuel-subsidised industrial agriculture of the United States (bottom). The purpose of juxtaposing these systems was to show how the high yields of American agriculture were based on large inflows of fossil fuels which (1) replace the work formerly done by man and animals and (2) do away with the more diverse network of animals and plants preserved in the two non-Western systems.
(Howard Odum 1967: 72, 74, 76, 82)

coral reefs). Of the three explicit laws which constrained each energy diagram, the first two laws required a careful accounting of every kilocalorie taken into a system and used up in the process, while the third law required a smooth flowing system in which successful systems were shown to make the maximum efficient use of energy resources.

In addition to the three explicit energy laws, there were a number of tacit assumptions made by the Odums. The first tacit assumption went back to the Odums' coral reef study, where they had learned that a mature ecosystem tended to have production equal to respiration, with minor oscillations. Consequently, mature systems were modelled as relatively stable, self-maintaining units where production (i.e., photosynthesis) roughly equalled respiration. Following a period of unrestrained growth, maximum power and efficiency were assumed to arise from increasing complexity and stability within the ecosystem. In addition, the Odums' focus on the system level, rather than the individual, led to an analysis which, both implicitly and occasionally explicitly, described selection occurring at the system level. The maximum energy use and the steady-state of the mature ecosystem were often described in terms of natural selection occurring at this higher level. Likewise, when it came time for the Odums to investigate the agroecosystem, it was at the level of national systems that most of their examples were derived. They would frequently diagram a prototypical 'American farm', but the examples that followed were inevitably drawn from different nations and cultures so as to make the comparisons more striking. This larger system analysis became important as the Odums sought definitively to tie industrialised farms into the larger industrial society. Finally, both ecologists assumed that all inputs into the ecosystem could be converted into energy. In Howard's earliest diagrams in the 1950s, the cycling of nutrients and the cycling of energy had often been portrayed in separate models. But with the beginning of Howard's more symbolic energy circuit models in the 1960s, all inputs – nutrients, labor, respiration – were reduced to common energy units (usually kilocalories). By the 1970s, even more problematic inputs such as money were occasionally being converted to energy units (i.e., dividing the gross national product in dollars by the total energy production in a nation).[41] The cumulative effect of the Odums' tacit system assumptions was the increasing reduction of all important inputs to energy, the necessity of examining many conversions at the larger system level (i.e., GNP only had relevance at the level of a national agroecosystem), and the further removal of the diagrams from any observed data, as they sought to encompass more elements deemed crucial to understanding the system.

With this combination of explicit laws and tacit assumptions, the Odums proceeded to outline the key characteristics of human ecosystems in general, with particular attention being paid to the American agroecosystem. From their earlier work on stable ecosystems, it seemed clear to the Odums that the key to understanding the agroecosystem lay in discovering how energy flowed through the basic trophic levels of a farm, with crops as producers and urban humans as the ultimate consumers. If all inputs and outputs on the farm could be reduced to kilocalories then alternative forms of agriculture could be examined and compared. Both Odums were particularly interested in the comparative aspects of agroecology, in part because of dominant political interests in the 1960s. The International Biological Program (1968-1974) and various food panels convened by the United States and the United Nations during this period were preoccupied with the question of how to feed the world's burgeoning population.[42] The Green Revolution offered one possible answer to fears of overpopulation.[43] However, Howard, through his work in Puerto Rico and with the White House Panel on World Food Supply, had become increasingly convinced that developing nations' agricultural systems were poorly understood and might contain hidden efficiencies unknown to American experts.[44] In particular, Howard was struck by the stability of millennial-old cattle raising practices in Uganda and monsoon agriculture in India. Never one to evade a telling catch phrase, Howard quoted Gandhi's statement that in India 'cows are sacred because they are necessary' to frame his own analysis about the protein and manure returns provided by cattle in India.[45] While experts were just beginning to study the systems of agriculture in the developing world, both Odums felt that the American agricultural system had also been largely unexamined from an energy perspective and had been widely misunderstood as a result.

According to the Odums, the myth of American agriculture was predicated on a common belief that increasing crop yields were the result of more efficient use of solar energy. This belief, the basis of the Green Revolution, claimed that fertilisers and pesticides had allowed any natural disasters and nutrient bottlenecks to be overcome, while hybrid seeds allowed for increasingly efficient use of all the solar energy received in a field. Both Odums referred to these common conceptions as 'energetic fallacies' for, like perpetual motion, they seemed to imply that something could be constantly increased without any compensatory costs.[46] The only way to combat these misconceptions involved detailing all the agroecosystem's energy inputs and outputs.

That is not to say the standard beliefs about American agriculture did not hold some truths. American agriculture was magnificent at increasing crop yields and all the diagrams of international agroecosystems tended to have the United States or Japan at the top of the list for crop yields per acre.[47] The misconceptions arose from equating yields with efficiency. In economics, the equivalent would be equating the gross national product with each individual worker's efficiency, regardless of other inputs such as mechanisation. Similarly, the Odums felt that the actual energy costs of American agriculture had been significantly underestimated, due to a lack of accounting for fossil fuel subsidies. Every ounce of fertiliser, pesticide, and diesel fuel used to run a piece of farm machinery was a fossil fuel energy input, and had to be accounted for in the energy ledger in accordance with the first law of thermodynamics (and ecoenergetics) – that energy is neither created nor destroyed. That this had not been previously attempted was partially due to difficulties in conversions. The units of energy for food were traditionally kilocalories, and those for fossil fuel were joules or watts. However, the conversions did exist and could be made increasingly with calculators and computers in the 1960s and 1970s. Another reason for the lack of analysis was the result of United States Department of Agriculture (USDA) being primarily interested in yield per acre, regardless of energy inputs.[48] The Odums argued the most important figure was not the *gross energy* produced (as measured in total crop yields of American agriculture frequently touted by the USDA) but rather *net energy* was the figure that counted in evaluating the agroecosystem. Net energy was a fairly slippery commodity in the Odums' hands, but a common sense one. Since solar energy was a free and continuous input into an open system, one could ignore it for the purposes of determining net agricultural energy.[49] But any energetic inputs – such as fossil fuel, labour, and chemical additives – had to be added up and compared to the amount of energy made available in kilocalories of food. When defined this way, there was an interesting evolution in the history of agriculture toward declining net energy yields. Hunting and gathering and subsistence farmers tended to have small net energy yields, while highly industrialised farming of the Western variety tended to have a net energy loss. In fact, there was a strong correlation between auxiliary energy inputs (that is anything but solar energy) and decreasing net yields.[50] The fact that human cultures had moved away from net energy yields over their history might have made it appear evolutionary advantageous. However, Eugene explained this phenomenon as being a result of humans being stuck in a type of 'pioneering stage' in which it was advantageous to maximise the use of energy resources, even while the human population and its complex modern civilisation required a transition to a steady-state climax stage.[51] The job of ecologists, such as the Odums, was to increase the human understanding of their current agroecological stage and help humans modify their policies in accord with it.

The first stage in increasing public understanding involved laying out the details of American agriculture. A distressing tale of the decline of American agriculture, as regards net energy, emerged from the Odums' breakdown of all the inputs that went into that agriculture. The Odums, in a rather uncharitable assessment of the general public, believed that most people saw sunlight and hybrid seeds yielding larger crops and never went beyond this type of superficial analysis. The ecologist's job was to pull back the curtain and expose, via energetic models, that the Great Oz of American agriculture was hardly as powerful and stable as it appeared. The key to the Odums' analysis of American agriculture revolved around what they termed *auxiliary* sources of energy. As opposed to a natural system, human systems could modify evolutionary tendencies by adding auxiliary sources of energy. A stable ecosystem in nature was forced to expend a fair amount of the system's energy providing for the maintenance of the flora and fauna. In fact, many ecosystems enjoyed a noticeable shift over time from pioneer species (those that used energy resources with profligacy), to more stable species (those which were more economical in their demands). Clements referred to these generalised species as pioneer and climax species; MacArthur and Wilson, in their 1967 work *Island Biogeography*, referred to them as r-selected and K-selected species; and the Odums referred to them as C4 and C3 plants.[52] Although the names changed with the changing ecological paradigms, they all referred to two classes of species: the former being adapted to quick, fast growth, the latter being adapted for stability and maintenance of a complex system.

According to the Odums, this natural evolution towards climax species was precluded in agriculture because farmers sought to maximise production via quick-growing species of plants (such as C4 varieties). The use of pioneer species and monocultures prevented the appearance of a well-developed climax community and, by all ecological laws, these stands of plants should have been prone to large boom-and-bust cycles. However, it was painfully obvious that the opposite had occurred in the United States, where crop yields kept on consistently increasing, with only periodic fluctuations during particularly adverse weather conditions. According to the Odums, American farmers had achieved this feat by using fossil fuels to bypass the evolutionary cycle. Plants that should have evolved to be more stable and self-sufficient (or been replaced by others that were) persisted in agricultural fields because virtually all maintenance functions were replaced by fossil fuels. Pesticides killed predators, fertilisers decreased the need for deep and efficient roots, feedlots decreased any energy expenditures cattle might require, and air-conditioned hen houses reduced the need for chickens to maintain their own temperature-regulating mechanisms. Farmers had been able to maintain highly productive pioneer-type species, that were inherently unstable, by releasing all the pioneer's energies for the production of increasing yields. These fossil fuel farms were populated by seemingly monstrous hybrids, bred not for any efficiency as regards sunlight, but rather for their symbiotic potential with fossil fuel auxiliaries:

> We now have chickens that are little more than standing egg machines, cows that are mainly udders on four stalks, and plants with so few protective and survival mechanisms that they are immediately eliminated when the power-rich management of man is withdrawn. Such varieties are complementary to the industrialised agriculture and cannot be used without it.[53]

With this analysis, the Green Revolution came across as a 'cruel hoax' since these hybrid species would have little hope of survival in a non-industrialised society without the requisite fossil fuel subsidies.[54] Equally important, this analysis boded ill for the sustainability of American agriculture. According to Howard: 'Nothing about man's present system [of agriculture] is balanced, for his inputs come from geological storage and from energies that used to go to balanced systems'.[55] Considering that one of the primary attributes of mature ecosystems was stability, American agriculture seemed at odds with all natural systems laws.

The Odums, in re-examining the American agroecosystem, had moved themselves to the forefront of a contemporary political debate on Third World development. Paul Ehrlich's 1968 polemic *The Population Bomb* had largely stigmatised people of the developing world as breeding themselves into poverty.[56] Ehrlich and population ecologist Garrett Hardin both raised questions about the efficacy of the Green Revolution to meet the needs of a burgeoning world population.[57] The Odums had previously entered the debate on the Green Revolution, yet they approached the problem from the opposite end. The Odums argued that American agriculture had been a vast experiment, replacing natural ecosystems with heavily dependent monocultures. The result of these policies had been a costly and unstable agriculture whose export to the rest of the world seemed dangerous and misguided. This confluence of political and scientific issues of the 1960s caused the Odums to examine international agroecosystems, but their primary interest remained the United States. They felt that a necessary first step involved understanding the domestic agricultural situation, before beginning a heedless export.

This analysis of the domestic agroecosystem led to a critique of the American economy ranging far beyond the Green Revolution. Once begun, the analysis of human agroecosystems called into question every aspect of human ecology and economy. As their ecoenergetic diagrams began to encompass cities (as the ultimate processors and consumers of agricultural products) both Odums began to question the energetic foundation of American society. Eugene, in an important 1969 article for *Science* entitled 'The Strategy of Ecosystem Development', attempted to explain two centuries of American history using the laws of ecosystems succession:

> In the pioneer society, as in the pioneer ecosystems, high birth rates, rapid growth, high economic profits, and exploitation of accessible and unused resources are advantageous, but, as the saturation level is approached, these drives must be shifted to consideration of symbiosis (that is, 'civil rights', 'law and order', 'education', and 'culture'), birth control and the recycling of resources. A balance between youth and maturity in the socio-environmental system is, therefore, the really basic goal that must be achieved if man as a species is to successfully pass through the present rapid-growth stage, to which he is clearly well-adapted, to the ultimate equilibrium-density stage, of which he as yet shows little understanding and to which he now shows little tendency to adapt.[58]

According to Eugene's prescription, the human ecosystem needed to approach more closely a mature equilibrium system where production and respiration are nearly equal. Some of the solutions Eugene directed at American agriculture included: the use of more 'detritus agriculture' (including oyster beds along the entire East Coast), less dependence on fossil fuel inputs on the farm, and a halt to the Green Revolution. In addition, Eugene argued for diverse homesteads, both for aesthetic reasons (he felt most heterotrophs prefer to have both grass and trees) and because they offered more stability from ecological perturbations.[59]

Howard went beyond these rather timid goals to envision more broadly what the entire human ecosystem would look like if it followed all natural ecosystem laws. First, Howard reaffirmed the equivalence of all natural and human systems by diagramming history and religion as ecological feedback loops in *Environment, Power, and Society* (1971) – his first attempt to make accessible to the general public the explanatory, predictive, and control aspects of his new energy diagrams. From this analysis, Howard proceeded to offer a prescription for bringing the entire human ecosystem into a more stable accord with natural laws. For a human ecosystem to approach a natural mature ecosystem required achieving an equilibrium between production and respiration, and growth must give way to self-maintenance. Obviously, the large energy subsidies for agriculture would have to cease and instead, more efficient use of decomposers, animals, and human labor would have to occur to maintain production. Howard's speculations in *Environment, Power, and Society* were relegated to the final chapters in an otherwise rather sober scientific treatise; however, the future outlined in the final chapters of that work was revisited in Howard's later, post-OPEC oil crisis writings, including his prototype of a textbook for the three E's – *Energy Basis for Man and Nature* (1976). In this later work, an emboldened Howard envisioned American agriculture in the near future as approaching the steady-state of traditional Asian agriculture, 'where the nutrients are recycled and most yields reinvested in work back on the land' and where the primary energy inputs involved human and animal labour.[60] Howard predicted that travel agents, luxury dealers, international traders, and other occupations at the top of the fragile, production-driven, fossil fuel economy would have to be transferred to the farm economy to meet the needs of a newly labour-intensive agriculture.[61] The resulting human ecosystem would be self-sufficient, regional, and less prone to non-local fluctuations. The loss of Florida oranges available year-round would be compensated for by the sustainability of the new system and the end of agricultural booms and busts. Howard presented this future as 'a happy place' and an optimistic alternative to the unnatural *Future Shock* (1970) previously described by Alvin Toffler.[62] Like a natural ecosystem, the human ecosystem at that point would be geared toward the sustaining of diverse regional units at a steady-state.

CONCLUSION

The OPEC oil crisis of 1973 seemed to vindicate the Odums' predictions about the instability of American energy use. Both Odums took the crisis as proof of their analysis and claims and prepared, in vain, for a more general acceptance of their ecoenergetic analysis of society. The Odums did succeed in attracting some adherents to their beliefs; unfortunately they were the wrong sort of disciples. The Odums' work began to attract attention from some of the same amateur groups whom systems ecology was meant to supersede. Organic farmers and environmentalists, in particular, found many points of intersection between the Odums' solutions and their beliefs. Organic farmers found an unlikely ally in the Odums' agroecology.[63] The organics' shunning of herbicides and pesticides need no longer be based on vague misgivings about industrialised farming, but could now be justified as the avoidance of unstable expenditures of geological resources – using the Odums' analysis. The organic farms and ideology also seemed to fit best into the mode of a low-energy and high-diversity ecosystem, toward which mature natural communities were constantly striving.[64] Even the rhetoric of the Odums at times seemed to echo that of organic farmers and alternative agriculturists when, for example, Howard claimed that 'agribusinesses' will be replaced by 'agrihumanity'.[65] The Odums' discussion of local food producing regions echoed the organic Rodale Institute's later work on 'foodsheds' – both ideas self-consciously building on SCS concepts of ecologically important agricultural watersheds. The organics could now find agroecological support for their agrarian nostalgia and health concerns.

In a similar manner, the type of society envisioned by Howard Odum in the steady-state future struck an appealing note with many environmentalists. Howard's 'Ten Energy Commandments' to replace anthropocentric religion, reflected the growing environmental belief in 'the rights of nature'. The final commandment declared 'Thou must find in thy religion, stability over growth, organisation over competition, diversity over uniformity, system over self, and survival process over individual peace.'[66] The fiercely biocentric 'deep ecologists' of the 1980s found much to admire in the Odums' human ecology.[67] In addition, certain pseudo-scientific organisations like Biosphere 2 (the ill-fated attempt to re-establish a sustainable and self-sufficient human ecosystem in southern Arizona) quoted approvingly from both Odums to support their research.[68] The Odums predicted a future steady-state human ecosystem that would be more agrarian, regional, stable, and diverse – all terms that have become buzz words of environmentalists in the last twenty-five years. In addition, the Odums, un-like most ecologists, were firmly committed to bringing the human ecosystem into line with ecological laws, since they saw no clear demarcation between them. This fitted in well with the modern American environmental movement's attempts to ensure a political role for ecologists and ecology.[69]

Unfortunately, scientifically suspect organic farmers and environmentalists had not been the intended audience for the Odums' agroecology. The Odums had hoped to move beyond amateur groups to create a working theory of human ecology that would follow its laws as inextricably as the coral reefs followed their ecoenergetic laws. The Odums assumed that ecologists would lead the way in the rationalising of the human ecosystem. The Odums needed to appeal to the ecology community and the federal government to achieve their goal of rationalising the human ecosystem. With these two groups, however, they were not particularly successful.

Ecologists were the first to pull back from some of the more unorthodox speculations of the Odums. These speculations were a natural outgrowth of the Odums' type of ecosystem analysis which constantly spoke of natural ecosystems having 'strategies' and 'goals' which were achieved via 'feedback loops' in the system. If natural ecosystems did indeed have such strategies then it made perfect sense for human ecosystems to be brought into accord with similar goals – especially agriculture, which was perceived by the Odums as an ecosystem in which all natural feedback loops and controls had been eliminated in order to maximise productivity. Unfortunately, the Odums' rather injudicious use of such anthropomorphic terms tended to scare off a new generation of ecologists. By the late 1960s, fewer and fewer ecologists were finding the search for ecosystem strategies and goals a particularly fruitful area of research. Instead of the Odums' purposeful ecosystems and system strategising, a new generation of ecologists was becoming more interested in the interworkings of selfish individuals to fulfill their Darwinian strategies. The next generation of ecologists, in the 1970s and 1980s, found their most fruitful paradigms in the evolutionary transformations of populations. The effect of the Odums' attempt to blend the natural and social sciences was that their work was attacked as anthropomorphic and teleological by the very ecologists they sought to empower. Agroecology was gradually marginalised as a hybrid anthropological ecology, with no effective base or adherents in American agricultural colleges throughout the 1970s and 1980s. In spite of its origin as an attempt to analyse American agriculture, agroecology was pushed to the edges of academia, as a tool to study 'other' systems or to understand the effects of the Green Revolution. Although their values were often sympathetic to developing nations, agroecologists largely took on the colonial role of studying other peoples from a Western perspective. Ecologists, by and large, gave up the human ecology work of the Odums, and restricted agroecological analysis to an academic niche most often found studying developing nations or primitive tribes, not the agricultural system that supported their universities.

Scientists associated with land-grant institutions either ignored the Odums or actively attacked their ideas. One of the great advantages of the diagrams pioneered by Howard was their ability to reduce all inputs to an interconvertible unit of energy. The Odums felt that this allowed meaningful comparisons to be made between human and natural systems for the first time. Unfortunately, the Odums' theories were vulnerable to charges of privileging energy above all other causal factors. This became especially obvious in the Odums' one sustained analysis of a human systems – the agroecosystem – which did not in fact revert to the type of nineteenth century farm communities predicted by energy flow models. Yet both Odums continued to make these type of generalised predictions about human systems well into the 1990s, based largely on their all-encompassing energy analysis. Agricultural economists were rather quick to attack this type of analysis, on the basis of two claims. First, agricultural economists had their own definition of energy as equivalent to its cost in dollars as an input. This was at odds with the Odums' conversion of all elements to energy. The issue among agricultural economists was whether you wanted to reduce everything to energy or money – a debatable point.[70] More problematic was the claim (again led largely by aggrieved land-grant economists) that the Odums' models did not mirror what actually occurred on American farms.[71] Here the critics seized upon the relatively small data base available to the Odums and the rather sweeping generalisations arising from this data.[72] The economist critics' general claim was that the agroecologists had not effectively demonstrated how their analysis was more 'useful' in policy decision-making than more traditional economic analysis. The economists argued that the agroecologists' scientific credentials and powerful use of physical laws effectively shrouded some of the utilitarian flows of their analysis. By exaggerating the utilitarian flaws in net energy analysis, the agricultural economists largely sidestepped the issue of whether net energy analysis held any explanatory power. Because flaws could be discerned, the argument went, agroecology need not be considered as a serious policy tool.

In part, the Odums' ineffective policy efforts were the result of their own personal and occupational limitations. The Odums' reductionist techniques and holistic claims allowed for an exciting research agenda in the 1950s and 1960s, while largely remaining above the mundane level of everyday practicalities in American agriculture. This allowed the Odums to examine American agriculture in a wholly new way and to detect heretofore unknown long-term instabilities in American agriculture. Unfortunately for the Odums, agricultural policy has always been formulated in the mundane present. Since the advent of large federal agricultural subsidies in the 1930s, American agricultural policies have been driven by a combination of historical, political, and economic factors. None of these translated particularly well into energy units. The Odums were, from the beginning, more interested in rewriting the history of agroecology than in studying the history that gave rise to American agriculture. This prevented the Odums from enlisting support from outside groups who had long sympathised with the types of dire agricultural warnings presented by agroecology. The Odums' political naiveté and disinterest also limited effective federal implementation of their programme. They could not reduce to kilocalories the effect of having federal subsidies controlled by powerful Senators with strong rural constituencies. In addition, the *long-term* nature of the 'predictions' implied by the Odums' energy diagrams made them politically problematic and a less than perfect policy instrument. Economically, crop values, with their historical, social, and political contingencies, were not necessarily the equivalent of their energy value as related to a barrel of crude oil. The Odums were relatively uninterested in why energy values did not function on the farm as they did in the ecoenergetic diagrams. The Odums always focused on two of the three E's (energy and environment always took precedence over economics), and the Odums' economist critics pointedly seized upon the gap between the energy diagrams and what actually was happening on American farms. Not surprisingly, recent environmental transformations in agriculture – such as no-till farming – owe more to the development of new pesticides and herbicides than ecoenergetic diagrams. The Odums' broader programme of a planning role for agroecology faltered on the irreducible human constructs of history, politics, and economics which drove, and continue to direct, American agriculture.

The Odums' legacy was ultimately a narrower one than they would have liked. Although their policy goals largely failed, the Odums' diagrams proved a useful heuristic device for comparing differing agroecosystems by reducing them to their most basic energy inputs and outputs, while allowing the viewing of the whole and the parts simultaneously. Their diagrams and personal popularity persisted, while their importance within the ecological community waned after the 1960s. The power of the Odums' combined work lay in equal measure upon their own scientific prestige and the usefulness of their diagrams. As a dominant force in the American ecological community in the 1950s and 1960s, their work was understood and replicated across the nation and hence was profitable as regards publications and research programmes. Both Odums were quite successful in setting up a number of ongoing ecological institutions at their home universities. In addition, their models were an important tool for the burgeoning field of ecology after the 1950s. Although Howard understood the mathematics probably as well as any ecologist in the field, his diagrams did not require specialised mathematical knowledge. The Odums (when they listed them at all) placed the equations for their models far away in the appendices of their papers and texts. As Howard noted, the energy circuit diagrams were 'a form of mathematics with emergent theorems and perceptions that extend the capacity to see the wholes and parts simultaneously'.[73] The models may have been mathematical in derivation, but their comprehension required no special mathematical skill and this may have accounted, in part, for their continuing popularity in introductory texts and among certain mathematically averse ecologists. Finally, the visual appeal of the models themselves almost certainly played some role in the continuing Odum legacy. As opposed to mere numbers or graphs, the Odums' models actually seemed to represent the way nature works and hence were reminiscent of an older tradition of natural history. In a field such as ecology, whose roots were largely observational, it is not surprising that the visual ecology of the Odums' models enjoyed a lasting popularity in texts and presentations – long after the research programme of most ecologists had abandoned the Odums' energy reductionism and system dynamics. The science of agroecology itself retained a large share of both the Odums' methodology and modeling techniques, while initially narrowing the scope of its focus largely to 'other' agricultural systems beyond the developed world's boundaries. As the science of agroecology began to define and delineate itself more strongly along these lines, it moved further from the Odums' social role and left the way open for a new generation of political ecologists to take over the social half of the Odums' programme.

The political ecologists largely took up the more extreme programme of a critique of American society that had been inherent in many of the Odums' more encompassing energy diagrams. Rachel Carson was the most important early populariser of some newly discovered agroecological dilemmas. Her environmental wake-up call, *Silent Spring* (1962), helped to present American agriculture for the first time as a system dangerously at odds with nature and capable of endangering the broader populace. Barry Commoner went on to re-evaluate the entire American agricultural system in the 1970s and concluded that the system was dangerously unstable both economically and environmentally.[74] Commoner attributed this to the cheap inputs of energy and its by-products which, he claimed, made it pay to pollute. Agriculture played an important role in Commoner's broader critique of capitalism as anti-environment. Yet, reminiscent of the Odums, Commoner was likewise unable to expand the domestic constituency for agroecology. Certainly, part of the domestic failure of agroecology can be attributed to the politics of the farm bloc. Since the 1930s, a coherent political interest group has emerged based on federal subsidies of agriculture and consisting of land-grant institutions, farmers, and federal agricultural agencies. This powerful force has consistently and effectively mobilised counter-attacks to any critique of current agricultural policies. But the failure to mobilise a larger share of environmental activists or ecologists surely cannot be attributed solely to the farm bloc. Rather, both environmentalists and ecologists seem unwilling to accept human ecology as a *natural* field of study or activism. Until human environments are incorporated into the same privileged realms of study and protection which natural systems enjoy, programmes such as agroecology must surely falter on the prejudices of their chosen constituency.

NOTES

1.   Howard T. Odum 1971: 116.

2. Howard W. Odum's two seminal works on regionalism were *Southern Regions of the United States* (1936) and *American Regionalism* (with Harry Estill Moore in 1938).
3. As quoted in Eugene Odum, 'The Attitude Revolution', 1970: 14.
4. Biographical details about the Odums are from: Hagen 1992: 101-145; Taylor 1988: 213-244; McMurray 1995: 1501-1503.
5. Eniwetok was a two-square-mile circular atoll in the Marshall Islands chain in the western central Pacific. Captured by the U.S. in 1944, it had been an atomic test site since the end of World War II.
6. The best treatment of the importance of the Odums' work in the 'atomic age of ecology' is found in Hagen 1992: 101-121.
7. H.T. Odum and E.P. Odum 1955: 291-320.
8. Eugene Odum was well-versed in Clementsian theories, having studied zoology at the University of Illinois in 1937 under Victor Shelford, who was then collaborating with Frederic Clements on *Bio-Ecology*. As regards evolution at the level of systems, Howard defined it as such: 'By systems selection, Lotka, Tansley, and the theories in this volume refer to self-organisation choices that contribute to the systems resources for meeting contingencies. Darwinian selfish selection is regarded as a secondary priority for survival.' (H.T. Odum 1983: 453)
9. H.T. Odum and E.P. Odum 1955: 291. The introduction went on to ask about the connections between productivity, energetic efficiency, and a steady-state equilibrium.
10. The most common SCS techniques to hold soil and water in their proper places included: plouging along contour lines, terracing steeper hillsides, intercropping strips of grasses or trees to retain soil and moisture, crop rotations to restore soil nutrients, and various drainage projects designed to better control water runoff and prevent soil erosion. The heyday of these projects occurred from 1933-1937, when an influx of youthful Civilian Conservation Corps labour allowed large-scale implementation of these projects on endangered farmlands.
11. The SCS in 1937 began to organise all the nation's farmers into Soil Conservation Districts. The SCS was committed to providing technical support and land-use maps to all farmers participating in the districts. In spite of the districts, it became apparent during World War II and immediately afterward that farmers felt free to drop out of any soil conservation plan whenever market conditions made it lucrative to do so. Faced with a growing dropout rate and a farmer-based agenda within the districts, the SCS after World War II began to focus more on land drainage, flood prevention, and efforts to increase crop production. For a more detailed discussion of the Soil Conservation Service during and immediately after World War II, see Madison 1995: 157-172.
12. E.P. Odum 1993: 140. Eugene in the 1950s originally thought the SCS might play a useful role as a provider of technical support along the natural boundaries of watersheds. However, the SCS's inability to develop a coherent theory-based science, its dubious reclamation efforts, and its political ties to the farm bloc all helped to disillusion Eugene about the potential of the SCS.
13. E.P. Odum 1964: 16; Howard Odum and Elisabeth Odum 1976: 7.
14. The group I labelled as 'agrarian romantics' all had close ties with the Soil Conservation Service. For example, one of the founders of the Friends of the Land was Hugh Hammond Bennett, Director of the SCS. The editor of the Friends of the Land journal, *The Land*, was a former writer of SCS pamphlets, Russell Lord.
15. E.P. Odum 1969: 262-270.
16. In 1942, J.I. Rodale began publication of *Organic Farming and Gardening* magazine. For the first two decades of its existence, American organic farming was largely dominated by Rodale's publications.
17. Olds 1969: 66-71. In this article, Barry Commoner speaks patronisingly of organics stumbling onto ecological laws. The Odums for their part rarely, if ever, mentioned the organics although

many of their proposals – such as low-input farming, compost heaps, and detritus agriculture – could likely be traced to them. Instead, the ecological histories produced in Eugene's textbooks and articles scrupulously avoided mentioning any organic predecessors.

18. In an interesting historical footnote, the official SCS publication, *Soil Conservation*, actually discussed a new science of 'agroecology' at least twenty years before the Odums first became interested in the topic. The article discussed the new European science of agroecology which sought to quantify appropriate agricultural plants based on the environment and particular farm culture of a region. The agroecology described in the article resembled a mixture of plant botany and social anthropology. It did not involve energy or any ecological concepts nor was it an important factor in shaping the American version of agroecology. Characteristically, the technical and engineering-based SCS paid no attention (besides the one article) to the new science. Bensin 1938: 138-141, 152.

19. Bensin 1938: 138-141, 152. Bensin was an agricultural scientist who occasionally taught at the USDA Graduate School in Washington, D.C.

20. Bensin 1938: 138.

21. Bensin 1938: 138.

22. Transeau 1926: 1-10. Transeau was a scientist at Ohio State University and President of the Ohio Academy of Science.

23. Transeau 1926: 1.

24. Transeau 1926: 10. Transeau did his energy conversion by converting glucose to calories and using calories as the standard for comparing various energy efficiencies. Transeau, unlike the Odums, did not consider the inputs of fertilisers, pesticides, mechanical fuels, or labor into his equations, in part, perhaps because they played a much smaller role energy-wise in the American agriculture of 1926 than they did in 1966.

25. Transeau seems to have also fallen out of later histories of agroecology. The only agroecologist to mention Transeau as a pioneer in the field was Stanhill 1984: 3.

26. Sir Albert Howard, an agricultural scientists in colonial India, introduced the Indore method of composting to American farmers in the 1940s through J.I. Rodale's magazine *Organic Farming and Gardening*.

27. E.P. Odum 1968: 11-18.

28. E.P. Odum 1964: 16.

29. H.T. Odum 1971: 9-11.

30. H.T. Odum and E. Odum 1976: 13. Every chapter in this book had questions at the end to provoke either the high school or undergraduate readers to envision the links between the three E's. In reality all of Howard's books after 1971 had this goal of uniting the three E's.

31. E.P. Odum, *Ecology: The Link Between the Natural and the Social Sciences*, 2nd ed. (New York: Holt, Rinehart and Winston, 1975). The first edition came out in 1963 and although the theme was the linkage between the two sciences, the subtitle was only added to the second edition. The Odums' book titles were often an explicit call for a link between the natural and human sciences: Howard Odum, *Environment, Power, and Society* (1971), and Howard Odum and Elisabeth Odum, *Energy Basis for Man and Nature* (1976). Even Eugene's earliest textbook, in 1953, strongly supported this linkage in its content.

32. E.P. Odum 1968: 16.

33. Disclimax stood for 'disrupted climax community.' Clements also used the term proclimax to describe agricultural systems. Clements and Chaney 1936: 48.

34. H.T. Odum 1971: 10.

35. H.T. Odum 1983: 579.

36. Hagen 1992: 142.

37. Howard's presentation before the ESA was reprinted in: Howard Odum, 'Ecological Potential and Analogue Circuits for the Ecosystem,' *American Scientist*, 48, No 1 (1960), pp. 1-8.

38. Norbert Wiener's influential book *Cybernetics* came out in 1948.

39. H.T. Odum 1967: 55.

40. Howard's laws were laid out quite early in the chapter on energetics he wrote for his brother's textbook. E.P. Odum 1959: 43. Originally these thermodynamic laws were supposed to be superseded by something called the 'ecoforce,' which Howard introduced along with his energy circuit analogues in 1959. But the power of the ecoforce to 'force' energy through the various trophic levels was quickly abandoned and the thermodynamic laws returned to the forefront of Howard's model-making.

41. H.T. Odum 1971: 297-298.

42. E.P. Odum 1968: 17. In this overview of systems ecology, Eugene explicitly tied the new energy flow diagrams to the International Biological Program and other attempts to grapple with the food problem.

43. By the Green Revolution I refer to attempts, beginning in the 1960s, to dramatically increase food production in the developing world using high-yield hybrid seeds, pesticides, herbicides, and large-scale mechanical farming.

44. Howard was given funds from the Rockefeller Foundation in 1957 to study the Puerto Rican rainforests and their response to a massive input of energy. From this natural system, Howard began to devise models for humans in response to massive inputs of fossil fuels.

45. H.T. Odum 1967: 60.

46. H.T. Odum 1971: 128.

47. Examples of such diagrams include: E.P. Odum 1971: 412; E.P. Odum 1993: 85; H.T. Odum 1971: 132. These diagrams were generally presented as straw men to be followed by analysis demonstrating that gross yields were not the crucial measure for agroecology, but rather net yields (in which the United States and Japan did not fare so well) were the important figure.

48. The USDA still does not have any breakdown for the energy costs of crops.

49. H.T. Odum and E. Odum 1976: 79. Howard succinctly defined net energy as 'the energy yielded over and beyond all the energy used in processing it.' According to his diagrams and analysis, solar energy was not a 'processing' energy but all other energy inputs were tabulated.

50. H.T. Odum 1983: 253.

51. E.P. Odum 1969: 262-270.

52. H.T. Odum 1983: 373.

53. H.T. Odum 1971: 128.

54. H.T. Odum 1971: 128.

55. H.T. Odum 1971: 288.

56. Ehrlich 1968.

57. Hardin's two most provocative discussions of the issue appeared in 'The Tragedy of the Commons' (1968) and 'Lifeboat Ethics: The Case Against Helping the Poor' (1974).

58. E.P. Odum 1969: 269.

59. E.P. Odum 1969: 267.

60. H.T. Odum and E. Odum 1976: 145. In looking to Asian agriculture, Howard was following a long tradition in American agrarian conservation going back to F.H. King, a turn of the century USDA employee, and culminating in Howard's contemporary the organic farming publisher J.I. Rodale.

61. H.T. Odum and E. Odum 1976: 250-251.

62. H.T. Odum and E. Odum 1976: 250-251. Howard presented his 'steady-state future' as a retreat from future shock back to an earlier simpler life. In many ways this was reminiscent of the

way certain agrarian Romantics (Louis Bromfield comes to mind) nostalgically recalled an earlier agricultural era.

63. In the 1960s the organic farming journal was taken over by J.I. Rodale's son, Robert Rodale. Robert had a keen interest in keeping up with the latest ecological trends (including agroecology) and he created the Rodale Research Center to bring together ecologists and agronomists to create new agroecological studies. R. Rodale 1968: 19-21; R. Rodale 1982: 22-26; R. Rodale 1983: 15-20.

64. Olds 1969: 66-71; Cox 1973: 90-94.

65. H.T. Odum and E. Odum 1976: 7.

66. H.T. Odum 1971: 253. The first nine commandments were: '(1) Thou shall not waste potential energy; (2) Thou shall know what is right by its part in survival of thy system; (3) Thou shall do unto others as best benefits the energy flow of thy system; (4) Thou shall revel in thy systems work rejoicing in happiness that only finds thee in this good service; (5) Thou shall treasure the other life of thy natural system as thine own, for only together shall all survive; (6) Thou shall judge value by the energies spent, the energies stored, and the energy flow which is possible, turning not to incomplete measure of money; (7) Thou shall not unnecessarily cultivate high power, for error, destruction, noise, and excess vigilance are its evil wastes; (8) Thou shall not take from man or nature without returning service of equal value, for only then are thee one; (9) Thou shall treasure thy heritage of information, and in the uniqueness of thy good works and complex roles will thy system reap that which is new and immortal in thee.' As to the inevitable question—was Howard serious about this?—the answer seems yes. As part of his attempt to outline a means of bringing human systems more into line with natural systems, Howard specifically targeted anthropocentric religion and proposed teaching this energy ethic (with its ten commandments) in schools and churches as an alternative.

67. For an example of American deep ecologists' treatment of the Odums see: Devall and Sessions, *Deep Ecology*, 1985: 85. In this popular American book, Eugene Odum and several other ecologists are praised as a small group of scientists who 'were to develop in their own philosophies some version of a biocentric perspective on the equality of all nonhumans and humans.'

68. Allen 1991: 3, 57. In spite of finding support for their project in the Odums' writings, Howard, at least, was quite skeptical of the project predicting only 20% of the species in the habitat would survive. Howard had been advocating experiments resembling Biosphere 2 since 1971, because he sought to determine how to create the perfect steady-state human system for the future. In contrast to the current project (with its eyes on the stars) Howard sought to use such an experiment to radically rework the human systems on earth so as to make them better adapted to the environment. H.T. Odum 1971: 287.

69. Both Eugene and Howard explicitly reached out to a broader popular audience. Howard, as already noted, published a series of books aimed at a more general audience. Eugene contributed articles to popular magazines. A particularly interesting example of the latter is Eugene's 'The Attitude Revolution' an introduction to a series of articles on ecology collected by The *Progressive* magazine. In his introduction Eugene described systems ecology as uniting the disparate forces of the day including: Howard W. Odum's technological 'lag time,' the youth-oriented ecology movement, and the need for a scientifically rigorous national environmental policy. E.P. Odum 1970: 13-15.

70. Some of the earliest volleys in the critique of agroecology came after a 1974 Congressional act mandating the use of 'net energy analysis' in the assessment of new energy sources. Many agricultural economists attacked the net energy enthusiasts both for their dismissive attitude toward traditional economics (often derided as 'contingent') and for their attempts to replace economic values with energy values. For a typical example of this critique see: David Huettner, 'Net Energy Analysis: An Economic Assessment,' *Science*, 192, No. 4235 (1976), pp. 101-104.

71. An important survey by Stout et al. in 1984, seemed to confirm the relatively minor role of energy in domestic agricultural policy. Whether this was the result of the failure of the Odums' programme to educate the public and transform agricultural policy, or the inevitable working of agricultural economics was left unexplored—though the implication was toward the latter. Stout et al. 1984: 167-189.
72. Connor 1977: 669-681; Pasour and Bullock 1977: 683-693.
73. H.T. Odum 1983: ix-x.
74. Barry Commoner's most sustained examination of agroecological issues occurred during his study of agrochemical pollutants in the water supply of Decatur, Illinois. From a careful examination of the conditions that led up to this disaster, Commoner concluded that the economies of agriculture made it efficient to overload the soil with fertilisers and pesticides. See: Commoner 1971: 81-93.

REFERENCES

Allen, John 1991. *Biosphere 2: The Human Experiment.* New York: Viking.

Bensin, Basil 1938. 'Agroecology as a Basic Science of Soil Conservation'. *Soil Conservation*, December: 138-141, 152.

Clements, Frederic and Ralph Chaney 1936. *Environment and Life in the Great Plains.* Washington, D.C.: Carnegie Institute.

Commoner, Barrry 1971. *The Closing Circle: Nature, Man, and Technology.* New York: Alfred A. Knopf.

Connor, Larry 1977. 'Agricultural Policy Implications of Changing Energy Prices and Supplies'. In William Lockeretz (ed.) *Agriculture and Energy*, pp. 669-681. New York: Academic Press.

Cox, Jeff 1973. 'Factory Farming is Not Efficient'. *Organic Gardening and Farming*, 20, No. 6: 90-94.

Devall, Bill and George Sessions 1985. *Deep Ecology: Living as if Nature Mattered.* Salt Lake City: G.M. Smith.

Ehrlich, Paul 1968. *The Population Bomb.* New York: Ballantine Books.

Hagen, Joel B 1992. *An Entangled Bank: The Origins of Ecosystem Ecology.* New Brunswick, N.J.: Rutgers University Press.

Hardin, Garrett 1968. 'The Tragedy of the Commons'. *Science* 162: 1243-1248.

Hardin, Garrett 1974. 'Lifeboat Ethics: The Case Against Helping the Poor'. *Psychology Today* 8, No. 4: 38-43, 123-126.

Huettner, David A. 1976. 'Net Energy Analysis: An Economic Assessment'. *Science* 192: 101-104.

Madison, Mark 1995. 'Green Fields: The Agrarian Conservation Movement in America, 1890-1990'. Ph.D. diss., Harvard University.

McMurray, Emily, ed. 1995. *Notable Twentieth-Century Scientists*, pp. 1501-1503. Detroit: Gale Research.

Odum, Eugene P. 1959. *Fundamentals of Ecology.* 2nd ed. Philadelphia: W.B. Saunders Company.

Odum, Eugene P. 1963. *Ecology.* New York: Holt, Rinehart and Winston.

Odum, Eugene P. 1964. 'The New Ecology'. *BioScience* 14, No. 7: 14-16.

Odum, Eugene P. 1968. 'Energy Flow in Ecosystems: A Historical Review'. *American Zoologist* 8: 11-18.

Odum, Eugene P. 1969. 'The Strategy of Ecosystem Development'. *Science* 164: 262-270.

Odum, Eugene P. 1970. 'The Attitude Revolution'. In *The Crisis of Survival*, pp. 9-15. Glenview, Ill.: Scott, Foresman and Company.

Odum, Eugene P. 1971. *Fundamentals of Ecology.* 3rd ed. Philadelphia: W.B. Saunders Company.

Mark Glen Madison

Odum, Eugene P. 1975. *Ecology, The Link Between the Natural and the Social Sciences.* 2nd ed. New York: Holt, Rinehart and Winston.

Odum, Eugene P. 1989. 'Input Management of Production Systems'. *Science* 243: 177-182.

Odum, Eugene P. 1993. *Ecology and Our Endangered Life-Support System.* 2nd ed. Sunderland, Mass.: Sinauer Associates, Inc.

Odum, Howard T. 1960. :'Ecological Potential and Analogue Circuits for the Ecosystem'. *American Scientist* 48: 1-8.

Odum, Howard T. 1967. 'Energetics of World Food Production'. *The World Food Problem.* Vol. III, pp. 55-94. Report of the Panel on World Food Supply. Washington, D.C.: The White House.

Odum, Howard T. 1971. *Environment, Power, and Society.* New York: Wiley-Interscience.

Odum, Howard T. 1983. *Systems Ecology: An Introduction.* New York: Wiley.

Odum, Howard T. and Elisabeth Odum 1976. *Energy Basis for Man and Nature.* New York: McGraw-Hill.

Odum, Howard T. and Eugene P. Odum 1955. 'Trophic Structure and Productivity of a Windward Coral Reef Community on Eniwetok Atoll'. *Ecological Monographs* 25: 291-320

Olds, Jerome 1969. 'The Worries of a Nebraska Farmer and a Missouri Biologist'. *Organic Gardening and Farming* 16, No. 12: 66-71.

Pasour, E.C. and J. B. Bullock 1977. 'Energy and Agriculture: Some Economic Issues'. In William Lockeretz (ed.) *Agriculture and Energy,* pp. 683-693. New York: Academic Press.

Rodale, Robert 1968. 'How Big is the Organic Idea'. *Organic Gardening and Farming,* 15, No. 10: 19-21.

Rodale, Robert 1982. 'A New Challenge for our Research Center'. *Organic Gardening,* 19, No.. 12: 22-26.

Rodale, Robert 1983. 'Breaking New Ground: The Search for a Sustainable Agriculture'. *The Futurist,* 1, No. 1: 15-20

Stanhill, G. 1984. *Energy and Agriculture.* New York: Springer-Verlag.

Stout, B.A., J.L. Butler, and E.E. Gavett 1984. 'Energy Use and Management in U.S. Agriculture'. In G. Stanhill (ed.) *Energy and Agriculture,* pp. 167-189. New York: Springer-Verlag.

Taylor, Peter 1988. 'Technocratic Optimism, H.T. Odum, and the Partial Transformation of Ecological Metaphor after World War II'. *Journal of the History of Biology* 2: 213-244.

Transeau, Edgar 1926. 'The Accumulation of Energy by Plants'. *The Ohio Journal of Science,* January: 1-10.

Pesticides and the British Environment:
An Agricultural Perspective

John Sheail

1. INTRODUCTION

There will always be debate over how the private/public policy interface functions but, as Ross and Amter have written, in their study of 'the making of our
chemically altered environment', 'one can at least hope for more empirical
investigation and less falling back on abstract reasoning'. Early struggles over
environmental control offer striking case studies of that relationship. Politics,
pollution and science came together in a way that foreshadowed the technological complexity of today's governance.[1] Such exploration of the pesticidal
contribution to 'our chemically altered environment' began early with Thomas
Dunlap's endeavour, in the late 1970s, to put the controversy arising from the
usage of DDT in historical context, not only for its material impact but for the
light shed on the complex relationships arising within government agencies,
private industries and the relevant sciences.[2]

There has been well-founded criticism of how environmental histories
have focused too closely upon the self-professed environmental groups and
their respective peripheral bureaucracies. As Thomas Zeller observed in a German context, there were other 'social actors who, in their activities and their
greater State support, made a considerably greater impact upon environmental
wellbeing'.[3] That wider perspective may be especially relevant in assessing
the manifold responses to the post-war 'chemical revolution' in agriculture
and, more particularly, why Rachel Carson's *Silent Spring* found so responsive
a readership in Britain. Conservationists acclaimed the book for its reinforcement of their worst fears. The Ministry of Agriculture's Chief Scientific Adviser
(Agriculture) saw nothing in the British experience 'to justify Miss Carson's
gloomy assertions'. Agriculture ministers, namely the Minister of Agriculture,
Fisheries and Food in respect of England and Wales and the Secretary of State
for Scotland, had already put in place such safeguards as were required in
pesticide usage.[4] How justified was such a remark in terms of safeguarding
humans and game- and wild- life? How pro-active, as opposed to begrudging,
were ministers in taking the necessary precautions? What were the respective
bureaucrats saying to one another, within the bureaucracies which determined
the scale and manner of usage?

Environment and History **19** (2013): 87–108

It was obvious, as the first accounts came to be written of the British experience, how narrow was their historical database. Such accounts were necessarily drawn from published material and the documentation of the game and nature-conservation bodies and, therefore, largely conveyed their perspective. As one author rightly anticipated, it would be many years before any substantial agricultural source was opened up to the historian.[5] The Ministry of Agriculture, Fisheries and Food continued to rebuff requests to see its relevant files, claiming that there was nothing more to learn than what had already been published.[6] Only more recently has the Public Records Act made it possible to reconstruct in any detail the role played by ministers, officials and their 'expert' advisers, in their endeavours to reconcile the respective sectoral interests of food production with what they perceived to be the wider public-interest. The present paper draws on the files of the Ministry's Agricultural Labour, Wages and Farm Safety Division which, from 1966, was principally concerned with pesticide-safety matters.

2. BEFORE *SILENT SPRING*

James Whorton's volume, *Before Silent Spring*, focused on pesticides and public health in pre-DDT America, emphasising how there was already an awareness of the side-effects of insecticides. It was Rachel Carson's achievement to raise that concern so dramatically.[7] Edmund Russell has used previously classified documentation to illustrate the assertive, as opposed to reactive, role of federal government. The Second World War provided both incentive to develop, and the means to assert, regulation of DDT in taking close heed of the unintended side-effects of its use. Such a role was dissipated with peace, control passing from military to civilian hands which, for all practical purposes, meant from government to manufacturers and users of the product. It was not until the 1960s that scientists and activists had gathered the knowledge, mobilised citizen groups, and perfected such use of the legal system as 'to return the chemical to virtually the same position as it had occupied during the war'.[8]

Eric Ashby has recounted how a comparatively few people had similarly voiced anxieties in a British context, such as the distinguished entomologist, V.B. Wigglesworth in *Atlantic Monthly* as early as 1945, when DDT was first released for civilian use.[9] Crucially however, there remained such a degree of government involvement that, by the time *Silent Spring* was published, there was a degree of pesticide regulation, however inadequate, 'to defend the countryside against the dangers'. The book's effect in Britain was to transmit such misgivings, through the mass media, to what Ashby characterised as 'the common man'. This in turn made 'politicians more sensitive to the need to deal with the issue; it stimulated research; it helped to give popular meanings to words like "pollution", "environment", "conservation"'.[10]

As a Deputy Secretary in the Ministry of Agriculture, Fisheries and Food had already briefed the Minister, in July 1958, there had been 'quite a few important and difficult problems' in the regulation of pesticides. As a flourishing export-industry, four of the 35 manufacturers of agricultural chemicals in Britain accounted for 90 per cent of total output. However essential the various weed-killers and insecticides, they posed, by their very nature, risks to those using them. Seven agricultural workers had died of DNOC (Dinitro-ortho-cresol) poisoning between 1946 and 1950. Over 200 were reported to have died of the side-effects of the highly toxic organophosphorus insecticides in the United States. A Ministry Working Group, chaired by the distinguished biologist, Sir Solly Zuckerman, provided the basis for an Agriculture (Poisonous Substances) Act of 1952, whereby Agriculture Ministers might make regulations requiring operatives applying dangerous chemicals to wear protective clothing. There was report of only one further death shortly afterward, with very few cases of serious illness.[11]

Two further Zuckerman Working Groups found it considerably harder to decide whether and how pesticides should be regulated, so as to minimise any residues in human foodstuffs and, secondly, to avoid impact upon game and wildlife. There was so little known about the presence, movement and significance of such residues. Zuckerman's Groups accordingly recommended, in reports of 1953 and 1955, that the Ministry should join with the Association of British Manufacturers of Agricultural Chemicals (ABMAC) in establishing a non-statutory Notification Scheme, analogous to the Agricultural Chemicals Approval Scheme, administered by the Ministry's Plant Pathology Laboratory, which adjudged the efficacy of crop-protection products. Pesticide manufacturers would similarly require approval before marketing any new chemical for use in agriculture or food storage or recommending a new use for an existing product. They would provide such data as were required to assess the chemical, its behaviour and any possible risks.[12]

The Minister, in adopting the Working Groups' recommendations, appointed an Advisory Committee on Poisonous Substances used in Agriculture and Food Storage to assist in both administering the 1952 Act and the Notification Scheme (which became the Pesticides Safety Precaution Scheme). Where the Advisory Committee was satisfied by the research data provided by the manufacturers that they had taken sufficient safeguard, bearing in mind the circumstances in which the pesticide would be applied, ministers might approve usage. In the event of fresh evidence of some hitherto unsuspected risk, ministers could require withdrawal, or modification, of the dossiers and, therefore, the product.[13]

The Advisory Committee comprised government officials, both administrative and scientific, and 'independent members', one of whom was chairman. Its Scientific Sub-Committee was made up of government scientists, with the Director of the Plant Pathology Laboratory as chairman. Whilst there was no

John Sheail

industrial representation on either body, the Ministry's sensitivity as to its views was highlighted by a two-year delay in appointing a chairman, following Zuckerman's brief tenure. Only after close consultation with ABMAC was Sir Charles Dodds (the Courtauld Professor of Biochemistry in London University) eventually appointed. The acting chairman (the head of the Ministry's Labour, Machinery and Seeds Division) wrote of how, whilst such consultation might have seemed unduly deferential, it was essential for manufacturers to have confidence in the body to which they voluntarily submitted such highly-confidential material. [14]

In appraising the challenge which confronted the Advisory Committee and its 'expert' Sub-Committee, the Ministry of Agriculture's files provide a more complete and intimate account of the scale and functioning of what Robert Rudd called 'the compromise zone in which political decision lies', as was devised for the regulation of large-scale pesticide use in post-war British farming. [15] Relevant insight may similarly be obtained for another contemporary framework of explanation, namely Eric Ashby's discernment of a complex chain-reaction between the disclosure of an environmental hazard and the political action required for its control, comprising an ignition stage, a second stage of objective assessment and a third, when such information was combined with 'the pressure of advocacy and with subjective judgment to produce a formula for political decision'.[16] The present paper points both to how the initial impacts of pesticide use emphasised the inherent diversity of agriculture, and to the particular instances of usage which caused such outcry outside the farming industry. It thereafter explores the attempts made by ministers to identify a publicly more-impartial 'umpire' within Government, in adjudicating the impact and, more particularly, attributing responsibility for minimising any deleterious effects and in taking further hesitant steps in devising some kind of comprehensive statutory regulation of both agricultural and non-farming uses.

3. CONFLICTING PRESSURES

Uppermost in the mind of the National Farmers' Union (NFU), the largest representative body in agriculture, was the damage caused where the pesticide sprays used upon arable land drifted onto neighbouring crops and, more particularly, horticultural land. Such animosity further exposed the divisions inherent within a Union whose membership encompassed so wide a range of activity and organisation. As the Union's president observed to Sir Richard Mantelow, the Ministry's Deputy-Secretary, at the Royal Agricultural Show of 1958, the situation was bound to become 'more and more troublesome with the rapid developments in spraying operations from the air'. Spraying had become so extensive as to make it even harder, if not impossible, to identify the specific cause of any particular damage. The NFU pressed the Ministry to appoint a committee to consider establishing a compensation fund. At a meeting convened by the Ministry with Union representatives, to discuss a levy from which compensation would be awarded, ABMAC representatives refused to be drawn into what they saw as a conflict between the two food-growing sectors. The NFU representatives claimed it was impossible for the Union to raise a fund, in as much as farmers in those districts where there was little or no horticulture would refuse to make any direct contribution to horticulturalists in other parts.[17]

The Minister compromised to the extent of appointing a committee of his Legal Adviser, an Under-Secretary, and a senior officer of the National Agricultural Advisory Service, to advise in the strictest confidence its chairman, Lord Waldegrave, the Parliamentary Secretary to the Minister. It both confirmed the destructiveness of such spray-drift upon horticultural crops and the unfairness of any general levy, given how such incidents were so localised to those parts of South and East of England where arable farming and horticulture were so intermixed. But most critically, the number of reported incidents had declined by the early 1960s, from some 113 in 1959 to 67 in 1961, such reduction being attributed to the impact of the various voluntary 'codes of practice' disseminated by the Ministry and Union. However hard it might be on the individual grower, the rarity of such incident and its insignificance in terms of the gross turnover of the horticultural industry, reaffirmed the Ministry's view that it was a matter for individual insurance, rather than some central compensatory fund.[18]

4. WILDLIFE AND SEED DRESSINGS

Sir John Winnifrith, the Permanent Secretary and the Ministry's most senior official, was, in his own words, 'ambushed' in May 1961, when he appeared before a standing parliamentary-committee charged with scrutinising the administrative efficiency of government departments. Winnifrith found himself completely unbriefed as to whether the Ministry should exert closer regulation of pesticide use. Worse still, as he minuted, the Ministry was shown to have been 'caught napping', in as much as potentially the most-dangerous compounds to game birds and wildlife had been introduced just before the voluntary Notification Scheme and had consequently never been submitted to its scrutiny.[19]

The Ministry had, in short, taken insufficient cognisance of the ramifications of two of the most significant post-war developments in insecticidal use, namely the introduction of dual-purpose seed-dressings containing a fungicide and, since 1956, the organo-chlorine insecticidal compounds aldrin, dieldrin or heptachlor. The second had seen the switch from merely dusting the seed to liquid dressings. Some 80 per cent of wheat, and 66 per cent of oats and barley, were so treated by 1958. There had, similarly since 1956, been increasing report of unusually large numbers of bird-deaths during the spring-sowing, their manner evoking 'humanitarian feelings'. Major John Morrison, chairman of the governing-party's powerful committee of backbenchers had first brought such deaths to the Minister's attention. As chairman of the British Field Sports Society, he relayed, in June 1957, complaints as to the toxic effects of Dieldrex 15. Whilst the manufacturer, Shell Chemicals, denied any association, Morrison pointed to a paper published in March 1957, describing the deaths of pheasants, as well as wood pigeons, from ingesting dieldrin-dressed grain in trials at the Royal Veterinary Laboratory.[20] The Minister, Derick Heathcoat-Amory, assured Morrison that 'the possible harmful effects of chemicals on birds and animals' were being kept under close review. Morrison regarded such a reply as 'thoroughly unsatisfactory', further writing of how 'you and I know how difficult it is to get direct evidence of the damage done'. In so far as all the reported deaths had arisen since Dieldrex 15 was first marketed, there was surely need to control its use.[21]

The scientist in charge of the Ministry's Pest Infestation Control Laboratory warned against taking hasty decisions. Whilst confirming reports of unusually heavy losses of game and wild birds, the Laboratory believed the dressings were so valuable to farmers that there could be no question of their withdrawal until equally efficacious substitutes were developed. The Scientific Sub-Committee of the Advisory Committee on Poisonous Substances used in Agriculture and Food Storage appointed a 'Bird and Seed Dressings Panel', whose eventual report of March 1959 found that, whilst there was fairly strong circumstantial evidence of the deaths caused by consumption of the dressed seed, there was no evidence of a major hazard. When the Nature Conservancy (charged by Royal Charter as a research council with the safeguard of wildlife) protested the need for a more systematic field-survey, in order to assess the 'true losses', others on the Advisory Committee argued that, if the Conservancy felt so strongly, it should itself undertake such survey from its own resources. As the director of the Ministry's Plant Pathology Laboratory (and chairman of the Scientific Sub-Committee), W.C. Moore, wrote 'I do feel rather strongly that we should not have to chase every alarm and excursion that is raised, but be allowed to get on with the jobs that really matter'.[22]

Protests at the even higher bird-casualties during the dry spring-sowing of 1960, when so much dressed-seed was left exposed, caused the Minister of Agriculture, Fisheries and Food, John Hare, to agree with officials that everything must be done to discover the truth in preparation for discussions with manufacturers before the next growing-season. Officials again warned of the impossibility of over-riding the farmers' natural reluctance to revert to less-effective treatments.[23] 'A holding programme' was called for, which both protected the Minister from charges of inaction from naturalists and claims of such over-reaction as might cause manufacturers, merchants and farmers to withdraw from voluntary regulation and, thereby, force the Minister 'back to legislation'. The Minister appointed a Research Study Group to review the issue more generally, with the Ministry's Chief Scientific Adviser (Agriculture), Professor H.G. Sanders, as chairman. Whilst the head of the Medical Research Council's Toxicology Research Branch, J.M. Barnes, protested at such surrender to ill-founded clamour, in as much as present safeguards were second to none in the world, Basil Engholm (an Under Secretary in the Ministry) emphasised how, far from showing any lack of confidence in what was a first-class example of co-operation between the Government and industry, the Study Group's authoritative reassurance would uphold the voluntary character of present safeguards.[24] Some ABMAC members had meanwhile become 'extremely perturbed' by the convincing evidence of such carelessness and deliberate poisoning of birdlife as would bring all uses of pesticides into disrepute. Lord Waldegrave readily conceded the need for even greater publicity to ensure the grain was properly sown, but insisted that there should be only the most guarded reference to deliberate poisoning, lest it encouraged further abuse.[25]

The 'holding programme' appeared to succeed in as much as officials were gratified by the success of 'a lively meeting', of December 1960, attended by all the interested organisations to discuss precautions for the coming spring-season, on the now accepted-premise of there being sufficient circumstantial evidence to show the dressings could cause appreciable casualties. The chairman, an Assistant Secretary in the Ministry, described the meeting as remarkable for the reasonableness and co-operation shown, with only the NFU foot-dragging for more evidence of bird casualties. Most importantly, the acknowledgement by the Nature Conservancy and voluntary wildlife-bodies that some use of the seed-dressings must continue until effective substitutes were found, had averted any serious campaign for an outright ban. Yet casualties remained as high as ever. Whilst Ministry officials emphasised, in April 1961, how long it would take for 'a big education campaign' to take effect, regional staff described farmers (other than those with game-preservation interests), as either apathetic, in their denial of any causal link, or deliberate in their concealment of the evidence. As an Under-Secretary, Henry Button, wrote, in March 1961, neither manufacturers nor farmers were 'at heart interested in wild life'.[26]

With ministers coming under increasing parliamentary pressure to intervene, officials conceded there must be a ban upon aldrin, dieldrin and heptachlor seed-dressings on cereals, except in the case of autumn-sown seed where the danger of wheat bulb-fly was significant. ABMAC insisted that any such prohibition was effected through 'the voluntary ban' and, therefore, shown to be a consequence of independent scientific assessment rather than public clamour. The challenge for officials was to prevent both Shell from breaking rank with the voluntary Notification Scheme, and 'the Nature people' from pressing for a complete ban. However shrill the company's protest, Winnifrith minuted how parliamentary and press criticism had made 'Shell's flesh creep'. There was also self-interest in as much as the company, in acknowledging that farmers must again rely on BHC (benzene hexachloride) -dressings against wireworm in spring, saw opportunity to market new forms of BHC-dressing, claiming them to be less phytotoxic to the actual grain-seed.[27]

The marked reduction in bird deaths soon caused public interest to wane, only reviving, as an Under Secretary, Claude M. Wilcox, later recalled, with the publication of the book *Silent Spring* and, ironically, the Ministry's own small booklet *Chemicals for the Gardener* in May 1963. The Ministry booklet was essentially a list of products covered by the Agricultural Chemicals Approval Scheme suitable for home-gardeners, identifying the most appropriate product for a particular use. The inclusion however of a comparatively few products containing aldrin and dieldrin provided, as the Minister himself pointed out, further leverage, particularly on the part of the Royal Society for the Protection of Birds and British Trust for Ornithology, for a campaign for the withdrawal of all chlorinated-hydrocarbon products. Officials warned of the most serious repercussions if such precedent were extended to agriculture generally. Ministry officials had become especially conscious of how the earlier seed-dressing 'troubles' had caused the Nature Conservancy, Game Conservancy and voluntary bodies to undertake (or more usually commission) analyses of bird and other animal-tissue, making full use of the highly-sensitive techniques becoming available. Attention was increasingly focused upon the possible long-term, sub-lethal effects arising from the build up of such residues in the food chain.[28] The research of the Nature Conservancy's Toxic Chemicals and Wild Life Section, at its new Monks Wood Experimental Station, focused almost entirely upon the presence and significance of what proved to be the highly-persistent chlorinated-hydrocarbons found in body tissue and the environment generally.[29]

5. PERSISTENCE IN PESTICIDES

There was ample time to prepare for the publication of the British edition of *Silent Spring* in February 1963. Its three-part serialisation in the *New Yorker* six months previously had stirred the political conscience, both in Westminster and Washington.[30] The Advisory Committee on Pesticide Safety (as the Advisory Committee on Poisonous Substances used in Agriculture and Food Storage had become more concisely known) had, at the Minister's request, assessed the book's claims at its 35th meeting in late November 1962. Whilst the chairman was in no doubt of the book's considerable political implications, the problems warned of by Rachel Carson had long been recognised in Britain. Lord Shackleton, who wrote the Introduction to the British edition, was impressed, following a meeting with Lord Waldegrave and officials, at the seriousness with which the Ministry treated the pesticides-issue, his only criticism being its lack of 'effective publicity'.[31]

Sanders had, with the Minister's consent, participated in a BBC television programme which, in his own words, had proved a 'shambles', the BBC being so determined 'to play up the suspicion of harm'. He was given no opportunity to speak of the particular agricultural need for pesticides. ABMAC, at a 'preparatory' meeting with Ministry officials, had shown them the dossier of material to be given to relevant agricultural and scientific journalists on the book's publication. Particularly prominent was a review of *Silent Spring* by Dr E.F. Edson of Fisons' Chesterford Park Research Station, critical of how, by its tone, the book had damned an industry and very large numbers of conscientious scientists and administrators. It had focused on every problem, real or hypothetical, without any recognition of the usefulness and life-saving properties of pesticides. By the Pesticides Safety Precautions Scheme, industry had given 'the Government powers of veto, of modification of proposals, powers of review, and powers of delay' in the marketing of its products.[32]

Within the Ministry, Wilcox minuted, in early January 1963, that the aspect which worried him most was not the risk of immediate death, but 'the possibility of insidious and chronic effects from constant exposure over a long period of time to possibly small concentrations of pesticides'. It was the kind of risk which those working on X-rays in the early days had unwittingly exposed themselves to, with serious effects on their health. Whilst Wilcox was reassured that nothing had been found to cause anxiety as to immediate human-health hazards in Britain, a leader in the newspaper *The Scotsman* of 21 January 1963 acknowledged 'the undoubted good of insecticides in controlling malaria but, remembering thalidomide, people were still apprehensive'. Like a drug, toxic chemicals might be admirable for some specific purpose, but nevertheless inflict quite unsuspected side-effects. Alongside the alarm first raised by naturalists, there was an understandable anxiety about humans, the plain truth being that 'scientists do not know enough about the vastly complex subject'. The sheer variety of the chemical substances used in industrial society, and the difficulty of tracing their effects, made it impossible, so the leader-writer continued, to pronounce with confidence on the dimensions of the threat to human wellbeing.[33]

The eventual report of the Sanders Research Study Group of autumn 1961, and indeed a later Research Committee, appointed by the Agricultural Research Council at the instigation of the Minister for Science, found no relevant research-field neglected. The recommendation of the earlier Group for closer study of pesticide-residue levels in human foodstuffs caused the Advisory Committee to appoint a Residues Panel under the Ministry's Chief Scientific Adviser (Food). The later Committee called especially for the study of the processes and implications of the reported sub-lethal effects of such residues on wildlife.[34] An Under Secretary, P. Humphreys-Davies wrote in June 1963 of how the Ministry was genuinely anxious to meet the wildlife-case, where it had *prima facie* validity. The evidence for any sub-lethal effect, say in the reproductive behaviour of raptor species, was 'entirely circumstantial'. Any complete ban on what Wilcox called 'the black three', of aldrin, dieldrin and heptachlor, would cost farming an estimated £3 million annually. Sanders guessed that dieldrin was used to control carrot-fly on 75 per cent of the crop.[35]

As the Ministry recognised, pressures for an outright ban would be strengthened by President Kennedy's recent acceptance of a report by his Science Advisory Committee, recommending an orderly reduction in the use of chlorinated hydrocarbons, with a view to their eventual elimination. Whilst the preliminary findings of the Advisory Committee's Residues Panel found no cause for disquiet at the levels ingested by adults, there should be the most careful consideration of those obtained by children, say, from cow's milk and butter. As the chairman of the Advisory Committee's Scientific Sub-committee, W.C. Moore (the Director of the Ministry's Plant Pathology Laboratory), acknowledged, such possibility of a build up of pesticide residues through the food chain was potentially the most important issue to have arisen since the inception of the Pesticides Safety Precaution Scheme. If true, it undermined the whole idea that a chemical was 'safe', if used in the recommended manner. Whilst it might take seven or eight years to prove such a causal-effect scientifically, there was suspicion on both sides of the Atlantic that the environmental persistence of chlorinated hydrocarbons, which made them so valuable as a pesticide, also posed a substantial, albeit future, hazard ultimately to human wellbeing.[36]

What had begun as an attempt to assuage wildlife interest had become an endeavour to protect humans from the self-same chemicals. The Minister of Agriculture, Christopher Soames, invited the Advisory Committee, in June 1963, to undertake a special review. Whilst the public pretext was the 'rumpus' over the Ministry's *Gardener's* booklet, it was the accumulating evidence of a potential risk to human health, as well as to wildlife, balanced against the considerable part played by such chemicals in modern farming, which called for wider study. Whereas previously the Committee had considered only individual products for clearance under the Notification Scheme, Soames called for an inquiry of unprecedented breadth, namely that

> in the light of existing information and that now coming to hand, generally to review the risks arising from the use of chlorinated hydrocarbon pesticides (more particularly aldrin, dieldrin and heptachlor) in agriculture (including gardening) and food storage and to make recommendations.[37]

The Chairman, Sir James Cook (a distinguished organic chemist and Vice-Chancellor of Exeter university), formally accepted the Minister's invitation at the 39th meeting of the Committee in July 1963. The Nature Conservancy's representative, Robert Boote, tried, in Wilcox's words, to bounce the Committee into recommending an immediate and drastic reduction in usage, pointing to how the Conservancy had previously been proved right to press for such a precautionary approach as to the harm caused by cereal seed-dressings. As Wilcox further recounted, Cook had not fallen for such a ploy. Any immediate prohibition would have risked the breakdown of the voluntary Notification Scheme.[38]

The Advisory Committee adopted what it acknowledged to have, in December 1963, a family resemblance to the American solution, but without the eventual elimination being made explicit. Its Scientific Sub-Committee had affirmed that, whilst such 'pretty-wide contamination' presented no immediate hazard of any significance to humans or wildlife, except certain species of predatory bird, it was concerned 'at the accumulative contamination of the environment by the more persistent organo-chlorine pesticides'. Their further build-up should be arrested and optimally reduced through selective restriction of use. The most pressing were a ban on dieldrin from sheep dips, and aldrin from insecticidal fertilisers, the former under the aegis of a Veterinary Products Safety Precautions Scheme, more recently established to scrutinise such products supplied for treating agricultural livestock and poultry. The Assistant Secretary responsible for the Ministry's Animal Health Division, Jock G. Carnachan, protested at the significant departure this would represent from the Advisory Committee's previous endeavours to balance the various sectoral interests. Such withdrawal would mean that sheep would again have to be dipped at least twice annually. Beyond the cost and practical difficulty, especially in upland country, there was bound to be an increased suffering from fly-strike and, therefore, conflict with the Ministry's animal welfare responsibilities. Carnochan pointed to how there was no scientific proof of the sheep dips being the source of such residues found in human body fat, further arguing that the Scientific Sub-Committee had been far too influenced by the alleged threat to carrion-eating birds, which were themselves enemies of the sheep farmer. As was emphasised at a further meeting of the Advisory Committee in January 1964, such recommendation for a ban on dieldrin dips was justified not by any direct evidence of an immediate threat, but by discovery of the possibility of a progressive, and perhaps dangerous, contamination of the environment.[39]

A point of political decision had been reached, requiring reference to the Cabinet's Home Affairs Committee. The Advisory Committee had assumed, in Sir James Cook's words, that Ministers would take 'a very conservative view' in providing safeguard before any insidious effects, as might arise, had begun to reveal themselves. It had recommended that the Government should take out 'an insurance policy, at a substantial premium, against the possibility of future damage to human life and wild life'. The Advisory Committee accordingly recommended 'fairly extensive voluntary restriction of the use of aldrin and dieldrin', and that such remaining uses, and that of DDT and certain other lesser-used organo-chlorine pesticides should be further reviewed after three years. Ministry officials feared such notice of immediate and further restriction would be 'the last straw that broke the back of the Notification Scheme'. The Board of Royal Dutch Shell had, in confidential pre-publication discussions with Ministry officials, violently objected to the proposed restrictions as unjustified by the scientific evidence adduced. Soames wanted to know, prior to negotiation, how far he might count on the backing of ministers in threatening legislation, if 'they won't come quietly under the voluntary scheme'.[40]

6. 'AN IMPARTIAL UMPIRE'

Although loath to appear to be messing about with the Advisory Committee, just as it was addressing its most exacting challenge, that very remit had made the scope and manner of surveillance, and, therefore, the composition of the Committee all the more relevant. There had been an obvious advantage in officials forming the Committee's 'backbone' in making almost routine, yet commercially urgent, pronouncements as to the safety of individual products. The Committee had latterly comprised five independent members and fifteen officials, including the representative of the Nature Conservancy and such heads of government research establishments as the Institute of Animal Physiology at Cambridge. Except for the chairman, the independent members had played little part in practice in the very tight timetable, whereby the chairman and members were told in effect that 'this is what we [the Ministry] propose to do, we hope you agree; if you have any reservations, however, please let us know immediately'.[41] Any general principles had emerged more as a by-product than from any basic study of the issue.

More fundamental enquiry was now sought. The more general advice required by the Minister, for whole suites of pesticide compounds, went far beyond the Committee's original purpose. Officials on the Committee were placed in an especially difficult position. As Robert Boote frankly confessed at that month's meeting, he found himself in the position of prosecutor and jury in as much as the Nature Conservancy, in fulfilment of its Royal Charter, had already gone on public record in calling for the eventual elimination of such pesticides. The Advisory Committee concluded that whilst the Minister of Agriculture could not abrogate responsibility, given his statutory responsibility under the Agriculture (Poisonous Substances) Act of 1952, and his executive interest in agricultural chemicals generally, there were obvious presentational advantages in the Minister for Science taking responsibility for the Scheme, acting in consultation with the Ministers of Agriculture and Health. Soames obtained the approval of the Cabinet's Home Affairs Committee in February 1964, both to his immediate object of wider and closer restriction of the use of organo-chlorine pesticides, and to the reconstitution of the Advisory Committee under a more obviously 'impartial umpire', who shortly became the Secretary of State for Education and Science, its comprising only 'independent' members. The outcome, as one Ministry of Agriculture official put it, in February 1969, was that the Secretary of State had nominal responsibility, but the executive burden continued to fall on the Ministry – 'we do all the work'.[42]

The new Advisory Committee on Pesticides and other Toxic Chemicals was an unmistakably different body in its requirement 'to consider and advise on any improvements and extension of present safety arrangements' and, 'in particular to consider what stricter criteria should be applied in the approval of new products and review of the approval of existing products'. Where previously there had been full discussion between the Scientific Sub-Committee and the manufacturer's scientists, prior to any final decision being taken upon the individual product under scrutiny, Wilcox wrote, in October 1964, of how the new Committee must necessarily act in a more detached manner. Such implication was highlighted by Shell's response to what it regarded as the Advisory Committee's arbitrary and scientifically unsound judgement of its products. Whilst officials acknowledged that the Board and directors of Shell must be very worried as holders of world patents for aldrin and dieldrin, Wilcox advised Soames to take a firm line. As a study of further evidence submitted by Shell emphasised, the Advisory Committee did not dispute in any way the statements of fact made as to the mammalian toxicity of dieldrin, but rather the judgement which Shell attached to their significance. The Committee's recommendations were based not on any proven hazard (at least to humans) arising from the use of the pesticides in isolation, but rather from concern as to the hazards which could arise, if growing contamination of the environment were allowed to continue unchecked.[43]

Soames brought what Wilcox called 'the difficult negotiations and discussion' to an end by announcing the Committee's findings to parliament, and at a press conference, on 24 March 1964. He particularly commended ICI on its earlier decision to cease production of heptachlor. It had shown how industry also subscribed to the maxim that it was 'Better to be sure today than, possibly, sorry tomorrow'. As Soames intimated, Shell had also agreed, with commendable public spirit, to abide by the Committee's recommended withdrawal of aldrin and dieldrin. Much to the Ministry's irritation, Shell published the next day large advertisements in the same newspapers as carried report of the Minister's press conference, challenging what was now essentially a Government decision. The further restrictions took effect in January 1965, with those on dieldrin sheep-dip postponed for a further year.[44]

7. A STATUTORY CODE

Wilcox wrote, in a brief for ministers in the new Labour Government of October 1964, that chemical pesticides had become an essential tool for improving both crop yield and the quality of produce. The Government had however, ever since the bird-deaths of 1960–61 and publication of *Silent Spring*, been under constant pressure to impose ever stricter controls over their use.[45]

An ABMAC press release of January 1964 reiterated the view, first voiced in confidence to Ministry officials the previous February, that there should be a statutory registration scheme, so as to ensure that all pesticides sold in the UK were cleared by an independent 'expert' committee. It was the only way of ensuring that every use, and every home and overseas manufacturer, trader and post-sales handler, was covered.[46] The Advisory Committee's 'Review of Present Safety Arrangements for the Use of Toxic Chemicals in Agriculture and Food Storage', which had taken two years to complete, came to a similar conclusion. The voluntary scheme had worked well despite the many theoretical loopholes, enabling Britain to have the best safety arrangement in the world. There had been 'no very serious incidents'. There was however a considerable argument for a statutory licensing scheme to control the supply and labelling of pesticide products. It would enable the 'backlog chemicals' to be tackled, namely those in use prior to the Notification Scheme, where neither the manufacturer nor user had any incentive to have them scrutinised.[47]

Following publication by the Department of Education and Science of the Advisory Committee's report in January 1967, and an ensuing six-month consultation period, the Ministry of Agriculture drafted proposals for legislation to control the supply and labelling of pesticide products used in agriculture and food storage by a mandatory licensing system. It would re-enact the 1952 Act, so as to require anyone applying such chemicals to wear protective clothing. There would be powers to make misuse an offence, regulate the storage of such products, and to require records of usage to be kept. Given the wide support for such a measure, the greatest obstacle was likely to be that of obtaining any priority in the legislative programme.[48]

The Advisory Committee had, in its Further Review of the residual uses of organo-chlorine pesticides (announced in November 1966), taken the unprecedented step of carrying out a survey of pesticide use in manufacturing industry, public health and throughout the home. It recommended that any Pesticides Bill should cover both agricultural and non-agricultural usage. An inter-departmental meeting, convened by the Department of Education and Science in March 1968, supported both the reconstitution of the Advisory Committee explicitly to cover the non-agricultural field, and that such usage should be brought as fully within the scope of a statutory scheme as appeared possible and desirable.[49]

The chairman of the Advisory Committee's Scientific Sub-committee soon complained of the consequent delay, particularly given how, according to confidential figures obtained from manufacturers, non-agricultural uses accounted for less than 10 per cent of total sales of such products as contained organo-chlorine pesticides. Although not opposed to control in principle, Board of Trade officials wrote of how there must be very good reason for adding to industry's burden. The Department of Employment and Productivity warned of potential overlap with the responsibilities of the Factories Inspectorate. Home Office officials believed the simplest and most economic arrangement would be for Agriculture Ministers to have executive responsibility over the entire range of pesticide usage.[50]

No other government department had come forward with sufficiently convincing reasons to justify a public consultation as to whether the Ministry's proposed Agricultural Pesticides Bill should be expanded. Rather than further delaying what was essentially an agreed agricultural pesticides Bill, for so contentious an enlargement, Agriculture Ministers were recommended to proceed simply upon the basis of the agricultural proposals. The ministerial submission for the Bill accordingly emphasised how it could 'never be certain that a voluntary scheme however well run or loyally supported will be comprehensive'. Both from the point of view of public safety and confidence, the Government should have adequate powers to act rapidly and effectively in unforeseen circumstances. Although not generally controversial, users were likely to press strongly for some kind of appeals procedure. The most that could be offered was the requirement that the licensing authority would consult an applicant before issue, amendment, withdrawal or refusal of a licence.[51]

Whilst an Agricultural Pesticides Bill was published in July 1968, and consultation completed by the following March, there remained so little prospect of room being found in the Government's legislative programme for what had become the 73-clause bill, that instructions were neither sought by, nor given to, parliamentary counsel actually to draft the bill. An obvious course, to save parliamentary time, would have been to promote a considerably shorter, enabling bill, which left the detail as to the scope and manner of regulation to administrative decision later. There would have been 'all-round rejoicing' but, as a Ministry official remarked in February 1969, the bill would have been so widely-drawn as to make it considerably more contentious among 'trade interests'. Such postponement of the regulatory detail would, however, have considerably facilitated extension to include non-agricultural usage. That dimension acquired considerably greater relevance following correspondence between the chairman of the Advisory Committee, Professor Andrew Wilson, and the Secretary of State for Education, Shirley Williams, in May 1969. Following further study, the Committee had 'concluded that it was right to take precautionary measures now rather than to wait for evidence of danger'. A general power should be incorporated whereby the Bill could be extended, in whole or part, and particularly as to labelling, to any pesticide product used outside agriculture and food storage. Mrs Williams acknowledged the wisdom of such a move.[52]

8. ENVIRONMENTAL POLLUTION

It was the increasing priority given by Government to combating environmental pollution which made the difference. As a Ministry of Agriculture official minuted, it provided 'our best chance of securing the new legislation in anything like a reasonable period'. The Secretary of State for Local Government and Regional Planning, Anthony Crosland, had identified such priority in December 1969. The Prime Minister, Harold Wilson, had, in a major speech marking the beginning of European Conservation Year in 1970, referred to how 'new legislation on pesticides will be introduced by the Minister of Agriculture'. As the Minister, Cledwyn Hughes, emphasised to Crosland, in correspondence, 'the Government stands to gain considerable credit for the earliest possible action'.[53]

A ministerial brief emphasised how, by this 'politically popular Bill', Agriculture ministers would be the first to respond to such heightened public interest in pollution issues. There was however parliamentary time for only a 'short' Bill – an essentially enabling Bill of some 20 clauses. Whilst there would be 'some parliamentary disquiet' at 'still more government by regulation', there were 'clearly potent political advantages for the Minister and the Government in striking while the iron is hot and the Sunday newspapers are full of pollution horror stories'. The Treasury had agreed that manufacturers should be charged a licensing fee sufficient to cover the additional costs of a mandatory system, estimated to be some £35,000 above those of the voluntary scheme of £80,000, of which 75 per cent represented the annual costs of such scrutiny as undertaken by the Ministry's Plant Pathology Laboratory and the Pest Infestation Control Laboratory.[54]

The Department of Education and Science welcomed the draft Bill for its inclusion of non-agricultural uses. Although they posed no immediate and obvious hazard, Ministers would find it difficult, in 'the present state of public concern over environmental pollution', to justify legislation 'which did not enable them to act in whatever way was necessary'. By way of compromise with the Home Office, the Ministry of Agriculture had, for example, conceded its being responsible for the licensing of pesticides about the home. Such concession caused Morris Cohen (the Director of the Plant Pathology Laboratory), to warn of the potentially high political price paid for 'the advantage to the Minister of being able to demonstrate that he is getting off the mark quickly in the general sphere of environmental pollution'. It might be 'the very thin end of a wedge' whereby the Ministry had foisted upon it the executive responsibility for regulating pesticides, wherever they were used. The Ministry of Technology (as successor to the Board of Trade) proved the most obdurate in inter-departmental negotiation, threatening to oppose the 'short' Bill at the Cabinet's Home Affairs Committee. Although acknowledging the political advantage of appearing to act quickly and decisively, there was no public pressure, or indeed public knowledge yet of the Government's interest in regulating non-agricultural usage. Ministry of Agriculture officials pointed to how such criticism missed the point. It was the over-riding purpose of such legislation to anticipate public opinion to the extent of enabling ministers to act appropriately in responding to what were presently entirely unforeseen hazards.[55]

The Ministry of Technology was on firmer ground in pressing the 'longer', and considerably more substantive bill, once there had been consultation with the non-agricultural user-interests and ministerial agreement had been obtained as to where executive responsibility lay within Government. There should be consultation of the kind that had caused the Ministry of Agriculture's relationship with agricultural pesticide manufacturers to be the best in the world. Most crucially to the argument, Basil Engholm (now the Permanent Secretary) was 'much disturbed', agreeing with the Ministry of Technology, in a minute of March 1970, that it would be silly to forfeit the co-operation and goodwill of industry for short-term political advantage. Engholm also recommended sticking with the longer Bill, the Minister taking the immediate step of obtaining the approval of the Cabinet's Home Affairs Committee to its being drafted and introduced before the parliamentary summer recess, as an integral part of the consultative process. There would be the political advantage of having at least a bill on the stocks. Cledwyn Hughes agreed. Further drafts of a paper to the Home Affairs Committee emphasised how Agriculture Ministers proposed taking the lead 'in this year of concentration on hazards to the environment' by tightening up control of agricultural pesticides. Any coverage of non-agricultural pesticides would however extend far beyond their departmental responsibilities.[56]

It was now the turn of the Department of Education and Science to protest, warning of how the Government would be severely criticised for a bill which so conspicuously failed to address the wider range of pesticide hazards so recently identified by its own Advisory Committee. There was concern too, from within the Ministry, that Agriculture Ministers, in taking charge of the Bill, were bound to be criticised for its omissions, however much it was the responsibility of others. It would, in any case, 'be a topsy-turvy world' if parliament's attention was focused entirely on agricultural usage, simply because the Ministry had 'all along been more conscientious than the other Departments, most of whom are at last coming into line'. It was not until 21 April 1970 that Engholm secured sufficient inter-departmental agreement for a paper to be put to the Home Affairs Committee.[57]

Cledwyn Hughes introduced the Paper to the Home Affairs Committee, signed by himself and the Scottish Secretary, William Ross, on 8 May 1970, pointing to how pesticides used in agriculture, home gardens and food storage had been controlled since 1957 through the voluntary Pesticides Safety Precautions Scheme. There was no similar scheme covering, say, general domestic purposes, commercial mothproofing, timber and cable treatments and rubbish tips, which might constitute as great a pollution hazard. In as much as there would be considerable public concern if statutory controls were introduced over 'agricultural and related sectors while doing nothing to limit the hazards that exist or may arise from the employment of pesticides in other fields', the Paper called for urgent consultation with 'manufacturers and user industries with the objective of including in it powers to enable appropriate Ministers to regulate these other uses of pesticides', even if that meant their insertion during the later stages of the parliamentary passage of the Bill. From the chair, the Lord Chancellor, Lord Gardiner, summing up the brief discussion, spoke of how ministers on the Home Affairs Committee generally approved the proposals, subject to further consideration as to whether there should be provision for any appeal against the refusal of a licence by a Pesticides Licencing Authority.[58] Whilst work on the Bill continued, following the dissolution of parliament, on the premise that the new Government would require the Bill, there was further uncertainty following the election of a Conservative Government.

9. THE POLITICS OF REGULATION

Britain would seem to have adopted a more incremental approach to pesticide regulation than that, say, of the United States.[59] Through the documentation of the Ministry of Agriculture's Labour, Wages and Farm Safety Division, the present paper has traced how statutory acknowledgement of a responsibility for the safety of farm-workers was extended, over some two decades, to encompass the protection of the wider human and environmental wellbeing, as affected by pesticide usage.

As Robert Rudd wrote in his book *Pesticides and the Living Landscape*, published a year after *Silent Spring*, there was very little chance of pleasing everyone on such a question of values, particularly where viewpoints might be so extreme.[60] Far from there being more accommodation to be found within the ever more expansive 'compromise zone', the various environmental groups had become so buoyed up by the success of European Conservation Year, the appointment of a Department of the Environment, and prospect of a United Nations' conference on the environment in Stockholm, as to wonder whether they had been far too modest in their political demands. As Robert Boote minuted, in June 1972, they might have 'gone along' too much with industry in 'the pioneering work of co-operation between government, industry and voluntary bodies'. Such assertiveness was encouraged by the first report of the standing Royal Commission on Environmental Pollution, appointed the previous year, which identified pesticides as one of the pollutants most seriously affecting the environment. The report supported the Advisory Committee in arguing that 'mandatory control is desirable and will in the end be inevitable'.[61]

Such reinforcement of the Advisory Committee's stance might be perceived as signifying arrival at the third stage of Eric Ashby's 'chain reaction'. *Silent Spring* had secured the ignition stage, where the concerns previously voiced by the comparatively few had aroused such public concern as to stimulate political unease. Rachel Carson had condemned not the principled use of pesticides, but something much more readily grasped – their trigger-happy, indiscriminate use.[62] The Advisory Committee's reports, and now the Royal Commission's investigation, had marked completion of Ashby's second stage of more objective assessment of the incidence and consequences of such alleged abuse. They had made possible the third stage whereby Government could bring together such information with the pressures of lobbyists to produce a formula for political decision.

It was the Royal Commission's report that provided the pretext for an inspired Parliamentary Question, in April 1972, asking the Minister of Agriculture whether he proposed to proceed with statutory control. The Minister, James Prior, responded in a Written Answer by describing how the Commission's report had caused the Advisory Committee to make fresh appraisal, and the chairman to advise him in September 1971 of how, whilst there were still 'some blemishes', which could be removed 'more speedily if statutory powers were available', those remaining weaknesses in the 'effective and well-tried' voluntary system no longer justified its replacement. The Royal Commission had accordingly intimated that it did not wish to press for the immediate introduction of legislation. It maintained however, as a matter of principle, 'that it should not be ruled out as the ultimate sanction to control those substances which, when misused can harm – and in its view have harmed – the environment'. The Royal Commission welcomed the decision to refer the question of non-agricultural pesticides to the Secretary of State for the Environment, in furtherance of that new department's responsibility for the control of environmental pollution.[63] The Agriculture Ministers had accordingly decided that they did 'not propose to introduce legislation at this stage'. Rather, they had asked the Advisory Committee 'to have special regard to two contingencies foreseen by the Government and the Royal Commission', namely where the voluntary scheme had lost 'any of its present effectiveness or if there were new technological developments for which voluntary control' was inappropriate.[64]

In further exploring what might be characterised as the staged-dynamics of Rudd's 'compromise zone', the Ministry's files illustrate how both those defined by Thomas Zeller as the self-professed environmental groups and those whom he called the 'other social actors', strove for an identity which gave them ownership of such policies as were pursued. The impression obtained from the documentation, is one of insularity, in the sense that each was focused upon, and aware only of, its particular need for survey, analysis and decision. In as much as Robert Boote spoke in later years of how Ministry officers had kept him personally fully briefed about Ministry thinking and activity, and of how he had endeavoured to reciprocate on behalf of the Nature Conservancy, the archives may well be deceptive in providing only the more formal record.[65] However much that may be the case, they are above all revealing for the way in which the different bureaucracies succeeded in projecting to themselves a sense of ownership of the issues raised, and the decisions made. Each believed it had played the significant part in what had been achieved. It was not until the Food and Environmental Protection Act of 1985 that the statutory regulation of pesticides was attained.

NOTES

1. B. Ross and S. Amter, *The Polluters. The Making of Our Chemically Altered Environment* (New York: Oxford University Press, 2010) pp. 2–3.
2. T. Dunlap, *DDT, Scientists, Citizens and Public Policy* (Princeton: Princeton University Press, 1981) pp. 4–6.
3. T. Zeller, *Driving Germany* (New York, Berghahn Books, 2007) p. 3.
4. The National Archives (TNA), MAF 284/18.
5. J. Sheail, *Pesticides and Nature Conservation: the British Experience 1950–1975* (Oxford: Clarendon Press, 1985) p. v.
6. Personal Communication, 20 March 1990.
7. J. Whorton, *Before* Silent Spring (Princeton: Princeton University Press, 1974) pp. viii–xii.
8. E.P. Russell, 'The strange career of DDT. Experts, federal capacity, and environmentalism in World War II', *Technology and Culture* **40** (1999): 770–796.
9. V.B. Wigglesworth, 'DDT and the balance of nature', *Atlantic Monthly* **176** (1945): 107–13; T.F. West and G.A. Campbell, *DDT and New Persistent Insecticides* (London: Chapman and Hall, 1950)
10. E. Ashby, 'Perspective', *The Times Higher Educational Supplement* (20 February 1987): 9
11. TNA, MAF 284/377
12. J. Sheail, 'The regulation of pesticide use: an historical perspective', in L. Roberts and A. Weale (eds.) *Innovation and Environmental Risk* (London: Belhaven, 1991) pp. 38–46
13. TNA, MAF 284/370
14. TNA, MAF 281/1.
15. R.L. Rudd, *Pesticides and the Living Landscape* (Madison: University of Washington Press, 1964) p. x.
16. E. Ashby, *Reconciling Man with the Environment* (London: Oxford University Press, 1978) pp. 14–15.
17. TNA, MAF 284/167.

18. TNA, MAF 284/268-9.
19. TNA, MAF 284/62 and 246; Select Committee on Estimates, *The Ministry of Agriculture, Fisheries and Food. 6th Report. Session 1960–61* (London: HMSO, 1964).
20. R.B.A. Carnaghan and J.D. Blaxland, 'The toxic effects of certain seed-dressings on wild and game birds', *Veterinary Record* **69** (1957): 324–325.
21. TNA, MAF 284/62.
22. TNA, MAF 284/170 and 254.
23. TNA, MAF 284/62 and 201.
24. TNA, MAF 284/203.
25. TNA, MAF 284/63-5.
26. TNA, MAF 284/63-5 and 245.
27. TNA, MAF 284/246-8 and 350.
28. TNA, MAF 284/350 and 377.
29. J. Sheail, *Nature Conservation in Britain: the Formative Years* (London: The Stationery Office, 1998), pp. 101–124.
30. R. Carson, *Silent Spring* (London: Hamish Hamilton, 1963); G. Kroll, 'The "Silent Springs" of Rachel Carson: mass media and the origins of mass environmentalism', *Public Understanding of Science* **10** (2001): 403–20.
31. TNA, MAF/284, 318.
32. TNA, MAF/284, 318.
33. TNA, MAF/284, 318.
34. TNA, MAF/284, 377.
35. TNA, MAF/284, 350 and 370.
36. TNA, MAF/284, 350, 354 and 370.
37. TNA, MAF/284, 350 and 354.
38. TNA, MAF/284, 354.
39. TNA, MAF/284, 354-5, 372 and 377.
40. TNA, MAF/284, 350, 370 and 377.
41. TNA, MAF/284, 370.
42. TNA, CAB/134, 2054-5, and MAF/284, 354, 370 and 474.
43. TNA, MAF/284, 370.
44. TNA, MAF/284, 370 and 377.
45. TNA, MAF/284, 377.
46. TNA, MAF/284, 370-1 and 377.
47. TNA, MAF/284, 374–5.
48. TNA, MAF/284, 469 and 471.
49. TNA, MAF/284, 471.
50. TNA, MAF/284, 472.
51. TNA, MAF/284, 472-3.
52. TNA, MAF/284, 473-4 and 476.
53. TNA, MAF/284, 475.
54. TNA, MAF/284, 475-6.
55. TNA, MAF/284, 475-6.
56. TNA, MAF/284, 476.
57. TNA, MAF/284, 476-7.
58. TNA, CAB/134, 2865.

59. Edmund Russell, *War and Nature. Fighting Humans and Insects with Chemicals from World War I to* Silent Spring (New York, Cambridge University Press, 2001).

60. Rudd, *Pesticides and the Living Landscape*, p. x.

61. Sheail, *Pesticides and Nature Conservation*, p. 192; Royal Commission on Environmental Pollution, *First Report* (London, HMSO, 1971, Cmnd 4585).

62. Ashby, *Times Higher Educational Supplement*.

63. Parliamentary Debates (Hansard), House of Commons, **834**, Written Answers column *172*.

64. Parliamentary Debates, Commons, *172*.

65. R.E. Boote – Personal communication.

POST-INDUSTRIAL

Towards Sustainable Agricultural Stewardship: Evolution and Future Directions of the Permaculture Concept

Jungho Suh

INTRODUCTION

Since Carson's (1962) landmark report on the dangers associated with DDT and various agrochemicals, a huge volume of environmental regulatory legislation and a variety of environmental-policy instruments have been introduced to prevent, monitor and solve the unintended environmental problems associated with conventional agriculture. However, state-centred, top-down, anthropocentric environmental-management approaches alone are only limitedly effective in controlling and changing money-oriented human economic behaviour.

Since the 1970s, various forms of alternative agriculture have emerged. The grassroots efforts of their proponents have been designed to curb unsustainable industrial agriculture. From the viewpoint of alternative agriculture, industrial agriculture is tantamount to monoculture, the genetic modification of crops and the widespread use of agrochemicals. Large quantities of crops are grown to feed factory animals rather than for direct use by humans. Moreover, concern has been expressed that the crops produced are unsuitable for animal feed. Toxic chemicals have degraded the land and contaminated water systems, thereby harming living species, including humans, through the food chain.

Beus and Dunlap (1990) enumerated several key characteristics of alternative agriculture: decentralisation; independence (reduced reliance on external sources); small rural communities (increased cooperation); harmony with nature; polyculture (integration of diverse crops and livestock); and restraint (internalisation of all external costs and increased reliance on renewable resources). These characteristics overlap substantially with the key features of agricultural post-productivism (Ilbery and Bowler, 1998) and agricultural-stewardship movements (Welchman, 1999; Paterson, 2003).

Permaculture emerged in the 1970s as a practical *in situ* approach to creating collectively sustainable human settlements (Beus and Dunlap, 1990; Halfacree, 2007). In contrast with industrial agricultural systems, which rely on annual market-driven monoculture and heavy inputs of fossil-based energy, permaculture involves small-scale polyculture and depends on soft technology and renewable energy sources (Gundersen and O'Day, 2009). The term 'permaculture' is short for 'permanent agriculture', which has been defined as follows:

- 'an integrated, evolving design system of perennial or self-perpetuating plant and animal species useful to [hu]man[s]' (Mollison and Holmgren, 1978, p. 1);

- 'the conscious design and maintenance of agriculturally productive ecosystems which have the diversity, stability, and resilience of natural ecosystems' (Mollison, 1988, p. ix); and

- 'consciously designed landscapes which mimic the patterns and relationships found in nature, while yielding an abundance of food, fibre and energy for provision of local needs' (Holmgren, 2002, p. xix).

Although the concept of permaculture is of great relevance to efforts to mitigate land degradation and global warming, limited scientific or socio-economic research on permaculture has been undertaken (Veteto and Lockyer, 2008). This paper examines the features of permaculture as a philosophical and practical approach to agricultural stewardship, reviewing seminal texts on permaculture (especially Mollison and Holmgren, 1978; Mollison, 1979, 1988; Mollison and Slay, 1991; and Holmgren, 2002). The paper's chief areas of focus are the contribution of permaculture to discussions of sustainability and the ways in which permaculture as a design system is intended to actualise sustainability. The future developments in the concept of permaculture necessary to ensure a sustainable permaculture movement are also discussed. The paper first traces the evolution of the concept of permaculture, defining it as an outcome of the intellectual interplay between modern agricultural science and traditional ecological wisdom. The core ethics and principles of permaculture are then presented, and their relation to Kaya's (1989) equation explained. Next, the paper provides examples of applications of the permaculture principles, based on participant observation of the Food Forest, a commercial permaculture farm located on the outskirts of the metropolitan area of Adelaide, South Australia. Finally, strategies promoting the widespread awareness and use of permaculture as a form of alternative agriculture are presented.

EVOLUTION OF THE CONCEPT OF PERMACULTURE

The origin of the concept of permanent agriculture can be traced to Hopkins (1910) and King (1911), both university professors in agricultural science. King's (1911) ideas arose from his fieldwork in China, Korea and Japan, where he observed Oriental agricultural practices closely.

Neither Hopkins (1910) nor King (1911) explicitly defined permanent agriculture. Paull (2011) noted that these authors invented the term 'permanent agriculture' to differentiate their view of agriculture from the concept of 'orthodox agriculture' promoted by the United States Department of Agriculture. A concise description of Far Eastern permanent agriculture can be credited to Hopkins (1910, p. xvii), who stated that 'every landowner should adopt for his land a system of farming that is permanent – a system under which the land becomes better rather than poorer'. King (1911, p. 13) provided a more comprehensive definition of permanent agriculture in his introductory chapter, as follows:

> Almost every foot of land is made to contribute material for food, fuel or fabric. Everything which can be made edible serves as food for man or domestic animals. Whatever cannot be eaten or worn is used for fuel. The wastes of the body, of fuel and of fabric worn beyond other use are taken back to the field; before doing so they are housed against waste from weather, compounded with intelligence and forethought and patiently laboured with through one, three or even six months, to bring them into the most efficient form to serve as manure for the soil or as feed for the crop. It seems to be a golden rule with these industrial classes, or if not golden, then an inviolable one, that whenever an extra hour or day of labor can promise even a little larger return then that shall be given, and neither a rainy day nor the hottest sunshine shall be permitted to cancel the obligation or defer its execution.

It is evident from the above quotation that Far Eastern traditional-farming systems in this period were characterised by resource recycling, small and slow solutions and labour-intensive farming. As the following statement shows, King (1911, p. 194) saw the recycling of organic materials (especially human manure) as key to maintaining the Oriental system of permanent agriculture (Parr and Hornick, 1992; Jenkins, 1994).

> One of the most remarkable agricultural practices adopted by any civilized people is the centuries-long and well nigh universal conservation and utilization of all human waste in China, Korea and Japan, turning it to marvellous account in the maintenance of soil fertility and in the production of food.

On learning that soil fertility in the Far East had been maintained for forty centuries, King (1911, p. 274) made the following observations:

> China, Korea and Japan long ago struck the keynote of permanent agriculture but the time has now come when they can and will make great improvements, and it remains for us and other nations to profit by their experience, to adopt and adapt what is good in their practice and help in a world movement for the introduction of new and improved methods.

Another stage in the intellectual evolution of permanent agriculture was Fukuoka's (1978) development of the concept of 'do-nothing' or 'natural' farming. The Japanese system of natural agriculture does not require cultivation, chemical fertiliser or prepared compost, weeding by tillage or herbicides or chemical pesticides. The ploughing of soil is not permitted because the earth cultivates itself naturally by means of earthworms and the penetration of plant roots. No weeding is required because weeds play their part in building soil fertility and balancing the biological community. There is no need of ready-made compost because the green and animal manure generated within the farm is sufficient to maintain soil quality.

Fukuoka (1978, p. 117) claimed that for many centuries before the end of World War Two, the principles of organic farming popular in the West barely differed from those of the traditional agriculture practised in China, Korea and Japan. It is true that the Far Eastern farming systems introduced by King (1911) into the English-speaking world inspired many agricultural scientists (e.g. Smith, 1929; Pfeiffer, 1938; Howard, 1940).[1] However, it cannot be said that Oriental permanent agriculture had a practical influence on organic farming in the West (Vogt, 2007). Human manure, for example, which was key to Far Eastern permanent agriculture, has certainly not been perceived as a recyclable resource in the Western history of organic farming (Jenkins, 1994).[2]

Mollison and Holmgren (1978) coined the contracted term 'permaculture', which encapsulates the key permanent-agriculture approaches to contemporary environmental crises (self-sufficiency, small-scale production, low energy intensity and resource recycling). In their terms, proponents of permaculture pursue sustainable farming methods, aiming to restore and maintain ecosystem services as the basis of human livelihoods.[3] In the 1970s, while travelling across India, Southwest Asia and peasant Europe, Mollison (1996) observed certain communities maintaining farming systems that were thousands of years old. He realised as a result that it is possible to create farming systems and improve their productivity based on analogies with natural systems.

Mollison (1979), Mollison and Slay (1991) and Holmgren (2002) identified Fukuoka's (1978) book *The One-Straw Revolution* as the best statement of the underlying philosophy of permaculture. Yet the strategies for promoting sustainable agriculture that underpin the practice of permaculture are not identical to those of the 'do-nothing' approach. On the contrary, the proponents of permaculture take a 'do-something' approach to sustainable agriculture, in that permaculture is the conscious design and maintenance of agriculturally productive ecosystems that mimic the patterns and relationships observed in nature.

However, Mollison (1979, 1988) and Holmgren (2002) criticised the system of permanent agriculture described by King (1911), referring to it as 'feudal permanence'. They argued that farmers living and working under the feudal system are bound to the land and in service to their landlords by unremitting toil; a situation that leads eventually to famine and revolution. They advocated instead a system of 'communal permanence', a permanent-agriculture approach with no risk of tyranny or feudalism. Mollison (1988, p. 6) described the benefits of such an approach as follows:

Without permanent agriculture there is no possibility of a stable social order.

Thus, the move from productive permanent systems (where the land is held in common), to annual, commercial agricultures where land is regarded as a commodity, involves a departure from a low- to a high-energy society, the use of land in an exploitive way, and a demand for external energy resources, mainly provided by the third world.

However, Mollison (1979, 1988) and Holmgren (2002) were too quick to equate the traditional Oriental practice of permanent agriculture with feudal society, which led them to conclude that permanent agriculture had ended with the feudal system. Indeed, Mollison and Slay (1991, p. 1) later described permaculture as the brainchild of the marriage between 'the wisdom contained in traditional farming systems, and modern scientific and technological knowledge'. Permaculture incorporated the Oriental principles of permanent agriculture, including the composting and recycling of human manure, into a science-based design theory for sustainable living environments. This vision of permaculture was well articulated by Holmgren (2002, p. 2), as follows:

In drawing together strategies and techniques from modern and traditional cultures, permaculture seeks a holistic integration of utilitarian values. By using an ecological perspective, permaculture sees a much broader canvas of utility than the more reductionist perspectives, especially the econometric ones, that dominate modern society.

In sum, the concept of permaculture has evolved considerably over the last century. Hopkins (1910) and King (1911) sought to teach the science of permanent agriculture chiefly by reporting facts in a narrative manner. In contrast, Mollison and Holmgren (1978) aimed to teach the science of permanent agriculture by offering theories rather than reporting facts. Permanent agriculture may have been practised for thousands of years in the Far East, but it was Mollison and Holmgren (1978) who rediscovered the value and significance of traditional permanent agriculture in the face of global environmental crisis.

PERMACULTURE PHILOSOPHY AND TRADITIONAL ORIENTAL WISDOM

It is worth noting that the farmers encountered and observed by King as he travelled through the Far East in 1909 were living in an agricultural society whose pro-ecological traditions were determined by Taoism, Buddhism and Confucianism (Jenkins, 1998). Fukuoka (1978), who inspired the co-founders of the modern concept of permaculture, was influenced by the environmental philosophies of Taoism and Zen Buddhism. Mollison (1979, 1988), Mollison and Slay (1991) and Holmgren (2002) repeatedly acknowledged that the idea of permaculture was inspired by traditional and indigenous agricultural wisdom, as described clearly by Mollison (1988, p. 3) in the following:

For the sake of [the] earth itself, I evolved a philosophy close to Taoism from my experiences with natural systems…it is a philosophy of working with rather than against nature; of protracted and thoughtful observation rather than protracted and thoughtless actions; of looking at systems and people in all their functions, rather than asking only one yield of them; and of allowing systems to demonstrate their own evolutions.

Taoists see all changes in nature as manifestations of the dynamic interplay between *yin* (feminine) and *yang* (masculine), and believe that the balance between *yin* and *yang* is maintained by a continuous flow of *qi* (vital energy) (Capra, 1975; Xu, 1997). This aspect of Taoist thought influenced the formation of the traditional Chinese system of geomancy known as *feng-shui* (wind and water) (Tuan, 1970; Bruun, 1995). According to the theory of *feng-shui*, the earth is a living organism, with mountains for bones, soil for flesh, water for blood and vegetation for hair (Xu, 1998). Proponents of this theory regard environmental degradation as synonymous with a disruption of the balance of *yin* and *yang* on the land. It follows that sites of human-made capital, such as royal palaces, houses, temples and graves, should be selected to harmonise with the flow of vital energy.

Unsurprisingly, *feng-shui* and permaculture have much in common. First, permaculture is a scientific method of designing agricultural systems to represent the patterns and relationships found in nature. On these grounds, Hale (1998) called *feng-shui* the ancient art of permaculture, implying that permaculture is the modern art of *feng-shui*. The proponents of both *feng-shui* and permaculture consider soil to be the interface between non-living mineral earth and the atmosphere, and thus the most important site of interaction for all terrestrial life, inclusive of humankind (Holmgren, 2002). Consequently, the concept of permaculture retains the *feng-shui* principle that the agricultural landscape should be designed in accordance with local energy flow and landforms. The implementation of permaculture requires thoughtful observation of the patterns and flows of the surrounding ecosystem to obtain the information necessary to design an energy-efficient and low-pollution farming system.

However, despite these similarities between *feng-shui* and permaculture, it should also be noted that the principles of permaculture as a design system are not limited to the landscape analysis and location theories of *feng-shui*. In other words, permaculture should not be construed merely as a theoretical framework necessary to locate a site suitable for agricultural development. From the perspective of permaculture, river flats might be the preferred location for establishing grain agriculture, and upland areas for tree agriculture. However, permaculture can be developed on any type of landscape, because permaculturalists ask what the land can give them rather than what they can compel the land to do (Mollison and Holmgren, 1978; Mollison, 1979).

Buddhist ecology, another type of Oriental wisdom, also profoundly influenced the formulation of the concept of permaculture. Its influence is evident in the following observations made by Holmgren (2002, p. 71):

> Traditional societies recognised that external negative feedback effects are often slow to emerge. People needed explanations and warnings, such as *the sins of the fathers are visited on the children onto the seventh generation* and *laws of karma which operate in a world of reincarnated souls...* John Lennon's song *Instant Karma* suggests that we will reap what we sow much faster than we think. The speed of change and increasing connectivity of globalisation may be the realisation of this vision.

Two key concepts underpin Buddhist ecology and economics: *dharma* (truths) and *karma* (the law of cause and effect) (Conze, 1957; Pryor, 1990; Harris, 1997; Daniels, 2005; Swearer, 2006; Thich Nhat Hanh, 2008). *Dharma* is predicated on the awareness that everything is in constant change. *Karma*, the law of cause and effect, assumes that everything is interconnected with everything else in an ever-changing world. As a result, to harm nature is to harm human beings, and vice versa.

The phrase 'mindful consumption' coined by Thich Nhat Hanh (2008) reinforces a core tenet of Buddhist economic thought: that overconsumption leads to the overexploitation of natural resources, leading to environmental degradation that in turn harms humans themselves. Buddhist economists argue that externalities should not be generated in the first place, and that people should thus reduce their use of resources, recycle resources and choose ecologically appropriate production technologies. Buddhist economics also emphasises the necessity of a self-sufficient economy in which food and other necessities are produced from local resources to meet local needs. Unlike neoclassical economists, Buddhist economists consider measures such as an increase in ton-miles of freight per capita to indicate a deterioration in self-sufficiency (Schumacher, 1973).

Permaculture is an agricultural application of the Buddhist ecological tenet that everything is connected to everything else. It thus involves making connections between elements such that 'each element in the system performs many functions' (Mollison and Slay, 1991, p. 6) and 'each function is supported by many elements' (Mollison and Slay, 1991, p. 8). For example, ducklings can be released into rice paddies to control weeds and pests (Furuno, 2001). Similarly, animal manure can be used to fertilise fish ponds, which can in turn be used to irrigate vegetable gardens (Mollison and Slay, 1991).

PERMACULTURE PRINCIPLES, SUSTAINABILITY AND THE IMPACT EQUATION

Permaculture is a system for designing and creating human habitats that are in line with nature's patterns and use natural resources sustainably. Permaculture as a design science is constructed on ethical principles (Mollison, 1988; Mollison and Slay, 1991; Holmgren, 2002); its strategies are thus underpinned by broad and generic moral maxims. Its central ethical principles are as follows: caring for the earth; caring for people; setting limits on population and consumption; and redistributing surplus.

'Caring for the earth' means developing a sense of place through the lens of ecocentric environmentalism. Permaculturalists consider caring for the earth to be our greatest responsibility, because the natural environment is the basis of our livelihoods. Consequently, they highlight the need for self-sufficient and small-scale farming that maintains soil fertility to ensure perennial production.

The maxim 'caring for people' calls for individual responsibility for the human community, because all individuals are connected to the collective society (de la Bellacasa, 2010). The paradox of individual actions and collective consequences in relation to the use of natural resources is often described as the 'prisoners' dilemma' (Luce and Raiffa, 1957) or the 'tragedy of the commons' (Hardin, 1968). The dilemma or tragedy arises from the lack of incentives for individuals to refrain from overusing and depleting resources, because nobody expects their fellow members of society to cooperate to make use of those resources in a socially optimal way. For this reason, permaculturalists recommend that humans cultivate harmonious and cooperative relationships with each other.

The principle of 'setting limits on population and consumption' reflects the fact that the consumerist behaviour of the human species is a root cause of the overexploitation of the natural environment. As a result, the global population must be stabilised. It is clear that another central tenet of permaculture, 'caring for the earth', cannot be successfully implemented without controlling the pressure of growth in both population and in consumption. This ethical principle reinforces a core precept of Buddhist economic thought: that overconsumption leads to overexploitation of natural resources, leading to environmental degradation that in turn harms humans themselves.

The principle of 'redistributing surplus' has two dimensions: contemporaneous interpersonal equity and intertemporal environmental justice. The latter concerns the environmental issue of sustainability and is particularly important in the context of ecocentric agriculture. From the viewpoint of permaculture, intertemporal environmental justice can be achieved by, for example, planting trees or reducing the consumption of non-renewable energy sources. The following remark from Holmgren (2002, p. 10) exemplifies this point.

> We [need to] make all reasonable efforts to increase, and even transform, the biological capacity of soil for the benefit of future generations. Planting of trees and other perennial vegetation to restore the health of the land, without the need to gain an economic benefit, has been a central activity of the permaculture and broader Landcare movements.

Although neither Mollison nor Holmgren mentioned the ImPACT equation (Kaya, 1989) directly, the components of the equation dovetail usefully with the ethical principles of permaculture. The equation is written as follows:

$$Impact = Population \times \frac{GDP}{Population} \times \frac{Energy\ consumption}{GDP} \times \frac{Impact}{Energy\ consumption}$$

Equation 1 postulates that environmental impact (I) increases proportionately with population size (P),[4] affluence (A) in terms of per-capita gross domestic product (GDP), energy consumption (C) per unit of economic activity and environmental damage (pollution) per unit of energy consumption inflicted by the available technology (T). This equation, also known as the Kaya (1989) equation, is a variant of the IPAT (Environmental Impact = Population × Affluence × Technology) equation.[5] It is evident that the factors on the right-hand side of Equation 1 are not separate but directly interconnected.

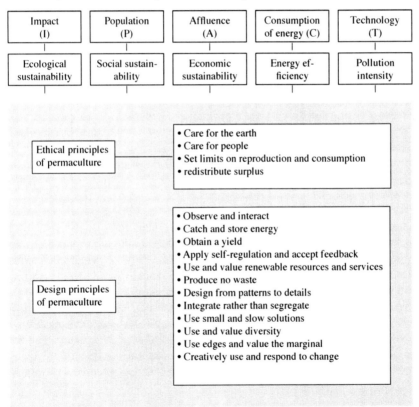

Figure 1. Conceptualising permaculture principles in relation to the ImPACT framework

Recalling the ethical principles underlying permaculture, the ImPACT formula implies that energy efficiency and low pollution intensity are the key feasible pathways to reducing humans' environmental impact. As illustrated in Figure 1, permaculture thinking is distinct from the ImPACT framework in that the former lacks a counterpart to the latter's 'equals' sign. The formulation of the ImPACT equation implies that the left-hand side of the equation is causally determined by the factors on the right-hand side of the equation. In contrast, the framework of permaculture thinking is clearly in line with the ecological teachings of Taoism and Buddhism, which describe more mutual or reciprocal causal relationships between factors. Not only is the level of environmental impact determined by population size, economic affluence, energy consumption and technology-related factors, but environmental degradation leads in turn to an unsustainable economy, and may threaten human settlements.

The design principles of permaculture are guidelines for putting its underlying ethical principles into practice. As listed in the box at the bottom right-hand side of Figure 1, Holmgren (2002) synthesised diverse aspects of permaculture thinking into a set of twelve design principles, based primarily on the work of Mollison (1979, 1988) and Mollison and Slay (1991). Holmgren (2002) noted that the logical sequence and consistency of these principles should not be considered complete, given the evolving nature of permaculture. The basis of the 'observe and interact' principle, in contrast to industrial agricultural practice, is that a given landscape should remain unchanged, because the nature of that landscape determines the best way in which the land can be used (Mollison and Holmgren, 1978). For this reason, the application of the 'observe and interact' principle involves intensive zone and sector analysis (Mollison, 1988). The 'apply self-regulation and accept feedback' principle asserts that negative social and environmental externalities should be prevented from taking place, or at least that their sources should be addressed to ensure that the living environment can continue to function. The permaculture approach characteristically fulfils this principle by creating mutually beneficial linkages and relationships between resources. Accordingly, fulfilling the 'integrate rather than segregate' principle entails choosing the right elements and putting them in the right places so that each element serves the needs of the others (Mollison and Slay, 1991; McManus, 2010). The 'use small and slow solutions' principle recalls a fundamental characteristic of small-scale traditional agriculture, wherein soil fertility is maintained by laborious resource recycling. It reminds those seeking to implement permaculture solutions that traditional agriculture has been destroyed by profit-oriented industrial agriculture, along with whole communities relying on permanent soil fertility.

The Food Forest: an on-the-ground application of permaculture principles

The Food Forest is a 15-hectare permaculture farm in the northern suburbs of Adelaide, South Australia. The northern side of the farm is bordered by the Gawler River. The buildings on the farm include a learning centre, a gallery, a visitors' lodge, compost toilets and a heritage-listed homestead built more than one hundred and sixty years ago during the first few years of white settlement in South Australia. The Food Forest is the largest commercial polyculture farm in Australia. It practises the principles of permaculture and grows many organic varieties of fruits and nuts, wheat, vegetables and poultry. When the Brookmans purchased the property in 1983, it was little more than a bare barley paddock. It took more than a decade to develop the property into a functioning permaculture farm, making the most of composted human manure as well as animal and green manure (Brookman and Brookman, 2005).[6]

The whole livelihood system of the Food Forest is characterised by intensive and sustainable bioenergy production and consumption in a closed cycle, because energy leaves the farm as food and returns as compost. Table 1 highlights the permacultural practices employed by the Food Forest. As noted by Holmgren (2002), an example of one permaculture principle may be used to illustrate several others. For instance, collecting rainwater from the roof and storing it for future use is an example of both the 'catch and store energy' imperative and the injunction to 'use and value renewable resources and services'.

The homestead was renovated to ensure its energy efficiency by installing devices for capturing solar energy and making use of strawbale construction. One of the environmental benefits of the latter is the superb insulation provided by strawbale walls, which make it unnecessary for dwellers to install electricity-hungry air-conditioning or heating systems.

Crop growing, livestock husbandry and biodiversity conservation are integrated in the Food Forest to form a permanent system that maintains if not improves soil fertility. About two hectares of the farm adjacent to its orchards have been reserved as a no-management zone, to maintain biodiversity within the farm. Brush-tailed bettongs (*Bettongia penicillata*) eat germinating seedlings and help to control a noxious weed known as soursobs (*Oxalis pes-caprae*), as well as revegetating the biodiversity zone. Cape Barren geese are free-ranged to exploit inter-row space in the orchard in spring and early summer. They are less selective feeders than European geese and thus provide more effective control of creeping weeds, including couch grass. Rabbit- and fox-proof fencing was built in 1993 to create a suitable environment for integrating native animal species into the farm to create a more productive agricultural system.

In response to the curiosity of smallholder farmers, gardeners and students, the Food Forest has begun to offer open days in the spring and autumn, along with a series of short courses on organic vegetable growing, agroforestry, build-

Table 1. Permaculture principles and examples of applications by the Food Forest

Observe and interact • Landscape analysis (e.g. wind directions, sun angles and frost pockets) undertaken as a crucial means of understanding and planning the placement of building structures and vegetation
Catch and store energy • Installation of energy-catchment and energy-storage devices (e.g. solar panels and rainwater tanks) • Construction of living soil from decayed plants, leaves, ash and animal manure to store mineral nutrients
Obtain a yield • Organically grown wheat and vegetables, fruit, nuts, free-range eggs, honey, carob beans and meat animals • Value-added products including organic wine, cider, vinegar and olives
Apply self-regulation and accept feedback • Jujube (*Ziziphus zizyphus*) trees planted, which are tolerant of warm and dry conditions • No or few agrochemicals sprayed
Use and value renewable resources and services • Use of a photovoltaic system to produce electricity and solar panels to heat water • Roof water captured for domestic and irrigation purposes • Use of groundwater for irrigation in the dry season
Produce no waste • Sewage water treated by reedbeds and reused to support tree growth • Utilisation of ash, composted pistachio hulls, grape marc and apple pressings from the property • Human manure composted and recycled into fertiliser to support tree growth • Legumes and animal manure used to provide nutrients for vegetable crops
Design from patterns to details • Selection of heat-tolerant vine varieties and dryland field crops (e.g. pistachios and walnuts) • Passive solar design
Integrate rather than segregate • Orchard floors grazed by Cape Barren geese and Dama wallabies • Brush-tailed bettongs (*Bettongia penicillata*) carrying and burying plant seeds • Bettongs, geese and wallabies loose in the same area
Use small and slow solutions • Fifteen years' worth of efforts to convert a bare barley paddock into fertile farming land, relying on ash, animal manure, green mulch and biodynamic farming • Emphasis on the optimisation of the farming system rather than short-term profit maximisation
Use and value diversity • Fifteen hectares of land devoted to vegetable gardens, vineyards, orchards, pastures, forestry and wildlife habitat (two hectares)
Use edges and value the marginal • Enormous volume of inputs to improve the fertility of top-soil • Annual planting in gardens and glasshouse at the side of the house • Windbreak trees planted along the eastern edge of the property
Creatively use and respond to change • Generation of income through agricultural education (short practical courses, open-day tours, • Permaculture Design Certificate courses and agrotourism) • Installation of fox- and rabbit-proof fencing covering eight hectares of the property • Sale of certified organic vegetables and fruit at a farmers' market at premium prices

ing with strawbales and free-range poultry management, among other topics. The Food Forest also holds a ten-day internationally accredited Permaculture Design Certificate course every year. Additional educational training is regularly provided in the orchards, gardens and bushlands of the property (Brookman and Brookman, 2005). The income from these educational activities accounts for a large portion of the annual income of the Food Forest (Brookman and Brookman, 2012).

The design system implemented by the Food Forest should not be interpreted as a model suitable for all places and circumstances. Notably, methods of applying permaculture principles can be adapted to suit landform, climate, local culture and other environmental conditions (Mollison and Holmgren, 1978; Holmgren, 2002). For instance, Mollison and Slay (1991) detailed different means of applying universal permaculture principles in tropical regions, aquaculture environments and urban communities.

FUTURE DIRECTIONS OF THE PERMACULTURE CONCEPT

Mollison and Holmgren (1978) formulated the concept and theory of contemporary permanent agriculture and presented it as a holistic, alternative form of agriculture. Permaculture calls for the integration of natural resources and food production into ecologically sustainable human settlements. Holmgren (2002, p. xix) described its uses as follows:

> Permaculture is not the landscape, or even the skills of organic gardening, sustainable farming, energy efficient building or ecovillage development as such. But it can be used to design, establish, manage and improve these and all other efforts made by individuals, households and communities towards a sustainable future.

As indicated above, organic gardening, energy-efficient building and ecovillage development are all components of permaculture. However, a part-whole fallacy needs to be kept at bay. For instance, permaculture does not merely entail growing organic food. Although organic farming is a fundamental aspect of permaculture, it should be noted that not all organic farming can be regarded as permaculture. In fact, some consider organic farming to be rather unsustainable (e.g. Esbjornson, 1992; Robinson, 2009). Interestingly, some of the produce of the Food Forest is not certified as organic because the farm grows fruit trees using recycled domestic grey and black water. Brookman and Brookman (2005) pointed out that the use of recycled sewage water and human manure is a key difference between permaculture and organic farming.

Information on permaculture practices has been rapidly disseminated to the general public. People in developed countries have access to private permaculture-consultation services, as well as permaculture magazines, guide books, websites, workshops and courses offered by permaculture-related decentralised organisations. Interestingly, permaculture theory has also been reimported with an injection of scientific thinking to China, South Korea and Japan, where farmers have historically practised permanent agriculture. The advocates of permaculture in these countries strive to promote sustainable rural development by recovering their forgotten agricultural traditions.

Despite the widespread practice of permaculture-based home and community gardening in cities and towns in the developed world, little socio-economic research has been conducted in the field of permaculture theory. Mollison and Slay (1991) and Holmgren (2002) attributed this paucity of academic investigation to the interdisciplinary nature of permaculture, which combines many fields of investigation, including agriculture, architecture, biology, chemistry, forestry and animal husbandry. However, the interdisciplinary themes of permaculture should not be blamed for the lukewarm academic response to the permaculture movement, because contemporaneous environmental problems such as global warming and land degradation are regularly approached from an interdisciplinary perspective. To increase the adoption of permaculture, therefore, especially among farmers, it is necessary to investigate and explore methods of implementing permaculture that diverge from the cross-disciplinary and holistic approaches.

Communitarianism rather than communalism

Mollison (1979, 1988) and Mollison and Slay (1991) argued that the formation of communal permaculture, wherein the land is held in common rather than regarded as a commodity, is crucial to mitigating the contemporary energy crisis and global environmental problems. With communal permaculture, the need for energy can be met more efficiently by on-site renewable energy sources. Holmgren (2002) saw the formation of communal permaculture as an execution of the 'integrate rather than segregate' principle.

Ecovillages have great potential to realise the vision of communal permanence. Some communal ecovillages (e.g. Crystal Waters in Queensland, Australia) have been developed specifically with permaculture in mind. Furze (1991) and Veteto and Lockyer (2008) provided two case studies of communal permaculture farms: Quindalup in Victoria, Australia and the Earthaven Ecovillage in the Appalachian Mountains of western North Carolina, respectively. Despite these examples, however, few communal ecovillages have been established to date worldwide. In fact, ecovillages are heterogeneous to the extent that no single model can describe all cases (Dawson, 2006). Moreover, it is debatable whether ecovillages can be classed as an environmental-stewardship movement (Pepper, 1996).[7]

Holmgren (2002) pointed out that weak property rights to land are one of the main impediments to communal permaculture. As a result, communalism (as opposed to feudalism) should not be regarded as a necessary condition for the successful implementation of permaculture. As a matter of fact, permaculture is often interpreted as red-green or eco-socialistic agriculture (e.g. Furze, 1992) owing to its radical approach to property rights. As the Food Forest shows, an individual farming household can successfully implement a permaculture system without the need for communal-property arrangements. Therefore, permaculturalists should focus in the future on revitalising the communitarian spirit of traditional farming villages instead of building intentional communal communities. As communitarianism (as opposed to individualism) sustained agricultural societies, with every individual bound to other individuals, it offers a means of actualising one of permaculture's core ethical principles: 'caring for people'.

Self-sufficiency with economies of scale

Mollison and Holmgren (1978, p. 1) argued that permaculture '[might] well be unsuited to a large commercial enterprise, or inapplicable to conventional farming, but has greater relevance to those who wish to develop all, or part, of their environment to near self-sufficiency'. Mollison and Holmgren (1978) may have regarded 'large-scale permaculture' as an oxymoron, because permaculture systems on a large scale tend eventually to import energy sources, resulting in the extensive use of machines that consume non-renewable fossil fuels.

A question is raised as to how much land is self-sufficient. According to Mollison (1979), this depends on whether one can manage the land without violating permaculture principles. This remark suggests that the physical limit of permaculture land can be flexibly defined. Indeed, the Food Forest's 15 hectares might be viewed as a large-scale system in many highly populated developing countries where the average size of farming land per household is approximately one hectare. Mollison and Holmgren (1978) and Mollison (1979) also advised permaculturalists to avoid subscribing to the 'isolated-fortress' mentality, commenting that pure self-sufficiency is a pointless goal. After all, it is questionable whether self-sufficiency really is an indispensible ingredient of permaculture.

A corollary is that self-sufficient farming is not necessarily incompatible with commercial production. Any surplus produce can and should be redistributed through markets, because not everyone can dedicate themselves to farming. Further, self-sufficiency can go hand in hand with economies of scale. It is not always the case that the average production cost can be reduced by expanding output (Curtis, 2003; Cheah and Cheah, 2005). Similarly, small-scale production systems are not always efficient or environmentally benign. Therefore, a balance must be found between larger- and smaller-scale farms to ensure that the principles of permaculture are upheld while the advantages of economies of scale are reaped.

Labour-intensive yet job-creating farming

Holmgren (2002) drew attention to the high labour intensity of permanent agriculture in China, Korea and Japan, as previously reported by King (1911). Even in the twenty-first century, permaculture demands the maximum possible use of human labour to secure a sustainable society and economy. Therefore, practising permaculture may require farmers to spend more time farming than their non-permaculture-practising counterparts. However, labour-intensive farming practices are not necessarily time-consuming. Fukuoka (1978) noted that although farmers in Japan in the distant past were probably poor, they still had the leisure to write poems.

The permaculture approach implements slow solutions to maintain soil fertility and environmental conditions. The 'slow-solutions' doctrine stems from the belief that producing more food crops more quickly than the land can do in its natural state has resulted in land degradation. Thus, the so-called green revolution is hardly a long-term solution to food shortage (Mollison, 1996). The non-market ecological benefits of permaculture practices such as slow solutions can compete with the time-saving benefits of profit-oriented industrial agriculture.

In addition to the ecological benefits of permaculture, it should be noted that labour-intensive farming systems can help to reduce unemployment in the urban sector and thereby contribute to the demographic sustainability of rural society (Robinson, 2009). A fundamental obstacle faced by capitalism is that urban industries need to keep producing goods and services to maintain employment, which often leads to overproduction as well as the overexploitation of natural resources. A labour-intensive yet low-pollution food-production system can break down this vicious cycle. Moreover, labour-intensive production does not necessarily mean low labour productivity. Even when labour intensity does make farms less labour-productive, the premium prices paid for organic produce may compensate for the reduction in labour productivity (Robinson, 2009)

Permaculture farms as educational centres

Education is a driving force in permaculture movements. Public awareness of permaculture education has grown massively, with an increase in the number of permaculture-design graduates worldwide from about 12,000 in the early 1990s (Mollison and Slay, 1991) to about 250,000 in the mid-2000s (Brookman and Brookman, 2005).

Regardless of their religious or ethnic backgrounds, urban dwellers are attracted to alternative agriculture. However, alternative agriculture is not adequately accommodated by the current system of formal education. Permaculture farms can fill the gap, playing the role of regional educational centres for traditional agriculture. Permaculture can be incorporated not only into agriculture-related subjects in school curricula, but also into a variety of topics in science, mathematics, history, ecology, biology, geography and economics courses. In this way, students can learn about the connections between farming and the whole world.

In reality, however, there are few functioning permaculture farms, particularly in less developed countries. In developed countries, permaculture is often understood as synonymous with organic farming or urban gardening. To rectify this insufficiency, governments could introduce accredited permaculture systems. Deserving permaculture farms could receive an accreditation similar to the National Park designation, which is intended to protect natural assets for educational, recreational and scientific purposes.

Urban and environmental planning in support of permaculture

Permaculturalists contribute to a sustainable future by organising their lives and careers around permaculture design principles (Holmgren, 2002). Given the increased public awareness of contemporary environmental crises, the ultimate aims of permaculture should not be treated as a mere agrarian utopian dream or as radical post-productivism. Rather, the advice of proponents of permaculture should be heard in urban-planning and regional development policy making contexts.

Permaculturists' approach to sustainability can be operationalised even at a national level as a path to ecologically sustainable economic growth (Daniels, 2010a, 2010b). An example is Bhutan's national approach to development via the 'Middle Path', a term originating in Mahayana Buddhism (Daniels, 2005; Uddin et al., 2007; Brooks, 2010). In Bhutan, economic, social and environmental values and performance are all considered with respect to gross national happiness (GNH). The calculation of GNH thus encompasses life safety and environmental degradation (e.g. air pollution and water pollution) as well as the conventional measure of national performance, GDP.

An appropriate national agricultural-policy orientation should be established to secure the non-market social gains of permaculture. For example, the contribution of permaculture to CO_2 emissions mitigation could be realised through carbon-based credit schemes and tax systems offer means of realising. In more general terms, a paradigm shift in policy-making from environmental managerialism to environmental stewardship is required. The demand for this paradigm shift is supported by some research findings showing that the distributional pattern of the environmental attitudes of the general public has changed since the 1980s. According to a multinational survey by Milbrath (1984), less than 15% of the population held ecocentric views in the early 1980s. Dunlap et al. (1992) reported that the proportion of ecocentrics to anthropocentrics increased considerably between 1976 and 1990, based on a longitudinal survey of Washington State residents. However, it should be noted that individual countries differ in terms of public awareness of, and attitudes towards, local environmental problems. Therefore, as Kilbourne et al. (2002) pointed out, each country's environmental policy needs to be designed in accordance with the dominant social paradigm of that country.

CONCLUSION

Alternative-agriculture movements emerged as a reaction to conventional or industrial agriculture. Due to the advantages of economies of scale and profit-maximising behaviour, a larger and larger proportion of agricultural and food industries has become concentrated in the hands of a smaller and smaller number of farmers and food manufacturers. The advocates of alternative agriculture blame the large-scale industrial farming system for various forms of environmental degradation observable today.

Permaculture has been recognised as an ideal alternative to conventional agriculture. The intellectual domain of permaculture is not confined to agriculture; it encompasses environmental ethics, design principles and farming methods. Central to the philosophy of permaculture is the need to care for the earth and for human beings, set limits on reproduction and consumption, enhance energy efficiency and utilise renewable energy. In a broader context, the permaculture movement is intended to contribute to developing and maintaining ecologically, economically and socially sustainable living environments. This aim should not be interpreted as empty rhetoric. The Food Forest, a privately owned commercial permaculture farm, provides evidence that the use of carbon-free or low-carbon energy sources, together with the reuse and recycling of much that is usually considered waste, is conducive to economic sustainability.

Permaculture was conceived at the interface between modern agricultural science and traditional ecological wisdom, particularly *feng-shui* and Buddhist ecology and economics. Extending beyond these Oriental teachings, permaculture is a design science that involves connecting resources strategically to ensure that the sum of their yields is maximised and also perennial. The permaculture principles informed by the ecocentric perspective of Eastern tradition provide new insights into the conceptual framework of the ImPACT equation: namely that humans' negative environmental impact eventually degrades the quality of human life, and that it is both fundamentally important and in our own best interests to contribute to ecological sustainability.

Despite regarding pre-industrial societies as a useful model, the proponents of permaculture do not urge a return to peasant economies (Mollison and Slay, 1991). Nevertheless, permaculture is often criticised as impractical and overly utopian, partly due to its dogmatic emphasis on communalism. Therefore, relaxing the 'communal' constraint would assist in the institutionalisation of permaculture without compromising its central theme and spirit. Similarly, self-sufficiency should not be considered to delimit the concept of permaculture.

Another main thesis of this paper is that a paradigm shift in public attitudes, values and behaviour should be reflected in government policy-making. Permaculture doctrines were first promoted well in advance of the general public's paradigm shift in environmental perceptions, attitudes and behaviour. Today, consumers are willing to pay premium prices for organically grown food. As government investment in bioenergy research and development is no longer an option in most countries, it is imperative to develop more aggressive policy measures to support permaculture and internalise the non-market social gains that come from reduced energy use and on-farm waste recycling.

ACKNOWLEDGEMENTS

The author would like to express his gratitude to the Food Forest. It would have been near-impossible to write the section of this paper on real-world application without participant observation and hands-on experience of this permaculture farm. Considerable gratitude is also extended to Mr Soon-Myung Hong, the former principal of Poolmoo Agricultural High School at Hongsung in South Korea, for his astute advice and warm encouragement. Finally, the author would like to thank Dr Steve Harrison, Principal Research Fellow at the School of Agriculture and Food Sciences at the University of Queensland, for reviewing an earlier version of this paper, along with two anonymous referees for their critical comments.

NOTES

1   Paull (2011) has provided historical and bibliological insights into King's (1911) influence on the intellectual development of Western agriculture.

2   It is notable that the 2004 Dover edition of King's (1911) study redacted the original title to *Farmers of Forty Centuries: Organic Farming in China, Korea and Japan*, despite the claim that not one word had been omitted or changed in the text (Paull, 2011). Paull (2011) suspected that the term 'permanent agriculture' in the book's title had been replaced with 'organic farming' for marketing purposes.

3   Other forms of sustainable-agriculture movements also emerged between the late 1970s and the early 1980s, including agroecology and regenerative agriculture (Dahlberg, 1991; Pepper, 1996).

4   Curtis (2003) noted that the economic theory of ecolocalism (inclusive of permaculture) has little or nothing to say about the population factor. However, this is not true of permaculture itself.

5   The IPAT equation was originally presented by Commoner (1971), who reformulated Ehrlich and Holdren's (1971) equation, Impact = Population × (Impact/Population). The IPAT equation has been extensively quoted in the literature as a conceptual framework and used as a tool for modelling and empirically analysing the nature and extent of humans' environmental impact. Chertow (2001) and York et al. (2003) provided an overview of the historical evolution of the IPAT equation and its variants.

6   Human manure is composted using the Clivus Multrum composting toilet system (Jenkins, 1994) and recycled for use in the production of timber for mulching. It is illegal to grow food crops using human excrement, composted or not, in Australia.

7.   Interested readers are referred to the debate between Fotopoulos (2000) and Trainer (2000, 2002).

REFERENCES

de la Bellacasa, M.P. 2010. 'Ethical doings in natureculture'. *Ethics, Place and Environment* **13**(2): 151–169. CrossRef

Beus, C.E. and R.E. Dunlap. 1990. 'Conventional versus alternative agriculture: the paradigmatic roots of the debate', *Rural Sociology* **55**(4): 590–616. CrossRef

Brookman, A. and G. Brookman. 2005. 'Commercial scale permaculture at the Food Forest'. A paper presented at the 2005 National Permaculture Convergence, Melbourne. http://www. foodforest.com.au/assets/pdfs/commercialscalepermaculture.pdf, accessed 10 November 2013.

Brookman, A. and G. Brookman. 2012. The owners of the Food Forest at Gawler, Adelaide, South Australia. Personal communication.

Brooks, J.S. 2010. 'The Buddha mushroom: conservation behavior and the development of institutions in Bhutan'. *Ecological Economics* **69**(4): 779–795. CrossRef

Bruun, O. 1995. '*Fengshui* and the Chinese perception of nature', in O. Bruun and A. Kalland (eds), *Asian Perceptions of Nature: A Critical Approach*, pp. 173–188. Richmond: Curzon Press.

Capra, F. 1975. *The Tao of Physics: An Exploration of the Modern Physics and Eastern Mysticism*. Boston: Shambhala.

Carson, R. 1962. *Silent Spring*. New York: Mariner Books.

Cheah, H.B. and M. Cheah. 2005. 'Small, diversified and sustainable: small enterprises in a sustainable production system', in C. Harvie and B.C. Lee (eds), *Sustaining Growth and Performance in East Asia*, pp. 28–71. Cheltenham: Edward Elgar.

Chertow, M.R. 2001. 'The IPAT equation and its variants: changing views of technology and environmental impact'. *Journal of Industrial Ecology* **4**(4): 13–29. CrossRef

Commoner, B. 1971. *The Closing Circle: Nature, Man and Technology*. New York: Knopf.

Conze, E. 1957. *Buddhism: Its Essence and Development* (3rd edn). Oxford: Bruno Cassirer.

Curtis, F., 2003. 'Eco-localism and sustainability'. *Ecological Economics* **46**(1), 83–102. CrossRef

Dahlberg, K.A. 1991. 'Sustainable agriculture – fad or harbinger?'. *BioScience* **41**(5): 337–340. CrossRef

Daniels, P.L. 2005, 'Economic systems and the Buddhist world view: the 21st century nexus'. *The Journal of Socio-Economics* **34**(2): 245–268. CrossRef

Daniels, P.L. 2010a. 'Climate change, economics and Buddhism – Part I: an integrated environmental analysis framework'. *Ecological Economics* **69**(5): 952–961. CrossRef

Daniels, P.L. 2010b. 'Climate change, economics and Buddhism – part 2: new views and practices for sustainable world economies'. *Ecological Economics* **69**(5): 962–972. CrossRef

Dawson, J. 2006. *Ecovillages: New Frontiers for Sustainability*. Devon, UK: Green Books.

Dunlap, R., K. van Liere, A. Mertig and R. Howell. 1992. 'Measuring endorsement of an ecological worldview: a revised NEP scale', a paper presented in the Annual Meeting of the Rural Sociology Society, The Pennsylvania State University, August 1992, PA.

Ehrlich, P.R. and J.P. Holdren. 1971. 'The impact of population growth'. *Science* **171**(3977): 1212–1217. CrossRef

Esbjornson, C.D. 1992. 'Once and future farming: some meditations on the historical and cultural roots of sustainable agriculture in the United States'. *Agriculture and Human Values* **9**(3): 20–30. CrossRef

Fotopoulos, T. 2000. 'The limitation of life-style strategies: the ecovillage 'movement' is not the way towards a new democratic society'. *Democracy and Nature* **6** (2): 287–308. CrossRef

Fukuoka, M. 1978. *The One-Straw Revolution*. New York: New York Review of Books.

Furuno, T. 2001. *The Power of Duck: Integrated Rice and Duck Farming*. Sisters Creek, Tasmania: Tagari.

Furze, B. 1992. 'Ecologically sustainable rural development and the difficulty of social change'. *Environmental Values* **1**(2): 141–155. CrossRef

Gundersen, D.T. and T. O'Day. 2009. 'Permaculture, a natural systems design approach for teaching sustainability in higher education: Pacific University's B-Street permaculture project', in S. Allen-Gil, L. Stelljes and O. Borysova (eds), *Addressing Global Environmental Security Through Innovative Educational Curricula*, pp. 165–177. Dordrecht: Springer.

Hale, G. 1998. '*Feng shui*: the ancient art of permaculture?'. *Permaculture* **18**: 3–5.

Halfacree, K. 2007. 'Trial by space for a 'radical rural': introducing alternative localities, representations and lives'. *Journal of Rural Studies* **23**: 125–141. CrossRef

Hardin, G. 1968. 'The tragedy of the commons'. *Science* **162**: 1243–1248. CrossRef

Harris, I. 1997. 'Buddhism and the discourse of environmental concern: some methodological problems considered', in M.E. Tucker and D.R. Williams (eds), *Buddhism and Ecology: The Interconnection of Dharma and Deeds*, pp. 377–402. Cambridge: Harvard University Press.

Holmgren, D. 2002. *Permaculture: Principles and Pathways beyond Sustainability*. Hepburn, Vic.: Holmgren Design Services.

Hopkins, C.G. 1910. *Soil Fertility and Permanent Agriculture*. Boston: Ginn and Company.

Howard, A. 1940. *An Agricultural Testament*. London: Oxford University Press.

Ilbery, B. and I. Bowler. 1998. 'From agricultural productivism to post-productivism', in B. Ilbery (ed.), *The Geography of Rural Change*, pp. 57–84. Harlow: Addison Wesley Longman.

Jenkins, J.C. 1994. *The Humanure Handbook: A Guide to Composting Humane Manure*. Grove City, PA: Jenkins Publishing.

Jenkins, T.N. 1998. 'Economics and the environment: a case of ethical neglect'. *Ecological Economics* **26**(2): 151–163. CrossRef

Kaya, Y. 1989. 'Impact of carbon dioxide emissions on GNP growth: interpretation of proposed scenarios'. Paper presented to IPCC Energy and Industry Subgroup. Geneva: Response Strategies Working Group.

Kilbourne, W.E., S.C. Beckmann and E. Thelen. 2002. 'The role of the dominant social paradigm in environmental attitudes: a multinational examination'. *Journal of Business Research* **55**(3): 193–204. CrossRef

King, F.H. 1911. *Farmers of Forty Centuries or Permanent Agriculture in China, Korea and Japan*. Emmaus, PA: Rodale Press.

Luce, D.R. and H. Raiffa. 1957. *Games and Decisions: Introduction and Critical Survey*, New York: Wiley.

McManus, B. 2010. 'An integral framework for permaculture'. *Journal of Sustainable Development* **3**(3): 162–174.

Milbrath, L.W. 1984. *Environmentalists: Vanguard for a New Society*. Albany: State University of New York Press.

Mollison, B. 1979. *Permaculture Two: Practical Design for Town and Country in Permanent Agriculture*. Hobart: Tagari.

Mollison, B. 1988. *Permaculture: A Designer's Manual*. Sisters Creek, Tasmania: Tagari.

Mollison, B. 1996. *Travels in Dreams: An Autobiography*. Tyalgum, NSW: Tagari.

Mollison, B. and D. Holmgren. 1978. *Permaculture One: A Perennial Agricultural System for Human Settlements*. Melbourne: Corgi.

Mollison, B. and R.M. Slay. 1991. *Introduction to Permaculture*. Tyalgum, NSW: Tagari.

Parr, J.F. and S.B. Hornick. 1992. 'Agricultural use of organic amendments: a historical perspective'. *American Journal of Alternative Agriculture* **7**(4): 181–189. CrossRef

Paterson, J.L. 2003. 'Conceptualizing stewardship in agriculture within the Christian tradition'. *Environmental Ethics* **25**(1): 43–58. CrossRef

Paull, J. 2011. 'The making of an agricultural classic: farmers of forty centuries or permanent agriculture in China, Korea and Japan, 1911–2011'. *Agricultural Science* **2**(3): 175–180. CrossRef

Pepper, D. 1996. *Modern Environmentalism: An Introduction*. London: Routledge.

Pfeiffer, E. 1938. *Bio-Dynamic Farming and Gardening: Soil Fertility Renewal and Preservation*. Translated by F. Heckel. New York: Anthroposophic Press.

Pryor, F.L. 1990. 'A Buddhist economic system – in principle: non-attachment to worldly things is dominant but the way of the law is held profitable'. *American Journal of Economics and Sociology* **49**(3): 339–350. CrossRef

Robinson, G. 2009. 'Towards sustainable agriculture: current debates'. *Geography Compass* **3**(5): 1757–1773. CrossRef

Schumacher, E.F. 1973. *Small Is Beautiful: Economics as if People Mattered*. New York: Harper and Row.

Smith, J.R. 1929. *Tree Crops: A Permanent Agriculture*. New York: Harcourt, Brace and Company.

Swearer, D.K. 2006. 'An assessment of Buddhist eco-philosophy'. *Harvard Theological Review* **99**(2): 123–137. CrossRef

Thich Nhat Hanh. 2008. *The World We Have: A Buddhist Approach to Peace and Ecology*. Berkeley: Parallax Press.

Trainer, T. 2000. 'Where are we, where do we want to be, how do we get there?'. *Democracy & Nature* **6**(2): 267–286. CrossRef

Trainer, T. 2002. 'Debating the significance of the global eco-village movement: a reply to Takis Fotopoulos'. *Democracy and Nature* **8**(1): 143–157. CrossRef

Tuan, Y.F. 1970. 'Our treatment of the environment in ideal and actuality'. *American Scientists* **58**(3): 244–249.

Uddin, S.N., R. Taplin and X. Yu, 2007. 'Energy, environment and development in Bhutan'. *Renewable and Sustainable Energy Reviews* **11**(9): 2083–2103. CrossRef

Veteto, J.R. and J. Lockyer. 2008. 'Environmental anthropology engaging permaculture: moving theory and practice toward sustainability'. *Culture, Agriculture, Food and Environment* **30**: 47–58. CrossRef

Vogt, G. 2007. 'The origin of organic farming', in W. Lockeretz (ed.), *Organic Farming: An International History*, pp. 9–29. Cambridge, Mass.: CAB International.

Welchman, J. 1999. 'The virtues of stewardship'. *Environmental Ethics* **21**(4): 411–423. CrossRef

Xu, P. 1997. '*Feng-shui* as clue: identifying prehistoric landscape setting patterns in the American Southwest'. *Landscape Journal* **16**(2): 174–190.

Xu, P. 1998. '*Feng-shui* models structured traditional Beijing courtyard houses'. *Journal of Architectural and Planning Research* **15**(4): 271–282.

York, R., E.A. Rosa and T. Dietz. 2003. 'STIRPAT, IPAT and ImPACT: analytic tools for unpacking the driving forces of environmental impacts'. *Ecological Economics* **46**(3): 351–365. CrossRef

In Search of Arcadia: Agrarian Values and the Homesteading Tradition in the Ozarks, USA

Brian C. Campbell

Elsewhere in Arkansas the latest blooming hippies have all cleaned up and moved back to the suburbs. Those who persist and endure in Newton County are the strong ones, fit survivors, like the real pioneers in the nineteenth century, who came as a kind of spillover of the mountain settlement to the east.

Donald Harington (1986: 98–9) in *Let Us Build Us a City: Eleven Lost Towns*

INTRODUCTION

Homesteaders come to the Ozarks because the region provides key ingredients for relatively self-sufficient living: cheap land, abundant water, and diverse flora and fauna for foraging. There have been waves of homesteaders who arrive in search of freedom, a piece of land that they can own outright to provide for themselves. This continuous homesteading tradition from the eighteenth century through the present distinguishes the region from others in the United States. Many such immigrants lasted only a short while in the rugged landscape (Blevins 2002); therefore those who remain receive minimal treatment in the academic literature. The extant back-to-the-land contingent represents a small percentage of the overall Ozark population and purposefully makes itself difficult to monitor and document, which contributes to the paucity of research on the movement. While the back-to-the-land phenomenon in the Ozarks has been overlooked, an examination of this long-standing trend provides insights into the cultural ecology of the region, and potentially a glimpse into the ethnobiology and motivations of the earliest settlers in the Ozarks.

Rather than exceptional or anomalous immigration events, the contemporary back-to-the-land movement and 'Old Stock' frontier settlement represent distinct but related waves of an Ozark homesteading tradition by peoples who share agrarian values, practices and motivations. These commonalities allowed traditional Ozark homesteaders (Old Stock) to survive and back-to-the-landers to endure in the Ozark Highlands, and served to conserve agricultural biodiversity (agro-biodiversity). I will present the media representation of Ozark culture and landscape that influenced these two Ozark homesteading populations and explore their shared agrarian values: self-reliance, frugality and ecological, physical (human) and spiritual health. I will conclude with discussion of how sustained agrarian values relate to the conservation of regional agro-biodiversity and agro-ecological knowledge. Archival and ethnographic research in multiple field sites in the Missouri and Arkansas Ozarks from 2002–2004 and 2006–2013 informs the discussion.

THE OZARK HIGHLANDS, USA

Geographers distinguish the Ozarks, situated predominantly in the southern half of Missouri and the northern half of Arkansas, by the general rugged verticality of the landscape caused by karst topography (Rafferty 2001). As groundwater filters through the hillsides, it dissolves the soluble rock (dolomite and limestone) that constitutes the majority of the Ozark landscape, simultaneously ushering many critical nutrients out of the reach of the roots of agricultural crops (Aley 1992). Land cover in the Ozarks relates directly to these karst effects; unlike the neighbouring Midwest's expansive agricultural monocultures, smaller diversified farms, cow-calf operations and forestry dominate the Ozark landscape (Campbell 2009). The Ozarks' native forests consist of oak-hickory-pine combinations with interspersed cedar glades, and 'high-quality water resources' abound in the form of streams, lakes and human-made reservoirs (Rafferty 2001: 1).

OZARK SETTLEMENT HISTORY

Ozark settlement history has been summarised in three general phases: 1) Old Ozarks Frontier from late eighteenth century to approximately 1860; 2) New South Ozarks from Post-Civil War until World War I; and 3) Cosmopolitan Ozarks from World War I to the present (Rafferty 2001). During the 'Old Frontier' settlement phase, Native American groups (Cherokee, Choctaw, Delaware and Shawnee primarily) and Euro-American frontier families from the Upper South Hill Country relocated from the eastern US to the Ozarks (Blevins 2002, Blansett 2010). The frontier families were economic generalists who cleared patches in the virgin forests to plant corn, legumes and cucurbits and relied on hogs and wild game for their meat. The propagation and foraging of wild and domesticated medicinal and culinary herbs, greens, berries, and shrubs constituted a major component of the early homesteader diet (Massey 1979, McDonough 1975). While considered destructive by some researchers (Otto and Burns 1981: 178), free range livestock, patch farming and foraging allowed self-sufficiency in a sparsely populated frontier region (Otto 1985).

Ozark immigrants during the second phase, the 'New South' (Reconstruction), differed from the previous settlers because they arrived along the newly constructed railways (sometimes with 'carpetbags'), established towns with economic linkages outside the region and engaged in commercial agricultural,

Figure 1. US Forest Service Historical Photo Collection. Photo Credit: J.M. Wait.

mining and lumbering enterprises unlike the more agrarian Old Stock homesteaders (Rafferty 2001). The traditional semi-subsistence economy continued in the more rural Ozarks during this period, and many new arrivals homesteaded in the same tradition of their Old Stock predecessors in regions previously unclaimed or now abandoned because of the agricultural marginality of the lands (Otto and Burns 1981). A poignant image in a US Forest Service historical photo collection from 1928 pictures a derelict house with a USFS photographer's caption: 'No, this home was not destroyed by a tornado. It was wrecked by poverty induced by an attempt to wrest from nature, lands never designed by her for ag[riculture]' (Wait 1928).

The 'Cosmopolitan' phase (WWI – Present) begins with the exposure of the isolated regions of the Ozarks to modern life (Rafferty 2001). Young men from remote communities, mostly unaware of the relative extravagance and technological complexity of urban centres, became exposed to the non-Ozark world through military service (Blevins 2002). Railroads and the automobile also established previously non-existent direct interaction as tourists begin exploring the region and Ozarkers travelled to cities for work (Rafferty 2001). Out-migration characterises the middle of this phase. As industrial agriculture began to dominate food production throughout the US, Ozark farmers lost access to their local agricultural economy. Therefore, locals began to emigrate to cities to find work.

Despite steady Ozarker out-migration, a continual flow of 'returnees, escapists, and opportunity seekers' moved to the Ozarks throughout the twentieth century (Rafferty 2001: 70). 'Returnees' refers to locals who moved away and came back to their Ozark birth homes to retire or take over an inheritance that would allow them to live. 'Escapists' is a somewhat ethnocentric reference to disillusioned young people who began moving into the Ozarks, presumably from urban areas, in search of an elusive agrarian lifeway, part of the early waves of back-to-the-landers. These 'escapists' represent the focus of this research, a sub-population that remains under-studied. My interviews with this population reveal a fascinating encounter with the traditional Ozark population; when these back-to-the-land 'escapists' arrived in the Ozarks, many traditional Ozarkers had succumbed to modern derision and consumerism, so their simple living ideology and behaviour was not well-received, except by the oldest, stubbornly traditional Ozarkers (Fellone 2010b). As one back-to-the-lander (BTTL) candidly relates:

> The Ozarks as a place was just modernizing when we came in here ... as home-steaders. We were interested in working in old time ways and growing stuff and heating with wood and not using money ... and bartering and canning and saving stuff, using old seeds. All of that stuff in the Ozarks was going out of fashion; people were moving to town, getting flush toilets, getting modern houses that were energy inefficient, getting ... new trucks and new cars and new stuff 'cause there was more money ... |Our group| primarily |was| constituted of people that had a homesteader mentality that was becoming passé in terms of fashion, lifestyle in the Ozarks, early '70s ... The people that we wanted to talk to, who were interested in us, were the Old Timers. The old people, in their 60s, 70s, and 80s ... those are the people that were there that thought we were ... kind of all right, even though we were weird. And we liked talking to them 'cause they knew stuff that we wanted to know.

ARCADIAN OZARK IMAGERY IN POPULAR MEDIA THROUGH THE MID-TWENTIETH CENTURY

Throughout the twentieth century, diverse media have disseminated images of backwoods Ozark homesteaders as representative of the general Ozarks, result-ing in a paradoxical mystique of freedom and Arcadian independence, but also backwoods (backwards) primitivism (Rafferty 2001, Blevins 2002). Neither of these extreme images fully or accurately depicts the Ozarks, but waves of immigrants have arrived with such Ozarkian myths in mind. Historical analysis stresses the inaccuracy of popular media (comics, films, television shows) ste-reotypes (Blevins 2002) and archaeological research debunked the 'traditional Ozark myth' (Stewart-Abernathy 1986) of a completely isolated, self-sufficient, distinct Ozarkian 'other' (Blevins 2002). Most Ozarkers of the late nineteenth and early twentieth century were not *completely* self-sufficient, but the isolation of rural Ozark homesteads necessitated self-reliance and ecological awareness.

The non-Ozarkian public's fascination with this particular subset of Ozark life says something. As researchers (and interviewees) indicate, the Ozark stereotype was paradoxical; it was simultaneous nostalgic romanticisation of an agrarian past and disparagement of the non-capitalistic (lazy), non-materialistic (barefoot and ragged), non-modern (backwards and poor), non-Northern (bushwacker, Confederate) lifeway of the traditional Ozarker (Blevins 2002, Brandon and Davidson 2005). While the latter (pejorative) was meant to display the superiority of the modern, urban, technologically advanced Yankee society over backwoods primitivism, the 'hillbilly's' laid-back, relaxing, egalitarian image appealed to many city people tired of the cosmopolitan rat race (Harkins 2004).

Many Ozarks promoters, folklorists, and travel writers celebrated the agrarian character of the region, exaggerating with their Arcadian evocations (Rayburn 1949, Blevins 2002). This literature attracted tourists and back-to-the-landers (BTTL) who sought beautiful wilderness and traditional Ozark culture (Blevins 2002). The first phase of back-to-the-land settlement in the Ozarks emerged out of influences from the Country Life and Arts and Crafts movements (Blevins 2002:128). As the Western frontier closed, academics and government bureaucrats echoed Theodore Roosevelt's Jeffersonian concerns about the moral fibre of the US population as more and more farmers abandoned the plough for the urban life (Shi 1985). The Arts and Crafts movement similarly oriented people towards a celebration of rural simplicity and wilderness (Cumming and Kaplan 1991). Images of impoverished Ozarkers, 'of log cabin homesteads inhabited by broad-brimmed hat-wearing, barefooted moonshiners and wrinkled women weaving homespun', circulated during the Depression and attracted writers to document this 'semi-arrested frontier' (Blevins 2002: 120). Donald Harington (1975: 255), an Ozark novelist who integrated extensive regional history and folklore into his works (the 'Gabriel Garcia Marquez' of the Ozarks), playfully references the flurry of back-to-the-land literature that appeared during the early and mid-twentieth century:

> These people had spent all their lives in the cities, laboring in business and industry, saving their pennies, and dreaming of a better life. There appeared on the newsstands of the cities a rash of new magazines, *Country Life in America, America Outdoors, Rural Digest, Arcadian Times, Hill and Dale, Silvan Weekly, Ladies' Bucolic Companion, Back-country Journal, Pastoral Pictures*, extolling the healthful benefits of a return to the soil.

Several key writers described the Ozark culture and region with Arcadian imagery in such publications. Charles Morrow Wilson, a relatively urban Ozarker from Fayetteville, Arkansas with family connections in the rural backwoods, wrote about the Ozark region for the *St. Louis Post Dispatch, New York Times, Reader's Digest, Saturday Evening Post, The Atlantic*, and *American Mercury*. His celebrations of the Ozarks as a unique culture and landscape, as seen in

Brian C. Campbell

Figure 2, gained wide audience. Wilson's 1934 article, *Ozarkadia*, introduces an idyllic agrarian paradise with a mix of back-to-the-landers and Old Stock Ozarkers who maintain neighbourly relations. He posits that the 'frontier temperament' of the region and the fact that four-fifths of the landholders live from the soil make it 'unique and outstanding' (Wilson 1934: 59). Otto Ernest Rayburn promoted the Ozarks' 'Arcadian' characteristics in a series of publications: *Ozark Life: The Mirror of the Ozarks*, (1925–1931), a column, 'Ozark Folkways', for the Sunday *Arkansas Gazette* (1927–1935), the column 'Ozark News Nuggets' in the *Tulsa Tribune*, *The Arcadian* (Magazine) 'A Journal of the Well-Flavored Earth' (1931–32), *Arcadian Life* (1933–1942), and *Rayburn's Ozark Guide* (1943–1960). Blevins (2002: 223) asserts that 'almost no one in the Ozarks – especially not Ozark natives' read *Rayburn's Ozark Guide*, but it

Figure 2. Charles Morrow Wilson's Ozarkadia.

'would lure more tourists and newcomers to the Ozarks than any other publication until Rayburn's death in 1960'. A 1949 edition lauded the agricultural possibilities in the Ozarks:

> There are numerous agricultural opportunities here, but the farmer should make a study of his land and possibilities for marketing before deciding on a specialty. It is well to consult the county agriculture agent of your county and to get material from the extension department of the State University. Because of the many cheese and milk plants throughout the Ozark region, dairying is now a very profitable industry. But there are many others to be considered ... 1. Karakul Sheep ... 2. Bees and Honey ... 3. Watermelons ... 4. Ginseng Culture ... 5. Roses ... 6. Mushroom Growing ... 7. Flowering Plants ... 8. Poultry ... 9. Strawberries

Marguerite Lyon wrote a column, 'Marge of Sunrise Mountain Farms', in the *Chicago Times* that reportedly attracted thousands of Chicago BTTLs to settle in the Ozarks (Blevins 2002: 479). In addition to her columns, she published several influential books (*Hurrah for Arkansas*, *Fresh from the Hills* and *Take to the Hills*) that recount her Ozark homesteading adventures, and these surely encouraged many back-to-the-landers to seek out the Ozarks.

BACK-TO-THE-LAND IN THE MID-LATE TWENTIETH CENTURY OZARKS

> There were so many people moving into the area in the early seventies that you could drive around and just see a Volkswagen bus with California plates or Wisconsin or somewhere just kind of driving around, and you would think, 'I bet they're looking for land.' And they'd maybe stop at a little country store, and you'd pull in and start talking to them and, yes, they just got here and they're looking for land. And it just got to be funny. (BTTL Male)

By the mid-twentieth century, the folk music and counter-cultural movements encouraged innumerable young people into rural areas to explore their agrarian roots. From the 1960s through the 1980s, the Ozarks became a key destination because of the 'authentic' folk music scene: local Ozark musicians who played town squares and folk festivals. After such outings in the Ozark hills, the low population density, agrarian culture, clean, clear streams and cheap land attracted many visitors to return to live on their own farmsteads or in communes with friends and acquaintances. The following quotations from back-to-the-landers who remain in the Ozarks provide insights into their motivating factors, thought processes and circumstances, and the agrarian values they share with other Ozark homesteaders that I will discuss in forthcoming sections:

Male: My becoming part of the back-to-the-land movement was quite intentional, which is not to say it was well-planned .... We were beginning to see *Mother Earth News* and ... all sorts of ideas about self-sufficiency and living appropriately on the land, and gradually it just seemed like that would be a good thing to do ... They [friends] ... knew, in terms of land price, that it was either Nova Scotia or The Ozarks, and the Ozarks sounded warmer ... I decided to throw in, and some other friends from the Iowa City area did also ... Our intention was to form a sort of community, not a commune in any way. But initially we owned this one-hundred and twenty acres together, equal shares, but we picked our own house site. There were no buildings on it. There was one electric pole. That was it.

Female: I was in college in the sixties in the Deep South, and we didn't get no hippies down there. I taught school for eight or ten years, and I realized at that point that I loved the kids, but I wanted to rip the heads off of the administration and spit in their necks ... I was traveling with a friend who was teaching on a reservation at Pine Ridge, and we picked up ... the first issue of the *East West Journal*, and it had a little paragraph in it about a community in Missouri that was looking for new members ... At that point I'd decided I wanted to move to [a] community. I'd read about communes. I thought they were fiction, but it sounded so cool. I'd never had brothers and sisters ... I called East Wind [anarchic intentional community in the Ozarks] and they said, 'Oh yeah, come right on.' So I did, and it was exactly what I had needed at that time. I stayed there for four years. [Then] I moved into a place that actually was back-to-the-land ... growing all of their own food ... We didn't even buy white flour. We bought wheat berries and ground our own. We didn't buy sugar. We had bees. We just didn't have enough land. We only had twenty-five acres, and it was not enough to support our dairy herd. We had five or six cattle, and so we looked on and off for several years for a bigger place. We wanted about one-hundred acres, we thought. Well, we found some more old back-to-the-landers who were selling real estate at the time ... five-hundred acres. That's not what we'd had in mind, but it was so wonderful. It had this huge spring. It had half a mile of river. It had *some topsoil* ... So, we bought it. Then I went to the oil fields as a ship's cook, and stayed there for five or six years just cranking the money machine, and during that time the community sort of fell apart because part of the people were here, part of the people were there, and nobody was really communicating with each other.

Male: When we moved to The Ozarks in 1973, we didn't know what we were doing or what exactly we were getting into. We didn't have big aspirations to live on the land and be self-sufficient. The truth is, we didn't have anything better to do ... We were able to stay with some new friends on a hill just outside of town. They were from Detroit, very relaxed about everything, and made us welcome. We were just sort of hanging out there about a week later, when a man drove up in a pick-up truck and said they were all going to a party. We piled into the back of the truck on a beautiful summer day ... That was the day my life changed. The party was ... a summer solstice party. I guess there were about two hundred people there. There was a big potluck meal; people jumping in and out of the creek in various amounts of clothing, and there were people playing music beneath a big shade tree. [A young man] sang Muddy Waters' tune, 'Five Long Years', and he put his heart and soul into it. The Bryant [Creek] was wild and beautiful. There was a big garden, and a hand-made stone house, and the most interesting people I had ever met just doing all sorts of things. I'd never seen anything like it ... About a year later we bought some land ... [and] put in a big garden and swam the creek and generally sat out to live the dream. We hardly had any idea what exactly we were doing, but we knew we wanted to be doing it. We were twenty-one and having fun.

OZARK ARCADIA AS BACK-TO-THE-LAND MYTH: RUGGED REALITY

In the heydays of the late twentieth century Ozark back-to-the-land movement many young people showed up for the party; some had serious intentions of trying to homestead in the region, but many were easily discouraged and departed. Blevins (2002: 200) emphasises that the 'vast majority stayed only a short time before being driven back to civilization by snakes, chiggers, heat, cold, and starvation'. A BTTL relates the ephemerality of many such ventures:

> On a very regular basis one of the [mail] carriers would come in and say 'well, so and so moved in to this place with their back-to-the-land movement. They're going to live off the farm.' Sad thing about it, nearly every one of them, when they moved in had nothing but themselves. They had no equipment, no livestock, and very little knowledge most of the time of what they were getting into, and as a result very few of them lasted more than two or three years, and sometimes they would just disappear. Sometimes they would just sell what little they acquired ... She [indicates his wife] worked in the library, and she said they just wore out the how-to books on how to farm. You'd be surprised how many we had move into this territory that had the idea that if I could just get back to the land, everything is going to be okay, and they were quite disillusioned when they found out, 'hey, it ain't so', found out they were having to work thirty-six hours a day just to exist, and most of them finally gave it up and said 'no.'

A 1975 *Mother Earth News* article attempted to disabuse other unprepared 'escapists' of self-sufficiency illusions in the Ozarks:

> Since offering a piece of land for sale, I've received scores of letters from sincere, good people who want to live simple, natural lives in the seclusion and beauty of a wild forest area. I agree with these folks' motives, and wish them all possible luck ... but the tone of their correspondence reveals so much unrealistic wishful thinking that I doubt their chance of success ... In most cases it's a myth that a person can settle in the depths of the country and maintain himself there ... unless he or she is willing to devote him or herself to continual hard work, with

Brian C. Campbell

little time to enjoy life or pursue whatever mental and cultural pleasures he or she favors. Only the rich can lead full, beautiful lives in a really wild area. Most others are enslaved to endless drudgery and penny-pinching. Think hard and realistically before you make the leap (Durand 1975).

BACK-TO-THE-LAND BIBLE: *MOTHER EARTH NEWS*

Mother Earth News published a series of reports from the Ozark region in the 1970s and 1980s about homesteading possibilities and showcased advertisements for land in the area. The publication embodied the ethos and motivations of many late twentieth century back-to-the-landers, encouraging them to seek out the Ozarks for their Arcadian homesteads, and providing practical guidance for self-reliant living. During research with back-to-the-landers in Ozark County, Missouri who initiated the Ozark bioregional movement, I learned of a vinyl 45 record that unambiguously demonstrates the influence of *Mother Earth News* on back-to-the-landers' preconceptions of the Ozarks. Ron Hughes, a talented musician and energy efficiency expert, wrote the song, *Ozark Mountain Mother Earth News Freak*, and played it with his band *Hot Mulch* on the record.[1]

Lyrics:

I'm moving to the country where everything is fine
 gonna live in a dome and drink dandelion wine
 when the collapse comes I won't get the blues
 I'll have all the back issues of the *Mother Earth News*.

I'll get myself a sweetie and a Volkswagen Van
 See the real estate man and buy me some land
 a few acres cleared with lots of trees and peace
 that we can fix up however we please

We'll get our eggs from chickens and milk from a cow
 a horse that plows and a book that tells how
 an organic garden growin' comfrey and peas
 we'll get honey from our bees and fruit from our trees

Self-sufficient that's the name of the game
 gonna get myself a system self contained,
 a wind mill to give me my electricity
 no phone in my dome I'll use ESP

No more Coca-Cola, gonna stop eating trash
 get into planting, gonna grow my own stash.
 Plant by the moon and talk to my plants
 consult the I Ching and learn to do a rain dance.

Get into harmony with Nature and the Universe
    I'll do Yoga in the morning if my back don't get worse
    Red Zinger Tea and Vitamin B-6
    One keeps me high, the other keeps off ticks.

I'll get a Solar Air Heater from New Life Farms
    in the Winter time it will keep me warm
    my methane digester's a great big hit
    It makes methane gas out of Chicken manure

If I keep my faith in the Organic ways
    maybe some day my tomb stone might say
    we shredded his body, it turned to compost in a week
    he was an Ozark Mountain *Mother Earth News* Freak.

The song's content demonstrates the influence of *Mother Earth News* and counter-cultural trends, but also BTTL adoption of (or at least familiarity with) traditional agrarian practices of the region: planting by lunar signs, ploughing with horses and localised self-reliant food production (eggs, milk, honey, fruit) (Campbell 2009).

BACK-TO-THE-LAND AND OLD STOCK INTERACTION

By the twentieth century, self-sufficient homesteading in the Ozarks was more difficult than ever, because the best lands were unavailable and timber and mining operations had devastated most inexpensive remote lands (Harington 1986, Blevins 2002: 128). Vance Randolph (1951: 27) documented a local folktale that presents a relatively common sight in the twentieth century Ozarks:

> a village realtor who was trying to sell a rocky little farm ... explained to the sucker that flint-rocks are necessary to productivity, and that land is no good without 'em. Stones retain moisture, prevent erosion, keep vegetables from getting dirty, and so on. Just then a man began loading some rocks into a wagon, to haul them off the field. 'Let's get out of here', said the land agent. 'We might get tied up as witnesses in court. That fellow is stealing those rocks!'

This local humour pokes fun at newcomers' naiveté regarding soil productivity, but also land speculators' unscrupulousness. Many BTTLs were misinformed and ill-prepared for self-sufficient living in a marginal landscape and they did not possess the traditional ecological knowledge necessary to survive on their own (Fellone 2010a). But not all BTTLs fled the Ozarks; some stayed and learned from their experienced, knowledgeable neighbours (Fellone 2010c). An Ozark BTTL eloquently reflects on his experiences with old-time Ozarkers:

> The people who lived around there regarded us with a curiosity. They were mostly

old people. The [nearest] highway ... was only about ten years old. The folks there had electricity for less than twenty years at that time, and possibly a lot less. Those older folks were the kindest, sweetest, and possibly the nosiest folks I had ever met. One old couple that lived a couple of miles away stand out in my mind as exactly the kind of folk anybody would want to grow up to be. They lived in a cute little house on their extensive cattle farm. They had a Willy's pick-up truck that they must have bought new from God at the creation day. Once a week the Mrs would drive into town while the mister would sit in the passenger seat with one foot on the dashboard with his slouch hat pulled low over his eyes. That was all the driving they ever did. Most days I could see the old man walking around his various cattle pastures spread out around the countryside. He would always stop and talk. He had clear blue eyes that kind of crinkled when he was pulling your leg. He reminded me of a Leprechaun, and not in any kind of Saint Patrick's Day way either, but in a very real and elemental way.

We had the good fortune to meet a number of people who were probably the last of their time. They were the real thing. They were old-timers. They grew up when the country around there was still pretty wild. For the most part they had been born in the last century. They walked to school when the century was young, and they worked Missouri Jacks [mules] and raised hogs and looked after one another. In winter, when everything was covered in snow, they went visiting and played checkers. They formed a tight but spread out community. They didn't have telephones most of their lives, or televisions, or roads or cars. They hitched rides into town on log trucks and often walked the fifteen or twenty miles home. They were more a product of the nineteenth century than the twentieth. They were *good ole* Ozarkers through and through. They were tough and hardy, stubborn and proud, religious and terrible sinners. I could get into trouble for making generalizations because every one of them was singular. I don't mean to romanticize them either. They were a crusty bunch, and if they didn't take to you, you knew it. We got taken advantage of a few times, and considered it a part of our education, and I'm sure that's how they looked at it too.

They were all very interested in our little band of beautiful hippies. They stopped by to visit and wanted to see the garden and brought fruit pies and canned goods to our shacks. They worried about us. They invited us to church, but all of our neighbors went to either one of two churches, and we thought it best not to take sides. Most of their adult children had moved away to find work. I think the old-timers were touched by our unabashed love of the land, the creeks, the rivers, the deep woods, the rolling hillsides, the bluffs, the caves. They always waved when they saw us out picking blackberries. They could see we were a little better than fools, but they never seemed to judge us. They accepted us, and they loved that we gardened, and played music, and had time to talk to them. They didn't always have a lot of patience with us, but they always had time for us. They taught us a lot about community, about survival, and about living in the Ozarks, most of all just by example. That was a privilege getting to know those folks. They were a link to another time, and we won't see their like again

… Every back-to-the-lander I've ever met has a similar story to tell about old timers who took these odd young under their wing, who were their bridge to the future, just as they were our bridge to the past.

AGRARIANISM AND OZARK HOMESTEADING

Agarianism plays an important symbolic, mythological role in the history of Western life, especially in the United States (Freyfogle 2001). The Latin root *agrarius*, 'pertaining to land', anchors the values of an agrarian. The agrarian knows the land (ecology) must be taken care of to ensure the economic, spiritual, and physical health of the humans living on it. Smith (2003:17) characterises US agrarianism dichotomously: the democratic, characterised by hard work, frugality and self-reliance, and the aristocratic, leisurely, intellectual, poetic, even theoretical engagement with the natural world. The two are not mutually exclusive, however; many BTTL and traditional Ozarkers demonstrate a fluid engagement with both realms – working hard with their animals and in their fields by day, and then writing or singing about the beauty and passion inherent in their agroecological pursuits in the evening or in the winters when the workload lessens.

This distinction between threads of agrarianism provides insights into the differences between the back-to-the-landers who came and left, and those who remained. This research focuses more on 'democratic' agrarianism, whereas those who were 'driven back to civilization' subscribed more exclusively to 'aristocratic' agrarianism (Blevins 2002: 200). Those who left did not realise the hard work that would be required of them and the inhospitable conditions they would have to endure. They may have romanticised or desired to commune with nature, perhaps because of the exaggerated descriptions in 'Arcadian' publications, but did not understand the difficulties and requirements of true farm life. Some, through hard work, became agrarian Ozarkers because they were willing and enjoyed such work with nature. As Donald Harington (1986: 98–9) noted (introductory quotation), BTTLs who continue to farm and forage in the Ozarks share an agrarian foundation with Old Stock homesteaders.

SELF-RELIANCE IN A LAND OF MILK AND HONEY

When Old Stock Americans homesteaded the Ozarks they were looking for cheap or free land on which they could make a living. Frequently one or a few members of a family preceded the rest of the group and established a homestead. In letters sent back to family they described the Ozark land, frequently in glowing terms to entice family to join them. Such letters are likely some of the first written Arcadian descriptions of the Ozarks, predecessors to Rayburn's Ozark promotions. As one such letter from a nineteenth century homesteader sent back East related about the Ozarks: '[there is] … rich valley land that could be had for the taking, where wild game was so plentiful, and wild honey was in abundance' (Bromley 1916: 18). Later BTTL accounts sounded remarkably similar about the potential of the Ozarks for homesteading. Marguerite Lyon (1941: 39) wrote:

> Lucille and her husband, Clair, had just returned from the Ozark Mountains in Arkansas, where they had gone to buy a truckload of peaches to sell in Iowa towns. They had fallen head over heels in love with the wooded hills of the Ozarks. They talked of the Ozark region as the Land of Milk and Honey … the place where the climate is forever mild and pleasant, where chickens and sheep can find their food the year round, where millions of cords of wood can be had for the cutting, where taxes are too small to talk about, and where you can buy a farm for practically nothing. They were eager, yes, wild, to get back there to a farm and live forever in the hills.

Rafferty (2001) cites self-reliance as one of the foundational Ozarkian characteristics. A traditional farmer from Madison County, Missouri, who continues his family's agrarian traditions, explained his grandfather's Jeffersonian ideals:

> My grandpa thought farming was the only way to make a living. That's all he ever did and he thought that everybody ought to do it. Your dad could say; 'my son is going to college' … and he'd say, 'oh, that's nice, it's a shame he's not farming, though'. See, that was the way he looked at it. I have two cousins … They're real nice, educated, great guys. But my grandpa … said; (emphatically) 'Aahhh, they're soft. They couldn't do a days work if they wanted to.' And mom said, 'Well, grandpa, not everybody makes a livin' with your back, ya know. They're makin' their living with their education.' And grandpa says, 'Hummph.' That didn't mean nothin' to him. He thought farmin' was the only way to go. That's why he didn't want me to leave the farm. He would have provided whatever I needed to keep me on the farm. He always said, 'You never got anywhere carryin' a dinner bucket.'

FRUGALITY

As many BTTL and Old Stock Ozarkers have related, one must not live beyond one's means. Otto Ernest Rayburn (1937: 4) waxed philosophical about the interconnections between health and happiness and Ozarkers' frugality: 'The Arcadian philosophy … emphasizes the common sense of pastoral simplicity as a way of life … If we could reduce our luxuries one-half, this utopian dream would become a reality in our generation … America would sing a new song of health and happiness, an American song, dedicated to the Well-flavored Earth.'

James West (1945: 125), in an ethnography about the Ozark community of Wheatland, Missouri, documented a common perception held regarding the 'hill people': 'Them's the happiest, cheerfulest people in the world because they don't want anything more'n they've got.' While this comment sounds ethnocentric, and it very well may be, when studying traditional Old Stock descendants in the region, the characteristic frugality emerges in the affirmative attitudes and resourceful tendency to reuse everything possible. In an ethnographic work by L.L. Broadfoot (1944: 160), a female informant explains:

> The sort of material I use is old worn-out overalls, shirts, underwear, stockings, scarfs, and all kind of waste clothin' that people have throwed away; and I set up late at night and cut 'em into narrow strings and sew 'em together, and with this little hook you see in my hand I knit the rug, and here's one I'm jist finishin'. I have even made all the furniture I have in my house – bedsteads, chairs, clothes chests, flour chest – and every bit of my furniture is my own handmade stuff.

In an anthropological study, Parker (1992: 155) documents the similar frugality of the countercultural back-to-the-landers:

> As a group, these 'hippies' are often viewed by insiders with suspicion (as marijuana growers) and/or amusement/puzzlement (for subsisting with few modern conveniences 'when they could do better'). As 'real' individuals, they may be respected for their skills and sense of community, and accepted at least as well as other outsiders.

PHYSICAL HEALTH AND WELL-BEING

Early settlers worried about the 'health' of the land and their families, especially when it came to the presence of malaria (Valencius 2001). Elevation served as an indicator of health in many cases, as did the presence of natural springs (Valencius 2002). In the 1840s the name of Buncombe Township in Independence County, AR changed to Healing Springs in recognition of its perceived healthfulness. While some Independence County settlers chose valley land because of the relative fertility of the soil, they found it to be 'polluted with mosquitoes and malaria' (Clark and McGary 1987: 29). Jacob Kever wrote to relatives back in North Carolina in 1878: 'We live in a broken rough rocky part but is healthy as much so as where you live. I can tell you that the river bottom is for wealth and the ridges for health. We have good water here as any in these parts'. (Clark and McGary 1987: 29).

By the 1960s and 1970s the concern with health related to the industrial contamination of many urban areas and the Ozarks presented a relatively pristine environment for back-to-the-landers. An Ozark BTTL explains his motivations as

Brian C. Campbell

Figure 3. Ozark National Forest, 6/1914 'Typical valley farm. Cotton in foreground. House built on hillside to avoid malaria'. Image Credit: Huey, RC.

they relate to physical and ecological health and the political economic practices degrading US society and ecology:

> I hail from Cleveland, Ohio, born in the year 1957 [when] Cleveland was still home of the century long iron ore coke and chemical processing manufacturing base. In the year of 1969 the Cuyahoga River burned ... In 1974 I was moved to a Philadelphia, Pennsylvania suburb. Philadelphia was a century long steel and processing manufacturing base . In 1979 I moved myself to San Francisco, California, and became involved in the radical political movements at the time, which promoted land and nature based solutions as a guide to betterment of the planetary situation ...
>
> At the time I did not know how much these environmental, political, and social events would shape my future. Sometime in 1983 I moved myself to the Ozarks of Missouri. The air, water, and earth quality of the Oregon, Shannon, Howell, Texas, Douglas, and Ozark counties of Missouri was as environmentally clean as anywhere one might find in the contiguous United States of America. I left after two years not having fully the skills necessarily to live rurally. I continued to live mostly rurally for the next twenty years in various areas of the US. Once again in 1995 I returned to this environmental anomaly called the Ozarks to reside permanently in this region. I built a small passive solar cabin

from mostly recycled materials that are byproducts of the local forest industry and other scavenging efforts. Of course there was the perfunctory and necessary garden that usually gives me nine months of fresh produce, also the use of wood heat at the cabin which is plentifully abundant here. The outhouse is to recycle human waste instead of as usual adding it to water to contaminate the water itself.

SPIRITUAL AND ECOLOGICAL HEALTH

Traditional Ozarkers considered themselves more virtuous than their city counterparts; the absence of work with the natural world corrupted, causing excessive greed and human temptations (Randolph 1931). Vance Randolph (1931: 299) explains:

> The old-time hillman is impressed not a whit by the fine clothes and big motor-cars of the invaders, for he knows perfectly well that nearly all tourists are salaried people – 'nothin' but hired hands' – while he himself is a landed gentleman despite his rags, and 'don't take no orders from nobody'. Like most rustics, the Ozarker regards all cities as sink-holes of iniquity, and believes that all city-dwellers are grossly immoral.

BTTL Ozarkers believe that the natural world and local ecology deserve respect and in most cases perceive it as sacred. Many have sought alternative training or formal education in ecological sciences that provide them case studies about the interconnections between species and their services, functions, and uses. Similarly, West (1945: 192) says of traditional backwoods Ozarkers: 'knowledge of nature is more respected among this class than any other.'

Research conducted with traditional farmers in the Missouri Ozarks demonstrated that multi-generational Ozarkers who continue in traditional farming practices believe their biblical interpretations compel them to be stewards of the land (Campbell 2009). A farmer illustrates:

> Farming dates back to biblical days and most of your illustrations in the bible are based on farming and in an orchard or vineyard or something like that. That's the way they explain things: stewardship. I think a lot of the problems in our country nowadays is because people are raised in the city where they are out of contact with nature. God sets up nature to reproduce and cleanse itself and recycle; everything recycles. Well, the people raised in the city don't understand that because they've never had a chance to.

BTTLs leave the city for the Ozarks because they seek physical and spiritual health through direct connections with the natural world, which they perceive as unattainable in an urban environment. As a BTTL male recounted:

> We just wanted to get as far away from the city as we could. We looked at the map

and here was Memphis, and here was St. Louis, and here was Kansas City, and we looked for the place that was farthest away from all of those. It happened to be around here. You know at that point, late '60s–1970, looking again ecologically at what was going on in the economy and everything else, none of it made any sense. It didn't make a bit of sense and it still doesn't. It just seemed ludicrous; everything that was going on seemed absurd to me from that perspective, so I decided I had to get as far away from it as I could – somewhere where there was some ecological integrity left.

BTTLs and Old Stock Ozarkers share a disdain for urban lifeways and a desire to take care of a piece of the Earth and ensure that future generations have the opportunity to enjoy and be nourished by it. An Old Stock Farmer from Madison County, MO explained:

He was ... about 85 years old then and he was reading a couple magazines and I said to him, 'What are you reading?' He goes, 'I'm ordering some fruit trees.' I was looking at him and I said, 'Uncle Rod, you're 85 years old, why are you ordering fruit trees? Do you think you're gonna live long enough to eat fruit off of those trees?' And he said, 'I'm 85 years old and I've been eating fruit off of trees all of my life since I was a little boy and now it's my turn to make sure other little boys get to eat off of fruit trees.'

Many BTTLs express concern about the current mistreatment of our natural resource base, citing Native American concern seven generations ahead, which meshes quite seamlessly with the values of the Old Stock settlers of the Ozarks.

AGRICULTURAL BIODIVERSITY

I became interested in and aware of similarities between traditional and back-to-the-land Ozarkers through my research interest in seed-saving. When I first began investigating Ozark agro-biodiversity I stumbled upon an internet seed exchange developed by a BTTL who had been living off-the-grid for ten years in an extremely remote section of the Ozarks. She had an extensive collection of open-pollinated seeds that she wanted to distribute to interested parties, and she directed me to some Old Stock families in her county who had family heirloom seed varieties. Distinct populations of Ozark agrarians share common interest in locally adapted seed varieties. As we developed seed trading exchanges (seed swaps) in the area, a range of Ozarkers attended, BTTL, traditional (Old Stock), young, Hispanic, intermingled in a way that was atypical in the Ozark social environment. BTTLs acquired local heirloom seed varieties from their Old Stock neighbours and vice-versa, and they discussed the strengths, weaknesses, growing habits, preferred propagation methods, uses, and seed-saving strategies (Campbell 2012).

The commonality between contemporary back-to-the-land and Old Stock descendants is sustainable, inexpensive food production, the creation of a bio-diverse landscape that produces with minimal effort ... Arcadia! BTTLs operate organic farms and nurseries in the Ozarks and follow traditional practices. Some of the few farmers still ploughing with mules and horses in the Ozarks are BT-TLs. They want to conserve traditional agricultural practices and crop varieties because they allow the agrarian lifeway to continue. But there are significant differences. Old Stock homesteaders from the nineteenth century engaged in biodiversity conservation by necessity; their survival depended on it. They had to ensure a range of options were available in case drought or floods or myriad circumstances wiped out one of their food sources. Now, however, many Ozark agrarians have options, but choose for ideological purposes to live a frugal life and engage in conscious biodiversity conservation.

Contemporary homesteaders have much more information at their disposal to apply to self-sufficient living, and they utilise sources and networks outside of the region to acquire information about gardening, farming, energy produc-tion and other aspects of postmodern homesteading. In their research on self-reliance, many BTTLs read about and experience firsthand the marginality of Ozark lands: the infertility, verticality, climatic extremes, droughts and floods. Locally adapted open-pollinated seeds with unique genetics adapted to this landscape are extremely useful for homesteaders attempting to implement a low-energy lifestyle (Campbell 2012). While Old Stock homesteaders arrived without locally adapted seeds, BTTLs can access them through local connections with traditional Ozarkers. Old Stock agrarians continued to save seed because they understood the implications of buying hybrid seed every year. Buying seed rather than saving it was a slippery slope – eventually resulting in the loss of independence. Despite this awareness, most family heirloom seeds have disap-peared as the majority of the local population transitioned away from market agriculture and self-sufficiency (Campbell 2012).

HARD LIVING AND 'GRIT'

When BTTLs arrived in the Ozarks, they frequently had a relatively advantaged upbringing compared to their new Ozark neighbours, in terms of access to cos-mopolitan resources such as education, technology and diverse store-bought foods. Yet, in many cases, they did not realise their privilege until they attempted to wrest a living from the rocky Ozark soils. A female BTTL reflected about the toughness of traditional Ozark life: 'It was hard living here, a very difficult place to live; the elements are harsh, between the ticks and the chiggers, the snakes, the lack of soil, it can be a difficult place.' BTTLs who persevered had 'grit', a desire and ability to withstand difficult times. As another BTTL said in no uncertain terms: 'I discovered others who bear some of these traits: a combination of the ability to live in fantasy, which I definitely was doing there, and grit and tenacity under great hardship. If you've got those two things going, at least in my experience, that has helped to stay here.' After spending sufficient time in the Ozarks, BTTLs began to realise that, like their neighbours, they were poor. They had to reflect on the changes in their socioeconomic status, and the fact that their new 'grit' represented a shift in values. A female BTTL midwife and herbalist perceptively notes:

> It's poor here. It's funny to maybe have pride in that, but it's not easy here … The back-to-the-land movement suffered from [being] white drop-outs. I think a lot of us didn't really understand what that meant; we came from whatever level. I'm not saying that everybody was middle class, but I think from whatever level, we were people by and large who took our privilege for granted … because we'd always had it. And when you give that up, you don't realize that you're giving anything up. You try to get rid of the stuff you don't like; you don't realize that you're getting rid of stuff that made your life easy. But over time, if you stick it out – now some people say 'this is not for me, I'm going back!' – but if you stick it out, then you get it.

This illustrative quote bridges the gap between the traditional Ozark and the back-to-the-land subcultures by coming to grips with hands-on life in a harsh landscape and the 'attitude' changes that occur through such experiences. BT-TLs who remain in the Ozarks discard a detached, idyllic (aristocratic agrarian) myth of living with nature without getting dirty. Through their engagement with the place and its people, they emerge with a new notion of Arcadia, a reciprocal relationship with this beautiful, demanding, wild place forged through hard-won subsistence. Their new agrarian philosophy is grounded in the tangible, tick-infested Ozarks, infused with the wisdom learned from traditional Ozarker-neighbours and harsh and pleasant experiences with life and death (children, friends, dreams, livestock and wild animals) that come when trying to grow one's living.

CONCLUSIONS

The real Arcadia, a rugged mountainous district in the Peloponnese peninsula of Greece, became an idealised myth because Virgil's poetry celebrated it as a pastoral setting of agrarian harmony. This myth has persisted throughout Western history, from Renaissance to Romantic, as people tried to recreate their Arcadias, their dreams of rural happiness and tranquility. The Ozarks are a myth, a tall tale, a dream, just as Virgil's Arcadia was simultaneously a place and a vision of agrarian hope. Charles Morrow Wilson, Donald Harington,

Marguerite Lyon, Otto Earnest Rayburn and Vance Randolph's *Ozarkadia* persists both in reality, among the heirloom Whippoorwill peas, Arkansas Black apple orchards, Butterfly and Poke weed, Goldenseal, Ginseng, Persimmon and Pawpaw; and in myth, among the mineral-Rush ghost towns, White Oak-clearcuts, lead-contaminated superfund sites, chicken processing plants and the socio-economics of *Winter's Bone*.

Is the Ozarks the Tyson chicken house or the organic pick-your-own blueberry farm next door? Is it the Hispanic chicken processing plant workers, the Hmong market gardeners, or the vertically integrated Old Stock chicken man? Is it methamphetamines, moonshine, and marijuana or Old-Time Protestantism, quilting bees, and pie suppers? It's all, none and a little of each. Its hollers and hidden valleys continue to attract agrarian-minded humans who seek a refuge from conspicuous consumption in wilderness, and an opportunity to tap into a tradition of self-reliance. Charles Morrow Wilson (1959: 165) attests: 'people came to the Ozarks in great part because good food was available or readily attainable there. It is difficult, or impossible, to name a better reason for coming or for staying.'

NOTES

1.   Locals recount that a bar/saloon in Mountain Home, AR had the song on the jukebox, but only the words 'Ozark Mountain Mother' were legible, so people who selected it expecting a traditional country song experienced some chagrin.

REFERENCES

Aley, T. 1992. 'Karst topography and rural poverty'. *Ozarkswatch* 5: 19–21.

Blansett, K. 2010. 'Intertribalism in the Ozarks, 1800–1865'. *The American Indian Quarterly* 34: 475–497. CrossRef

Blevins, B. 2002. *Hill Folks: A History of Arkansas Ozarkers and Their Image*. Chapel Hill: University of North Carolina Press.

Brandon, J. and J. Davidson. 2005. 'The landscape of Van Winkle's mill: Identity, myth, and modernity in the Ozark upland South'. *Historical Archaeology* 39: 113 –131

Broadfoot, L.L. 1944. *Pioneers of the Ozarks*. Caldwell, Idaho: The Caxton Printers, Ltd.

Bromley, J. 1916. *Biography of John W. Morris*. Marshall, AR: Self-published.

Campbell, B. 2009. 'Ethnoecology of the Ozark highlands' agricultural encounter'. *Ethnology* 48: 1–20.

Campbell, B. 2012. 'Open-pollinated seed exchange: Renewed Ozark tradition as agricultural biodiversity conservation'. *Journal of Sustainable Agriculture* 36: 500–522. CrossRef

Clark, L. and J. McGary. 1987. *Wolf Bayou, Arkansas and Healing Springs Township*. Drasco, Arkansas: L. Clark.

Cumming, E. and W. Kaplan. 1991. *The Arts and Crafts Movement*. New York: Thames and Hudson.

Durand, P. 1975. 'Advice on Ozarks homesteading'. *Mother Earth News* 33: 93.

Fellone, F. 2010a. 'Still on the Land: Oh, pioneers: Back-to-the-landers from the 1970s still call Arkansas hills home'. *Arkansas Democrat-Gazette*, 14 December 14: 29.

Fellone, F. 2010b. 'Ozarks as counterculture lore'. *Arkansas Democrat-Gazette*, 14 December: 34.

Fellone, F. 2010c. 'Still on the land: Happy to be nappies'. *Arkansas Democrat-Gazette*, 15 December: 35.

Freyfogle, E. (ed.) 2001. *The New Agrarianism: Land, Culture, and the Community of Life*. Washington, D.C.: Island Press/Shearwater Books.

Harington, D. 1975. *The Architecture of the Arkansas Ozarks*. New Milford, CT: Toby Press.

Harington, D. 1986. *Let Us Build Us a City: Eleven Lost Towns*. New Milford, CT: Toby Press.

Harkins, A. 2004. *Hillbilly: A Cultural History of an American Icon*. Oxford: Oxford University Press.

Lyon, M. 1941 *Take to the Hills: A Chronicle of the Ozarks*. New York: Grosset & Dunlap Publishers.

Massey, E. 1979. *Bittersweet Country*. Garden City, New York: Anchor Press/Doubleday.

McDonough, N. 1975. *Garden Sass: A Catalog of Arkansas Folkways*. New York: Coward, McCann, and Geoghegan.

Otto, J. 1985. 'Migration of the Southern Plain Folk: An interdisciplinary synthesis'. *The Journal of Southern History* 51: 183–200. CrossRef

Otto, J. and A. Burns. 1981. 'Traditional agricultural practices in the Arkansas Highlands'. *Journal of American Folklore* 94: 166–87. CrossRef

Parker, J. 1992. 'Engendering identity(s) in a rural Arkansas Ozark community'. *Anthropological Quarterly*. 65: 148–155. CrossRef

Rafferty, M. 2001. *The Ozarks: Land and Life*. Norman: University of Oklahoma Press.

Randolph, V. 1931. *The Ozarks: An American Survival of Primitive Society*. New York: Vanguard Press.

Randolph V. 1951. *We Always Lie to Strangers*. New York: Columbia University Press.

Rayburn, O. 1937. *Rayburn's Arcadian Life, A Journal of the Well-Flavored Earth*, October.

Rayburn, O. 1949. *Rayburn's Arcadian Life, A Journal of the Well-Flavored Earth*, published by Otto Ernest Rayburn, 1934 (Arkansas Collection).

Shi, David E. 1985. *The Simple Life: Plain Living and High Thinking in American Culture*. New York: Oxford University Press.

Smith, K. 2003. *Wendell Berry and the Agrarian Tradition: A Common Grace*. Lawrence: University Press of Kansas.

Stewart-Abernathy, L. 1986. *The Moser Farmstead, Independent but Not Isolated: The Archeology of a Late-Nineteenth-Century Ozark Farmstead*. Arkansas Archeological Survey Research Series No. 26. Fayetteville, AR: Arkansas Archeological Survey.

Valencius, C. 2002. *The Health of the Country: How American Settlers Understood Themselves and Their Land*. New York: Basic Books.

Valencius, C. 2001. 'The geography of health and the making of the American West: Arkansas and Missouri, 1800–1860', in N. Rupke (ed.) *Medical Geography in Historical Perspective*. London: Wellcome Institute Trust for the History of Medicine. pp. 121–145.

Wait, J.M. 1928. U.S. Forest Service Historical Photo Collection, unpublished, Russellville, AR: U.S. Forest Service.

West, J. 1945. *Plainville, U.S.A.* New York: Columbia University Press.

Wilson, C. 1959. *The Bodacious Ozarks: True Tales of the Backhills*. New York: Hastings House Publishers.

Wilson, C. 1934. 'Ozarkadia'. *The American Magazine* 117: 58–61.

Index

camellone 79–91

Campbell, Brian C. xi, xii, xiii, xiv, 395–430

capital; capitalism 125, 126, 130–1, 137, 138, 141, 142, 179, 180, 186, 198, 266, 305, 311, 342, 378, 388

capsicum 85, 90, 91

Caribbean 231, 242

Carruthers, Jane 125, 147

Carson, Rachel xii, 342, 349, 350, 356, 367, 373

Cato 59, 60, 199

cattle vi, viii, x, xii, 5, 6, 12, 14, 15, 16, 22–56, 85, 87, 101, 133, 138, 140, 145, 177–203, 210, 283, 285, 287, 288, 290, 292 331, 333, 335, 358, 402, 404, 406

cattlemen 281–302

Cayambe, Ecuador 79, 80, 84

Central Menabe, Madagascar 205–14, 220–3

change v, vii, xiii, 13, 23, 25, 27, 32, 39, 44, 68, 104, 107, 111, 126, 133, 138, 142, 155, 158, 168, 169, 177–203, 204–29, 240, 246, 281, 291–3, 378, 379

cheese 19, 24, 26, 28–45, 401

chemicals *see also* agrochemical; DDT; fertiliser; herbicide; pesticide v, vi, ix, x, xi, xii, 23, 153, 158, 160, 250, 311–17, 323, 334, 349–70, 375

China 374, 375, 376, 386, 388

chinampa 77, 86, 90

Christian 62, 68, 71, 87, 88

classification 60, 110, 115, 119, 178, 181, 209, 326

clearing *see also* deforestation 27, 31, 71, 104, 106, 126, 180, 184, 186, 187, 198, 219, 221, 265–6, 272, 273

Clement, Vincent vii, viii, ix 57–76

Clements, Frederic 285, 291, 296, 321, 322, 327, 335

climate 5, 7, 8, 26, 27, 29, 34, 78, 79, 83, 102, 104, 125, 206, 207, 211, 268, 306, 308, 385, 408

change *see also* global warming 13

coast 26, 81, 86, 87, 88, 90, 91, 92, 101, 102–05, 107–10, 233, 239, 243, 265, 266, 267, 269, 308

coca 81, 83, 85, 88, 89, 90

coffee ix, 102, 105, 108, 111, 114, 116–18, 121, 243

Colombia 77, 81, 85, 86, 89, 90

colonialism v, vii, viii, ix, x, xi, xiii, 77, 78, 80, 83, 85–90, 92, 101–24, 125–50, 151–74, 205, 208–14, 216–17, 220–2, 231, 339

Columella 59, 109, 115, 119

commercialisation 25, 306

Commoner, Barry xii, 342

communality v, vi, vii, x, xiii, 3–5, 16, 33, 39, 376, 386–7, 390

compost xiii, 323, 375–6, 377, 383, 384, 405

concession (of land) 127, 130, 208, 211, 212, 213, 216–17, 221, 222

conflict xi, 12, 15, 17, 25, 63, 68, 125–8, 130–2, 136, 138, 147, 156, 157, 283, 286, 353, 359

Confucianism vii, 377

conquest 58, 61–2, 64, 67–9, 71, 77

conservation v, vii, viii, ix, xiv, 3, 14, 19, 22, 68–71, 113, 121, 125, 127, 141, 145–7, 204, 230–64, 268, 272, 273, 284, 285, 291, 322–4, 330, 349, 350, 375, 383, 395, 413

consumption; consumer xii, xiii, 43, 44, 45, 85, 86, 89, 128, 133, 146, 147, 269, 305, 306, 313, 317, 331, 333, 379–83, 390, 391, 398, 415

contour viii, 117, 237, 239, 241, 243–56 tillage 230, 236, 240, 241, 242, 245, 249, 252–3, 255–6

cooperative 31–3, 35, 37, 40, 46, 231, 380

Index

Lightning Source UK Ltd.
Milton Keynes UK
UKOW02f0633171116
287833UK00001B/282/P